Urban Water Resources Toolbox

Integrating Groundwater into Urban Water Management

Urban Water Resources Toolbox

Integrating Groundwater into Urban Water Management

Edited by

Leif Wolf, Brian Morris and Stewart Burn

With contributions from

 Universität Karlsruhe (TH)
Research University · founded 1825

 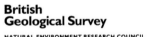 British Geological Survey
NATURAL ENVIRONMENT RESEARCH COUNCIL

 CSIRO AUSTRALIA

 FUTUREtec GmbH

 IRGO

 Robens

The GKW Group

 IWA Publishing

Published by IWA Publishing, Alliance House, 12 Caxton Street, London SW1H 0QS, UK
Telephone: +44 (0) 20 7654 5500; Fax: +44 (0) 20 7654 5555; Email: publications@iwap.co.uk
Web: www.iwapublishing.com

First published 2006, Reprinted 2007

© 2006 IWA Publishing. All rights reserved by the individual chapter contributors.

Printed by Ashford Overload Ltd, UK

Apart from any fair dealing for the purposes of research or private study, or criticism or review, as permitted under the UK Copyright, Designs and Patents Act (1998), no part of this publication may be reproduced, stored or transmitted in any form or by any means, without the prior permission in writing of the publisher, or, in the case of photographic reproduction, in accordance with the terms of licences issued by the Copyright Licensing Agency in the UK, or in accordance with the terms of licenses issued by the appropriate reproduction rights organization outside the UK. Enquiries concerning reproduction outside the terms stated here should be sent to IWA Publishing at the address printed above.

The publisher makes no representation, express or implied, with regard to the accuracy of the information contained in this book and cannot accept any legal responsibility or liability for errors or omissions that may be made.

Disclaimer

The information provided and the opinions given in this publication are not necessarily those of IWA and should not be acted upon without independent consideration and professional advice. IWA will not accept responsibility for any loss or damage suffered by any person acting or refraining from acting upon any material contained in this publication. Mention of a commercial company or product in this report does not imply endorsement by IWA. The use of information from this publication concerning proprietary products for publicity or advertising purposes is not permitted. Trademark names and symbols are used in an editorial fashion with no intention of infringement of trademark or copyright laws. We regret any errors or omissions that may unwittingly have been made.

British Library Cataloguing in Publication Data
A CIP catalogue record for this book is available from the British Library

Library of Congress Cataloging- in-Publication Data
A catalog record for this book is available from the Library of Congress

ISBN: 1843391384

For bibliographic and reference purposes, this publication should be cited as 'Wolf L, Morris B, Burn S, *AISUWRS: Urban Water Resources Toolbox*, 2006 IWA Publishing, London UK.'

To purchase copies of this publication, please contact:

IWA Publishing,
Alliance House, 12 Caxton Street,
London SW1H 0QS, UK
www.iwapublishing.com / publications@iwap.co.uk

Partner organisations in the production of this publication:

 COMMISSION EUROPEENNE
Research Directorate General
B-1049 Bruxelles
Belgium

 University of Karlsruhe,
Kaiserstr. 12
76128 Karlsruhe,
Germany

 British Geological Survey
Keyworth, Nottingham
NG12 5GG
UK

Contents

Dedication vi

Foreword vii

AISUWRS team members ix

Summary x

1. **Introduction** 1
   1.1 Problems in urban groundwater management 1
   1.2 The AISUWRS approach 8
2. **The Models** 15
   2.1 Urban Volume and Quality 16
   2.2 Network Exfiltration and Infiltration Model (NEIMO) 34
   2.3 Contaminant transport through the unsaturated zone: The SLeakI and POSI models ... 51
   2.4 Simple approach for balancing transient unsaturated soil processes in urban areas by the analytical model UL_FLOW 60
   2.5 Numerical Urban Groundwater Modelling using FeflowTM / ModflowTM HACCP approach to ground water modelling 73
   2.6 Integrating it all: the decision support system 91
3. **Application to real world problems** 100
   3.1 A matrix-flow aquifer: Rastatt, Germany 100
   3.2 A sandstone aquifer: Doncaster, UK 144
   3.3 A layered aquifer system: Ljubljana, Slovenia 191
   3.3 A karstic aquifer system: Mt Gambier, Australia 217
4. **Socio-Economics and Sustainability** 251
   4.1 Objectives, scope and approach 251
   4.2 Guidelines for data collection and analysis – the SEESAW Model 257
   4.3 Executive summary of the model applications in the case study cities 273
5. **AISUWRS Urban Water Resources Toolbox – A Brief Summary** 282
6. **References** 291

Dedication

✝

This book is dedicated to the memory of PD. Dr. Matthias Eiswirth who initiated and coordinated the AISUWRS project. Matthias died together with his 2-year old son in a tragic avalanche accident in the Swiss Alps on 30 December 2003. Matthias' death came as a great shock to those who had the good fortune to meet this exceptional person. He was the heart and soul of the AISUWRS project and he is remembered by his colleagues with gratitude for all the things he initiated and pushed to success with such energy and skill. His innovative ideas and science are a testament to his work as a scientist, but Matthias will also remain in our memory too as someone who cared about the people around him, providing support wherever possible and appropriate, and taking on many responsibilities without hesitation. We have lost a dear friend.

....All AISUWRS TEAM MEMBERS

Foreword from the European Commission

Back in 2002, the selection and funding of AISUWRS through the 5th Community Research Framework Programme (FP5, 1998-2002), demonstrated the importance attached to the need for developing a critical mass of knowledge capacity at European level to provide reliable responses to current and future challenges in the areas of urban water management, environmental technologies and pollution prevention in the 21st century.

With its commendable work on these strategic goals and its clear focus on challenging questions related to the mitigation of impact of diverse urban pollution sources on groundwater management, AISUWRS has developed illuminating scenarios with regard to sustainable practices for the handling of contaminants in urban water supply and disposal systems, providing more insight to one of the key elements of the urban water cycle. This book is designed to bring this topic to a larger audience.

Collective scientific work by recognised European partners coupled by qualified Australian participation, through the Commonwealth Scientific and Industrial Research Organisation (CSIRO), provided a peer-review context which guarantees the quality of produced knowledge.

AISUWRS has not acted in isolation but also as part of a wider cluster including another five related projects which collaborated under the CITYNET umbrella, an FP5 funded co-ordination action which worked on assuring the integrated dissemination of knowledge in all aspects of urban water management including both systems' related and key socioeconomic aspects.

We have to keep in mind that the implementation of integrated and adaptive future solutions is not only a challenge to our technological ingenuity and knowledge on pollutants behavior, but also to our 'water and material literacy', essentially our everyday habits and lifestyles. Public awareness and participation are fundamental to the success of integrated urban water management.

The continuous evolution of knowledge and deeper understanding of our changing ecosystems and societies' aspirations for better environmental management clearly demonstrates the value of projects like AISUWRS for formulating a paradigm shift in urban water management in Europe and worldwide based on key sustainability factors, adaptive management, and risk planning.

Still ahead of us lies the task of how to better manage the generated knowledge together with scientific and end-users communities, and streamline it to the everyday practice for improving the welfare of European and global citizens. The imminent launching of the 7th Community Research Framework Programme (2007-2013), and its strategic goals towards the promotion of a European knowledge-based economy provides a new framework for validating this kind of precious knowledge. Last, but not least, the provision of valid responses to the European Union's commitments and endeavours in international development and scientific cooperation for the Millennium Development Goals is an integral part of such validation effort.

This foreword on behalf of the European Commission and its Directorate General for Research is undeniably an expression of appreciation to the AISUWRS partners for their efforts but also a tribute to the sudden and unexpected loss of Dr. Matthias Eiswirth, the AISUWRS project co-ordinator, a European scientist whose vision and capacity should be always remembered and cherished by all of us who had the privilege to meet, work and exchange ideas with him.

Dr. Andrea Tilche, Head of Unit
ir. Zissimos VERGOS, AISUWRS Programme Officer
European Commission
Directorate General for Research
Unit I.3 'Environmental Technologies and Pollution Prevention'

Foreword from DEST

In light of the importance of the AISUWRS project to Australia and the International water community, financial support of the Australian Government through an International Science Linkages Competitive Grant from the Department of Education, Science and Training was provided.

Carley Rothnie, DEST

Foreword from the AISUWRS team members

This book is an outcome of the research project Assessing and Improving Sustainability of Urban Water Resources and Systems (AISUWRS). It was produced by the research partners as a means of disseminating to a wider audience the results of this 3-year international multidisciplinary project, which received funding support from 2002 to 2005 from the European Commission under the Fifth Framework Programme, the Department of Education, Science and Training of Australia and the UK Natural Environment Research Council. The organisations that participated in this research work were:

University of Karlsruhe, Germany (Coordinator)
British Geological Survey, United Kingdom,
Commonwealth Scientific and Industrial Research Organisation, Australia,
FUTUREtec Gmbh, Germany,
GKW Consult, Germany,
Institute for Mining & Geology (IRGO), Slovenia
University of Surrey (Robens Centre for Public and Environmental Health), United Kingdom

The AISUWRS project was designed to integrate knowledge on urban water supply and drainage systems with hydrogeological expertise on urban groundwater resources by means of a set of modelling tools. The project has developed some tools, linked them with existing model codes used routinely in groundwater resource evaluation and then applied them in a series of example case studies in urban areas overlying productive aquifers in Europe and Australia.

At the heart of the project was the recognition that while land and water use in urban areas is highly complex, cities are expanding. Municipalities and water utilities serving urban areas are increasingly recognising that they can no longer afford to neglect the water resource in an underlying productive aquifer just because it is difficult to quantify. The urban water balance *is* complicated, its components difficult to unravel, and the effects on water quality are only just starting to be understood in a way that could be used predictively. But the effort is worthwhile not only to avoid future problems like groundwater flooding of urban infrastructure as a consequence of water level rebound but also because the resource may be large and water is increasingly scarce. Cities fortunate enough to overlie a productive aquifer, the utilisation of which would take up a negligible amount of valuable space at the land surface, will increasingly value, and seek to redevelop, this sometimes neglected resource.

We would like to thank the end users who have participated in the AISUWRS project and all other institutions who provided invaluable support in sharing data and knowledge. Without them, it would not have been possible to complete this project in practice. We would especially mention the project partners to whom we are indebted; to the municipalities of Rastatt, Doncaster, Ljubljana and Mount Gambier, to Javno podjetje Vodovod - Kanalizacija, d.o.o. (Ljubljana Water Supply), EA RS - Hydrometeorological Survey of Slovenia, Yorkshire Water (UK), Environment Agency (UK), South East Water (Australia), SA Water (Australia), SA Environment Protection Authority (Australia), Dept Water, Land, Biodiversity and Conservation (Australia), City of Mount Gambier (Australia), South East Catchment Water Management Board (Australia), State Water Authority of Baden-Württemberg (Germany), Stadtwerke Karlsruhe (Germany) and Landesanstalt für Umweltschutz Baden-Württemberg (Germany).

We would also like to thank the European Commission and the administration of the Fifth Framework Programme that made AISUWRS possible by its financial support. The additional financial support of the Australian Government through an IAP-International S&T Competitive Grant from the Department of Education, Science and Technology and the UK Natural Environment Research Council is also gratefully acknowledged.

Heinz Hötzl & Leif Wolf
Executive Project Coordinators AISUWRS
On behalf of the AISUWRS team, Karlsruhe 2006

The AISUWRS project teams

The AISUWRS project formally started in November 2002 and ended in October 2005. During that time more than 60 individuals from 16 different agencies in 4 countries played a role in progressing the various components of an international venture whose many strands and innovative nature have made this a truly multidisciplinary project. Our thanks and acknowledgements to those colleagues on the joint voyage of discovery that has been AISUWRS.

Heinz Hötzl & Leif Wolf
Executive Project Coordinators AISUWRS
Karlsruhe 2006

All project participants in alphabetical order:

British Geological Survey	*Debbie Allen, Jenny Cunningham,* George Darling Daren Gooddy, *Rosemary Hargreaves,* Andrew Hughes, *Majdi Mansour, Ilka Neumann, Brian Morris,* Marianne Stuart, *Barry Townsend, Emily Whitehead,*
CSIRO	Karen Barry, *Stewart Burn,* Steve Cook, *Ray Correll, Dhammika De Silva, Clare Diaper, Peter Dillon, Scott Gould,* Leorey Marquez, *Grace Mitchell, Magnus Moglia, Ros Miller,* Tony Miller, *Mike Rahilly, Paul Sadler, Tara Schiller, Grace Tjandraatmadja,* Simon Toze, Gerardo Trinidad, *Joanne Vanderzalm.*
Department of Applied Geology, University of Karlsruhe	Christine Buschhaus, *Matthias Eiswirth, Inka Held, Heinz Hötzl, Jochen Klinger,* Michel Lambert, *Christina Schrage, Leif Wolf*
FUTUREtec Gmbh	*Andrej Druta, Klaus Hoering, Caterina Rehm-Berbenni*
GKW	*Uwe Arras, Ralf Bufler,* Stefan Doerner, *Hans-Peter Lammerich, Ulrike Voett,* Clemens Wittland.
Institute for Hydromechanics, University of Karlsruhe	Meike Bücker-Gittel, *Ulf Mohrlok.*
IRGO / University of Ljubljana / GeoSi, VO-KA	Branka Bračič-Železnik, *Barbara Čenčur-Curk,* Marko Gspan, Brigita Jamnik, Andrej Juren, *Ben Moon,* Darko Petauer, *Petra Souvent,* Nataša Šušteršič, Martin Tancar, Miran Veselič, Sandi Viršek, *Goran Vižintin.*
Robens Centre for Public and Environmental Health	Owen Baines, *Aidan Cronin, Joerg Rueedi,* Richard Taylor.

*Core team members in italics

Summary

The aim of the AISUWRS project was to develop assessment tools that would improve the sustainability of urban water resources and systems and to illustrate the practicability of their application through the medium of case studies. These tools comprise a chain of interconnected models that link urban water supply, urban drainage and urban groundwater resources in terms of quality and quantity. Parallel with the model development, AISUWRS teams undertook detailed field investigations on the impact of wastewater management on groundwater (a poorly quantified area of urban hydrogeology), which combined the quantification of the source (e.g. the leaky sewer network) with the monitoring and modelling of the groundwater as the receiving water body. The investigations were not constrained to popular parameters, but also contained a suite of promising specific marker substances, from pharmaceutics to enteric viruses. The modelling exercises were validated at specific test sites for sewer leakage under operating conditions, which are much closer to reality than laboratory tests documented in contemporary scientific literature.

The field investigations were also designed to test the practicability of populating the models with real data, collected under typical rather than idealised conditions so the resultant innovative but ambitious pilot studies apply the models not to small research-scale locations of a few hectares but instead to whole districts of several thousand population, the scale most typically used for water planning. These pilot studies have applied the model chain to a range of commonly encountered city and aquifer settings in Europe and Australia.

The uppermost model applied in the AISUWRS system is the Urban Volume and Quantity model (UVQ), developed by CSIRO, Australia. Its main input parameters are climate records, water consumption characteristics (e.g. water use for laundry or typical contaminant loads through toilets) and urban sealing coefficients. The model calculates water flows and contaminant loads through the wastewater and stormwater systems together with direct recharge to groundwater from green space. The pipe network information is fed into the specially developed Network Exfiltration and Infiltration Model (NEIMO), which estimates the amount of wastewater exfiltration from or groundwater infiltration into sewers. Leakage rates are based on defect distributions observed by CCTV investigations or inferred from pipe network design/construction criteria. The output is then forwarded to specially developed unsaturated zone models calculating water flows and travel times to the water table and the combined effects of sorption and decay of contaminants.

The concept was applied to the four case study cities of Rastatt (Germany), Doncaster (UK), Ljubljana (Slovenia) and Mount Gambier (Australia). Different water management scenarios were modelled, such as decentralised rainwater infiltration, sewer rehabilitation, demand change and climate change. The results of the modelling exercise were compared with monitoring activities at specifically constructed test sites and groundwater observation networks. Once set up, the application of the tools was straightforward and the effects of radically different urban water usage scenarios were easy to compare, demonstrating that urban aquifers could be managed in a sustainable fashion. Another conclusion from the studies themselves and a comparison across the cities is that urban aquifers do not appear to be under severe contamination pressure from the constant emissions of the urban drainage systems, provided appropriate protection measures are in place.

A key outcome of the AISUWRS project is the establishment of a software framework with user interface (AISUWRS DSS) that incorporates all major parts of the urban water and solute balance. Once set up for a case study, it demonstrates the causal directions in urban water management and allows the fast comparison of different scenarios. Within the project, scenarios of climate change, demand change, decentralised rainwater infiltration, greywater reuse and different strategies for sewer rehabilitation were assessed. Apart from the DSS front end, all process models are also available as stand-alone versions, along with separate manuals or documentation.

Due to the large number of processes considered in what is a complex hydraulic and hydrologic environment, the developed model chain requires a large number amount of input parameters, and uncertainty is inherent if these are not properly known or specified. AISUWRS has aimed to keep the parameters and processes as simple as possible in order to allow the entire water system to be understood and operated by a single user, but uncertainty is inevitable when unravelling the urban water balance; its constraint is undoubtedly a topic for the future.

While the established framework allows qualitative comparisons of the scenarios and indicates probable responses to a specific management option, much effort will also be required in future urban water research to increase the reliability of the quantitative results. The AISUWRS system could allow the sensitivity analysis of the individual parameters and this could serve as a guideline for future research on the most critical points.

The socio-economic analysis in the case study cities sometimes uncovered distinctively different problem perceptions and priorities, both in the groups of experts responsible for the water management and with the remaining stakeholders. The AISUWRS project has developed tools to foster these urgently required deliberation processes. Methodologies for formal sustainability assessment with a triple bottom line background were also elaborated and tested during the case studies.

This book provides a concise documentation on each of the models developed within the project. The application and use of the models and methodologies is demonstrated by the set of four case study cities, which have diverse conditions in terms of hydrogeologic setup, climate and data availability. These have shown that the approach is valid and constitutes an important step towards integrated water management for those cities fortunate enough to be located upon a productive aquifer.

1
Introduction

1.1 Brian Morris[1] and Leif Wolf[2]

1.2 Leif Wolf[2], Brian Morris[1], Stewart Burn[3], Heinz Hötzl[2]

[1.] *British Geological Survey (BGS), Wallingford, UK*
[2.] *Department of Applied Geology (AGK), University of Karlsruhe, Germany*
[3.] *CSIRO, Melbourne, Victoria 3190, Australia*

1.1 PROBLEMS IN URBAN GROUNDWATER RESOURCE MANAGEMENT

1.1.1 Setting the scene

By 2030 it is estimated that more than 60% of the world's predicted population of 8400 million will live in towns or cities (UNCHS 1997), with much of the increase concentrated in the developing world. Cities large and small depend on aquifers for their water supply; about half of the world's megacities are groundwater-dependent* (Table 1.1.1), as are hundreds of smaller cities worldwide.

Table 1.1.1. Population 1996 (and 2015 projected) of megacities dependent on groundwater (population statistics from UNDES 1994, 1996).

City	Population	City	Population	City	Population
Mexico City	16.9 (19.2)	Buenos Aires	11.9 (13.9)	Cairo	9.9 (14.4)
Calcutta	12.1 (17.3)	Jakarta	11.5 (13.9)	Tianjin	9.6 (13.5)
Teheran	6.9 (10.3)	Dhaka	9.0 (19.5)	London	10.5 (10.5)
Shanghai	13.7 (18.0)	Manila	9.6 (14.7)	Beijing	11.4 (15.6)

*Groundwater dependency definition. The city's water supply (public and private domestic, industrial and commercial) could not function without the water provided by a local urban or peri-urban aquifer system. Typically groundwater would provide at least 25 per cent of the water supply to such a city, and often much more.

In Europe, where urban piped water supply is generally provided by water utilities (as opposed to individual domestic boreholes), some indication of the important role of groundwater in urban water supply can be gauged from Table 1.1.2. It has been estimated that over 40% of the water supply of Western and Eastern Europe and the Mediterranean region comes from urban aquifers (Eiswirth et al. 2002). In Australia too, groundwater is a valuable urban resource, providing water to consumers in Perth, Newcastle and many smaller towns.

© 2006 IWA Publishing. *Urban Water Resources Toolbox – Integrating groundwater into urban water management* Edited by Wolf L, Morris B, Burn S. ISBN: 1843391384. Published by IWA Publishing, London UK

Table 1.1.2. Importance of groundwater in public supply in various EC Member States.

Country	% Groundwater in public water supply	% Urban population(1996)	Country	% Groundwater in public water supply	% Urban population(1996)
Denmark	99	85	France	56	75
Austria	99	64	Ireland	50	58
Italy	80	67	Greece	50	59
Germany	72	87	Sweden	49	83
Netherlands	68	89	UK	29	89
Belgium	52	97	Portugal	80	36

Groundwater's role in urban development is especially critical where key aquifers are located below the city or in the immediate periurban zone. Here the urban services of water supply, waste disposal and engineering infrastructure interact strongly with each other and with the underlying groundwater system (Figure 1.1.1).

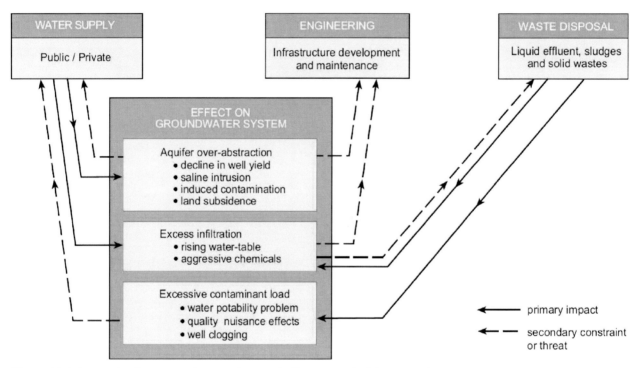

Figure 1.1.1. Interaction between urban services and effects on underlying groundwater resources (From Foster et al, 1998).

Yet in comparison with cities reliant on supplies from river intakes or from periurban reservoirs, such urban aquifers remain under-studied and under-protected.

In part this is a product of traditional water planning, which assumes future water demands will inexorably rise, must be catered for, and will inevitably involve the import of additional sources of water from the city's hinterland. An associated perception is that usage of urban/periurban aquifers is a relatively transient phase pending such development of more distant pristine sources, once the necessary capital investment can be found. However, urban planners are finding it increasingly difficult to replace a degraded urban aquifer by alternative water sources from the hinterland, where resources may already be fully utilised for agricultural or ecological purposes (Burke and Moench 2000). Moreover, sustainability principles, agreed to by more than 150 countries since the 1992 Earth Summit manifesto mean that in the future, a city fortunate enough to overlie a productive aquifer can no longer regard it as a discardable resource, an asset to be abandoned once dewatered or heavily contaminated.

The Water Framework Directive (WFD), which came into force in 2000, provides a clear sustainability message in this respect. As part of a comprehensive strategy for managing the water environment, Member States have environmental objectives for groundwater (Article 4.1[b] of the WFD) which require them to:

- Prevent or limit the input of pollutants into groundwater and prevent deterioration in status
- Protect, enhance and restore all groundwater bodies, with the aim of achieving good status
- Reverse upward trends in the concentration of any pollutant.

There is also an emerging awareness of the interdependence between the underlying aquifer system, the different urban land uses and the water infrastructure of a city, namely the water supply and sewerage system (Lerner et al. 1990, Foster et al. 1993, Eiswirth 2001) and urban reaches of rivers (Grischek et al. 2001). Cities like Venice, Mexico City or Bangkok have suffered from overexploitation of their aquifers, which has resulted in land subsidence (Morris et al. 2003). In contrast, several European cities have already experienced drastically falling water levels during early expansion followed by groundwater flooding at later development stages (London, Birmingham, Paris, Berlin, Hamburg, Moscow, Barcelona). A recent survey of communities in Germany has indicated that about 46% have experienced problems with flooded basements and properties (BWK 2003). A large number of private homeowners are affected as the houses were built according to the water levels prevailing at the time, before reduction in abstraction at nearby groundwater treatment plants caused groundwater rebound. These cases have become (2006) a major issue in German courts as the homeowners are suing the municipalities. As well as costly adverse effects on existing urban structures (tunnels, road cuttings, basements, foundations), rebound in aquifer water levels adds uncertainty to the design of future engineering projects, impeding redevelopment.

All this points to the need for:
- A better quantification of the urban water cycle for cities overlying aquifers
- An improved understanding of how new sources of recharge impact on underlying groundwater quality
- Application of that enhanced understanding to improve urban water management, by enabling the effects of deploying different water strategies to be predicted.

Against this backdrop the AISUWRS research project (**A**ssessing and **I**mproving **S**ustainability of **U**rban **W**ater **R**esources and **S**ystems) that is described in this publication was conceived and executed. The research partners in this project have sought to provide assessment and modelling tools for the better planning of existing city water supply, wastewater and pluvial drainage networks and for the evaluation of innovative alternative approaches to urban water systems.

The products of this research aim to improve the sustainability of those cities where local groundwater is a vital resource.

1.1.2 The Issues
1.1.2.1 Changes in recharge quantity

When assessing the impact of urban infrastructure on an underlying aquifer, the effects of quantity and quality of recharge are inextricably linked. Considering quantity first, the process of urbanization causes major changes in the frequency and volume of recharge compared with its rural equivalent. These can be readily predicted in a generalised way (Table 1.1.3).

Table 1.1.3. Effects of urbanization on subsurface infiltration and groundwater recharge (adapted from Foster et al. 1993).

Process	Effect on subsurface infiltration		
	Rates	Area	Time base
Modification to pre-urban condition:			
Surface impermeabilization	Reduction	Extensive	Permanent
* Stormwater/pluvial soakaways	Increase	Extensive	Intermittent
* Mains pluvial drainage	Reduction	Extensive	Intermittent to continuous
* Surface water canalization	Marginal reduction	Linear	Variable
* Irrigation of gardens and public open space	Increase	Extensive	Seasonal
Introduction of water service network:			
Local groundwater withdrawals	Minimal	Extensive	Continuous
Imported mains water supply leakage	Increase	Extensive	Continuous
* Mains sewerage leakage	Marginal increase	Extensive	Continuous
** On-site sanitation (septic tanks etc)	Increase	Extensive	Continuous
* ** Typical accompanying major (**) and minor (*) impact on groundwater quality			

The net effect for many cities is a rise in the total volume of recharge, the land-sealing effect of paving and building being more than compensated for by the enormous volume of water circulating through, and lost from, the water infrastructure of pipes and from soakaways draining the built area. This effect has been observed in a wide range of climatic and aquifer settings (Figure 1.1.2), although it appears to be most pronounced in cities where on-site sanitation or amenity watering is important, or in arid to semi-arid climates where the new sources may increase the total infiltration several times over the pre-urban situation.

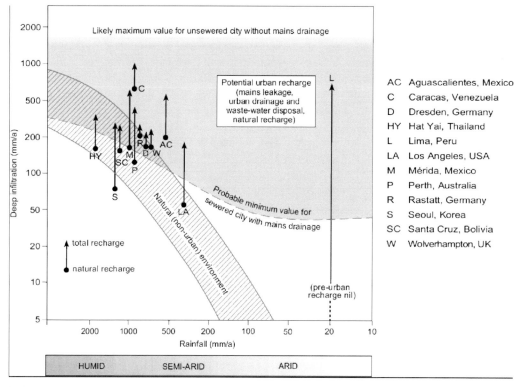

Figure 1.1.2. Increase in groundwater recharge due to urbanization (modified from Foster et al. 1993, Eiswirth et al. 2002, Krothe et al 2002).

However, in detail, the resultant urban water budget is complex (Figure 1.1.3). The changes vary from city to city, depending on climatic, geological and developmental setting, on national sanitary engineering practices and maintenance regimes (separate sewers, combined sewers or septic tank drainage) and on local water usage rates, to produce a water balance that is quite distinct from that occurring in adjacent rural areas.

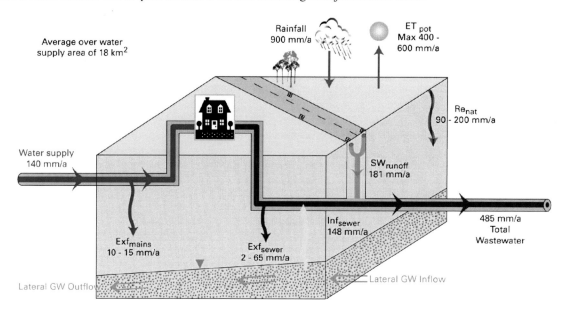

Figure 1.1.3. Overview of urban water fluxes, estimated for the entire City of Rastatt, SW-Germany. (ETpot = potential evapotranspiration, Renat = natural groundwater recharge, SWrunoff = stormwater runoff into sewers, Infsewer = Infiltration of groundwater into sewers, Exfsewer = estimated exfiltration of wastewater from leaky sewers, Exfmains = leakage from drinking water mains; redrawn from Wolf et al 2005b).

1.1.2.2 Changes in recharge quality

The changes in the origin, frequency and magnitude of the various components of urban groundwater recharge exert a corresponding influence on the net quality of recharge replenishing the underlying aquifer. The load generated at the land surface comes from a variety of potentially contaminating activities that is especially diverse if the city is industrialised (Table 1.1.4) and the mode of entry to the subsurface can be as point-source, linear or diffuse sources. The resultant pollutant range is enormous; microbiological, chemical, inorganic and organic, with each contaminant group having its own physicochemical and transport properties.

Table 1.1.4. Summary of activities that could potentially generate a subsurface contaminant load (adapted from Chilton in Chapman, 1996).

Activity/structure	Character of pollution load			
	Category	Main types of pollutant	Relative hydraulic surcharge	Soil zone by-passed?
Urban waste, wastewater, drainage:				
Unsewered sanitation	P-D	pno	+	✓
Land discharge of sewage	P-D	nsop	+	×
Stream discharge of sewage	P-L	nop	++	✓
Sewage oxidation lagoons	P	opn	++	✓
Sewer leakage	P-L	opn	+	✓
Landfill/solid waste disposal	P	osnh		✓
Highway drainage soakaways	P-L	soh	++	✓
Wellhead contamination	P	pn		✓
Urban Industrial development:				
Process water/effluent lagoons	P	ohs	++	✓
Tank and pipeline leakage	P	oh	+	✓
Accidental spillages	P	oh	++	×
Land discharge of effluent	P-D	ohs		×
Stream discharge of effluent	P-L	ohs	++	✓
Landfill disposal residues and waste	P	ohs		Sometimes
Well disposal of effluent	P	ohs	++	✓
Aerial fallout	D	a		×

Category: P point /   D diffuse /L linear
Pollutant types: 'p' faecal pathogens, 'n' nutrients, 'o' organic micropollutants, 'h' heavy metals, 's' salinity, 'a' acidification
Hydraulic surcharge: + to ++ increasing importance, relative volume or impact of water entering with pollution load

Unravelling the complexity of recharge in an industrialised urban catchment is thus an enormous task which can be quite site-specific, and so the AISUWRS research project has limited its water quality impact brief to the assessment of contaminant fluxes generated in predominantly residential catchments. Such residential districts typically comprise the largest proportion of a city's surface area. Recharge water quality in such areas is dominated by contaminant fluxes from domestic and public areas (infiltration through domestic gardens, amenity greenspace, diverted runoff from local roofed, and paved surfaces), and from sewer network leaks (foul, storm, combined). However, it must be accepted that there is inevitable overlap with industrial/commercial zone land uses e.g. suburban fuel filling stations, local workshops etc.

Table 1.1.5. Impact on groundwater quality from various sources of urban aquifer recharge (from UNEP, 1996).

Recharge source	Importance	Water quality	Pollutants/commonly used pollution indicators
Leaking water mains	Major	Excellent	Generally no obvious indicators
On-site sanitation systems	Major	Poor	N, B, Cl, FC, DOC
On-site disposal/leakage of industrial waste water	Minor-to-major	Poor	HC, diverse industrial chemicals, N, B, Cl, FC, DOC
Leaking foul and combined sewers	Minor	Poor	Diverse industrial & household chemicals inc HM, N, B, Cl, FC, SO_4
Stormwater/pluvial soakaway drainage	Minor-to-major	Good-to-poor	N, Cl, FC, HC, DOC, diverse industrial chemicals inc. HM
Seepage from canals and rivers	Minor-to-major	Moderate-to-poor	N, B, Cl, FC, SO_4, DOC, diverse industrial chemicals inc HM

B	boron	HC	hydrocarbons (fuels, oils and greases)
Cl	chloride and salinity generally	N	Nitrogen compounds (nitrate or ammonium)
DOC	dissolved organic carbon (organic load)	SO_4	sulphate
FC	faecal coliforms	HM	heavy metals (lead, zinc etc)

The net effect on water quality from the various contaminant loads in the different recharge sources (Table 1.1.5) can be mitigated by the beneficial effects of attenuation. The latter is important because it is the effectiveness of subsurface attenuation that is one of the principal reasons that groundwater remains so widely used for potable supply; even if complete contamination elimination is not possible, attenuation can reduce the extent and the cost of treatment of raw groundwater before it is re-used for supply. Nevertheless, the mode of attenuation in the subsurface is varied, as can be seen from Table 1.1.6, which summarises the relative importance of different mechanisms for just a few of the more commonly occurring contaminants/groups.

Table 1.1.6. Transport characteristics of the common urban contaminants/contamination indicators in groundwater (from Morris et al. 2003).

Contaminant	Source	Attenuation Mechanism				Permitted Drinking Water Concentration	Mobility	Persistence
		Biochemical Degradation	Sorption	Filtration	Precipitation			
Nitrogen (N)	Sewage	✓	✓*	×	×	Moderate (10–20 mg N/l)	Very high	Very high
Chloride (Cl)	Sewage, industry, road de-icer	×	–	×	×	High	Very high	Very high
Faecal pathogens (FCs)	Sewage	✓✓	✓✓	✓✓✓	×	Very low (<1 per 100 ml)	Low-moderate	Generally low
Dissolved organic carbon (DOC)	Sewage, industry (esp. food processing, textiles	✓✓	✓✓✓	✓	×	Not controlled	Low-moderate	Low-moderate
Sulphate (SO₄)	Road-runoff, industry	✓†	✓	×	✓	High	High	High
Heavy metals	Industry	×	✓✓✓	✓‡	✓✓	Low (Variable)	Generally low unless pH low (except Cr [VI])	High
Halogenated solvents (DNAPLs)	Industry	✓	✓	×	×	Low (10–30 μg/l)	High	High
Fuels, lubricants, oils, other hydrocarbons (LNAPLs)	Fuel station spillages, industry	✓	✓✓	×	×	Low ((10–700 μg/l BTEX§)	Moderate	Low
Other synthetic organic	Industry, sewage	Variable	Variable	×	×	Low (Variable)	Variable	Variable

KEY ✓✓✓ highly attenuated ✓✓ significant attenuation ✓ some attenuation × no attenuation
*Ammonia is sorbed † Can be reduced ‡ where occur as organic complexes §Aromatic compounds with health guideline limits

1.2 THE AISUWRS APPROACH

1.2.1 Scope

The scale and diversity of the problems of urban groundwater management described in Section 1.1 make it desirable to identify the key processes and to develop appropriate modelling tools to support the planning and decision process. The AISUWRS project has encompassed both aspects, firstly by direct field investigations to measure and describe the impact of city water infrastructures on urban groundwater resources and secondly the development of new models that can be applied to other city settings and other water management practices. This book is designed to explain and to provide guidance on the new tools.

The scope of the AISUWRS initiative was to assess and improve the sustainability of urban water resources and systems with the help of computer tools. In a diverse set of case study urban areas in Europe and Australia, the project analysed a range of existing urban water supply and disposal scenarios. A sequence of linked, process-related models simulating the passage of water through the urban area and its associated pipe infrastructure demonstrates how each scenario differs both in its handling of contaminants within different urban water systems and in the effects on urban recharge, both volumetrically and in terms of the potential for contamination. The sources of contaminants, their flow paths and the sinks were identified for different urban areas and quantification of the contaminant loads was undertaken. For the calibration of the models, detailed field studies were carried out in focus study areas in each of the three European and one Australian case study cities, a process that also proved invaluable in testing the robustness in use of the component models and the practicalities of linkage. In addition AISUWRS developed a management and Decision Support System (DSS) that makes use of these innovative pipeline and urban water system assessments to inform recommendations for the safeguarding and protection of urban water resources.

As there are a large number of possible urban water and groundwater contaminants, a selection of pollution indicators was used for the system development. These were chosen as common markers that can be easily analysed, are likely to be widely present in many urban settings and which could indicate incipient deterioration of underlying high quality groundwater. For such vital resources, it is important to identify contamination early, when the concentrations are relatively low, to locate and characterize the sources of contamination and to observe long-term trends. A multidisciplinary research effort was required in order to develop this method.

1.2.2 Modelling approach and data flow concept

The simulation of the various elements of the urban water balance in AISUWRS includes several stages before the resultant recharge can be passed down to a groundwater flow model for groundwater quantity and quality prediction. The stages are covered by different standalone models (Fig. 1.2.1), which are interlinked by a decision support system:

- At the land surface, a model is needed that schedules water inputs (from precipitation and piped mains water supply) and assigns it to on-site infiltration or to the piped drainage infrastructure (combined sewers or separate foul and pluvial drainage networks) via different urban land uses. This role is carried out by the urban water and contaminant balance model UVQ, which divides the city into a manageable number of hydraulic entities or neighbourhoods (within each of which water use is similar) so as to assign volumetric and contaminant fluxes through domestic, commercial or industrial premises and associated green space. Inputs include precipitation, mains water supply, stormwater and foul sewer inflows from upstream neighbourhoods; outputs are assigned to stormwater and foul sewer outflows, on-site sanitation units (septic tanks) if present, and water directly infiltrating to the subsurface via gardens, public open space or roof/paved area pluvial soakaways. UVQ has been developed within the AISUWRS project for this purpose.
- A system for assessing leakage from the pipe network to the subsurface. Mains leakage, typically assessed by nighttime minimum flow metering, is handled by AISUWRS within the UVQ model. For leaks from sewers (foul, stormwater, combined) the NEIMO model has been developed; this code uses pipe characteristics to estimate losses or gains (depending on the relative position of the water table). NEIMO uses generic curves that take account of pipe material, joint type, diameter, condition and trench substrate to calculate infiltration or exfiltration rates per asset that are then assigned using GIS to the whole area under investigation. CCTV-derived defect reports can also be employed (if available).

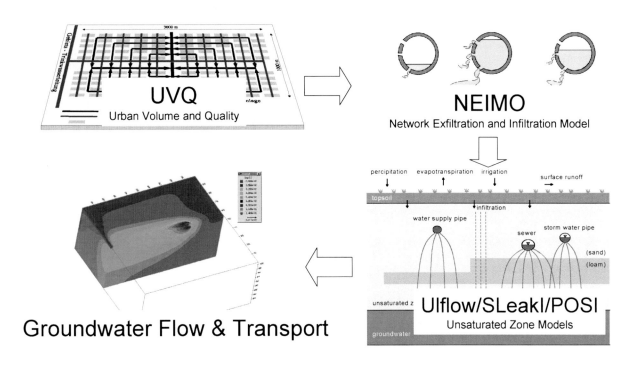

Figure 1.2.1. Major modelling compartments for the integrated approach in AISUWRS.

- Models that can track fluxes through the unsaturated zone help assess the role of contaminant attenuation and possible elimination as water percolates either directly through greenspace or indirectly via pipe leakage. The one-dimensional models PosI and SleakI have been developed to approximate the key processes, while the UlFlow model permits minimum and average residence times to be estimated for a given soil type and depth to water table
- The impact on the underlying aquifer is assessed using proprietary groundwater flow and contaminant transport models that can provide simulations of the effect on groundwater heads or on key contaminant concentrations, both spatially (across and through the aquifer system) and with time (as hydrographs or pollutant trend graphs). The Australian case-study utilises the data from models further up the chain in a different way, to inform hazard analysis and critical control procedures in an aquifer that is not yet conducive to numerical modelling
- A system is required to manipulate the complex array of datasets and facilitate the cascade of data from one model to another. Not only do the component models have their own data requirements (some quite extensive) but they also need to generate arrays of output files that subsequent models require, in the correct format, as input for their simulations. AISUWRS has developed a database structure and a decision support system to facilitate these complex and important tasks.
- The truism that there is no such thing as the universal pollutant is especially true of urban recharge, where a wide range of potential contaminants may be present, each with their own properties and propensities to attenuation. The AISUWRS project has started the process of contaminant flux estimation and prediction using a range of pollution indicators, selected initially from those that have been found useful in previous urban groundwater resource studies. Those applicable in the case study city settings where the model array has been applied are described in Chapter 4. The models referred to earlier track these pollution indicators, either in their own right in order to estimate final concentrations if the indicator has a health–related guideline limit, or because their movement and attenuation can be used as an analogue for contaminants with similar properties.

Thus the UVQ model schedules fluxes from precipitation and mains water through households to sewer, stormdrain, septic tank or garden, NEIMO transmits the leakage fluxes based on pollutant concentrations in the relevant pipe network and the unsaturated zone models apply appropriate attenuation factors. Thereafter the solute transport module of a groundwater flow model is used to assess the impact on the groundwater body as a whole in terms of aquifer contaminant concentration and down-gradient plume movement.

1.2.3 The urban water cycle and how the AISUWRS approach tracks the journey of water through the city

The AISUWRS modelling approach encompasses all urban water fluxes from source to sink (Figure 1.2.2). The cycle starts with water input from rainfall, and from public and private water supply (imported water). Volumetrically the principal water input in humid climates is precipitation, usually as rainfall. The rain falls either on roofs, paved areas, gardens or public open spaces. While most urban drainage calculations employ integrated runoff coefficients, these are very approximate, and do not permit differentiation of pathways that could have different water quality implications (e.g. runoff from de-iced roads versus roof runoff). The most upstream AISUWRS model-UVQ-provides the ability to schedule the actual area demand of each of these surfaces for a neighbourhood of similar water usage, providing more flexibility in simulation of alternative settings. For each surface type, the amount of runoff can be calculated and specified contaminant loadings added. In this fashion the quantity and quality of water that infiltrates on-site through green space or local soakaways and the stormwater component in the separate pluvial or combined sewer system can be calculated.

In combined sewer systems, the stormwater mixes with wastewater and leaves the area where it is generated via leaks, combined sewer overflows (CSOs) or the wastewater treatment plant. Rainfall percolating through unsealed areas like gardens or public open space enters the soil moisture store. Losses due to actual evapotranspiration can be calculated from climate parameters and the available soil moisture at a given time step. The excess water passes as seepage downwards to the groundwater body. For water entering the system via unsealed surfaces, contaminant contributions from sources like fertiliser application or atmospheric deposition are added. A separate model considers the degradation or sorption of these substances in the unsaturated zone before they reach the groundwater table.

On the water supply side the first diversion of flow occurs in the mains distribution system. Leakage from these pressurised systems is universal; the only question is the extent (varying from 5% in new, well-maintained suburbs to >80 % in some developing world cities). The water lost contributes to the soil moisture in the unsaturated zone or feeds directly into the groundwater. This contribution to the groundwater recharge, being of drinking water quality usually poses no direct pollution threat, although in some cities it can have other consequences if rising water tables approach the surface and cause flooding of engineered structures (basements, road and rail tunnels, underground car parks, septic tank drains). Also, an indirect effect in areas with significant soil contamination can be the enhanced downward transport of contaminants.

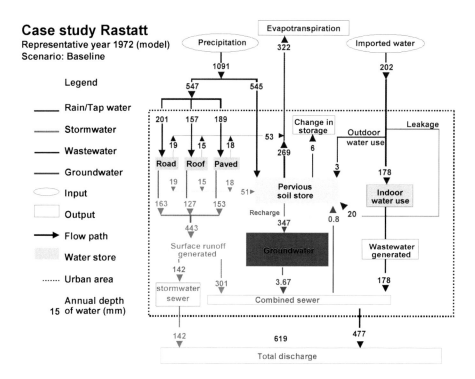

Figure 1.2.2. Example of urban water budget from the Rastatt case study.

The non-leaked water that actually reaches the customer is then used for a variety of purposes, depending on customer type. For manufacturing industry it may be used for process water, while in the service industry and public/government buildings it helps provide sanitary workplace conditions. In all districts, water may also be used for irrigation of public open spaces. Use in residential areas may be grouped into internal (kitchen, bathroom, toilet and laundry) and external uses (garden watering, car washing etc). Each of these uses adds both suspended and dissolved solids to the initial water composition. Standard rates reported in the technical literature can be used to describe the added loads per capita per day and calculate an average wastewater composition. These can if required be compared with on-site measurements.

The generated wastewater drains into a sewer system from which there are three outputs: sewer leaks, combined sewer overflows and the wastewater treatment plant. The recipient of the major part of the wastewater load is the wastewater treatment plant, where nutrients and some contaminants are removed or degraded before the release into surface waters. The wastewater that leaves the sewer through manifold leaks enters the unsaturated zone where, depending on the travel time, the degree of organic matter in the sediment, the availability of oxygen and the establishment of an adapted microbiological community, easily-attenuated contaminants are adsorbed or degraded en route to the water table. Once in the aquifer, the processes of contaminant removal and degradation slow down significantly. The third route for wastewater to leave the sewers is via CSOs. CSOs occur when the intensity or duration of a rainfall event causes a hydraulic overload of the sewer system, such that excess water volumes need to be released in a controlled or uncontrolled manner into the surface water environment.

In some countries separate foul and pluvial drainage systems operate, permitting stormwater runoff to be directed through a separate and parallel system, usually for direct discharge to the surface water drainage system. Such stormwater may also leave the piped drainage system via leaks.

Water, which has entered the aquifer, is then transported according to the local hydraulic gradient and will finally contribute either to a water supply well, or to natural system outflows such as springs or river baseflows.

1.2.4 Leaky sewers as a focus of the AISUWRS project

1.2.4.1 Relevance

Effective management of the urban water system is only feasible if the multitude of water and solute sources in the urban environment can be integrated with sufficient accuracy into the urban water balance. Leakage from defective sewers is believed to play a major role in this water and solute balance in some groundwater settings, where continuous contamination can occur from a permanently installed and widespread system by a complex mixture of substances (Barret et al., 1999; Dohmann, 1999; Eiswirth and Hötzl, 1997; Ellis, 2001; Sacher et al., 2002). The European Union standard EN 752-2 recognises this problem and therefore demands the structural integrity of urban sewer systems including their water-tightness (Keitz, 2002). Also, regulations in the Federal Republic of Germany request the water-tightness of the sewer system, a request that is impossible to fulfil in practice but towards which urban communities should strive.

The total length of the German sewer system amounted to ca. 486.000 km in the year 2001. According to the latest national German survey, based on CCTV inspections, 8.8 % (42,800 km) require immediate rehabilitation while another 10.8 % (52,500 km) require action in the short term (Berger and Lohaus, 2005). Awareness that almost 20% of the sewer network may leak under certain conditions gave rise to concern about the possible frequency and extent and consequent widespread detrimental impacts on soil and groundwater. Several major research initiatives were started at national (Dohmann et al 1999, Hagendorf 1996, Härig and Mull 1992) and international level (Rauch and Stegner, 1994, Vollertsen and Hvitved-Jacobsen, 2003, Ellis et al., 2003). Due to the complexity of the processes involved and the paucity of information about the actual sewer leakage condition, only limited attempts were done to employ physically based models on a real world scale. Existing calculations exhibit a high degree of uncertainty (typically only constrained by worst case and best case scenario) and have not been compared with measurements of the impact on groundwater quality. Direct evidence of groundwater contamination has been found only rarely and some authors have concluded that they have not been observed at all (Hagendorf 2004)

Recent events documenting the importance of sewer leakage for drinking water quality include the spreading of *Giardia lamblia* in the drinking water in Bergen, Norway during September-October 2004 (Hallström, 2005). Leaky sewers upstream of a drinking water reservoir have been identified as the most likely source for the bacteriological contamination. Rapid transport of wastewater into the lake was facilitated by the presence of coarse-grained sediments in the sewer surrounding and heavy rainfall. About 1700 people were affected by the outbreak. Examples of water quality deterioration in connection with excessive pipe leakage have been documented in the UK (Powell et al, 2003). Leaking sewers have caused public water supply contamination and associated gastric illnesses in Britain and Ireland (Misstear et al, 1996). In Germany, a likely example of sewage-derived contamination of drinking water has been reported from a waterworks near Cologne, where E.coli positives were repeatedly detected in nearby well water supply samples. The distance to the next settlement was less than 500 metres (Treskatis, 2003).

While the problem of sewer leakage is now widely known, major difficulties still exist in the quantification of the volumes exfiltrating from leaky sewers, and as a result, sometimes-contradictory statements can be found in the literature. Furthermore, estimates of the associated risk to urban groundwater resources are only sparsely available. Hagendorf (2004) concludes in his study that the sewer asset databases of all cities have to be updated with regard to ecological inspection in order to allow an appropriate evaluation of the need for rehabilitation works.

1.2.4.2 Exfiltration (sewer loss)

Härig & Mull (1992) attempted to quantify the effects of sewage exfiltration for the City of Hannover by water quality mass balances. However, a significant uncertainty remained as the substances considered (e.g. sulphate) could derive from a number of different urban contamination sources. In consideration of the multitude of sources for typical urban contaminants, Barret et al (1999) tried to use sewage specific marker species to detect the influence of wastewater on groundwater underneath the City of Nottingham, UK. Dohmann et al (1999) attempted to quantify exfiltration rates based on laboratory and field studies. However, the extrapolation to larger scales was based on very general statistical information on the structural condition of the sewer system. Extrapolations, which are based on CCTV-inspection data have been attempted by Wolf et al (2005a). An approach to validate the assumptions on sewer leakage by the measurement of marker species distributions in the urban aquifer has also been described in Wolf et al (2004).

Hagendorf (2004) concludes that emissions into the groundwater are very unlikely if clayey and silty sediments of more than 100 cm thickness are present below the sewer. He supposes that this is independent of the leak dimension. Emissions into the groundwater are likely if sand and gravel are present beneath the sewer and the distance to the water table is less than 100 cm.

1.2.4.3 Infiltration (sewer gain)

Sewers and groundwater are in close interaction in many cities worldwide. While this project has focused on the exfiltration of wastewater, sewer gain from shallow groundwater is a well-known problem for utilities as it increases the cost of wastewater treatment plant. In shallow water table situations, leaky sewer systems often serve as groundwater drains, preventing cellar flooding and related problems. Getta *et al* (2004) specifically addressed these problems for German communities and recommended a comprehensive prior assessment of the consequences of sewer network rehabilitation before construction. Numerical groundwater models that consider interaction with the sewer network are potentially powerful tools for this task.

Karpf & Krebs (2004b) have referred to the integration of leaky drainage systems into the flood protection plans for historical buildings in the area of Dresden, a problem also encountered in the Coptic quarter of Cairo where foundation subsidence and efflorescence problems have been noted (Amer *et al*, 1997).

A good correlation between the groundwater influenced pipe length and the dry weather runoff has been observed for the City of Dresden. A model based on the concept of a leakage factor has been developed which describes the sewer gainwater volumes reasonably well after model calibration. However, the structural condition of the pipes (e.g. cracks) is not taken into account and the calibration effort was quite high (more than 240 time spots used for calibration)(Karpf & Krebs, 2004b).

1.2.5 Field studies - The necessity of direct observations

1.2.5.1 Motivation

While the AISUWRS approach offers a range of models to describe the urban water cycle and to predict the future state of urban water resources, they need to be informed by direct field studies. The sampling of urban groundwater quality and the monitoring of groundwater levels is especially important. Standard information on these parameters is often already available from environmental agencies or similar public bodies but it typically does not (a) include a proper suite of marker species which are able to indicate the influence of sewer systems or septic tanks nor (b) originate from a sufficiently dense network of wells inside the city area. The case study examples described in this book should assist the choosing of an appropriate monitoring strategy for other cities and the minimisation of costs through the selection of appropriate marker species and well placements.

1.2.5.2 Consideration of the hydrogeological setting

The strategy for the urban groundwater sustainability assessment may vary significantly according to the prevailing hydrogeological setting, as one needs to know the general characteristics of groundwater flow in the aquifer system

before any sampling campaign can be designed. Four case study cities are covered and they demonstrate a range of different hydrogeological settings:
- A porous granular aquifer (Rastatt): The city is underlain by unconsolidated alluvial sand and gravel deposits in which matrix flow dominates. The aquifer is highly permeable, with some local heterogeneities. .
- A sandstone aquifer (Doncaster): The city is underlain by a sandstone formation, which contains both poorly consolidated and more cemented horizons. In the former, flow is intergranular, but in the latter water and mass transport occurs through fractures as well as through the matrix.
- A layered aquifer system (Ljubljana): Different aquifers are separated by layers of low hydraulic conductivity. Minor karstification processes result in locally very high permeabilities. Water and mass transport occurs through the matrix and karst conduits.
- A karstic aquifer system (Mt Gambier): The main aquifer body is located in a heavily karstified limestone. Water and mass transport occur rapidly through the often large karst conduits and only to a minor extent through the matrix.

1.2.5.3 Available and applied methodologies

Currently a variety of methods have been proposed at international level to investigate the urban impact on water resources. These include use of marker species, mass balances, vulnerability mapping and direct measurements to quantify leakage from pipe systems.

Marker species approaches (e.g. as documented by Barret et al, 1999) use chemicals that are typical of a certain part of the urban water system as tracers. Depending on the knowledge of the input concentrations, distribution of sources and possible attenuation mechanisms, mass balance calculations might be performed in order to quantify the respective flows to the groundwater body. Ideal marker species are those which, when present in groundwater, indicate recharge from a specific source. This requires that the species should be uniquely related to one recharge source, and be easily detectable in both the water of origin and in the groundwater receptor. Many of the currently known marker species and indicators have also been analysed in the different case study cities of the AISUWRS project and more detailed information can be found in the respective case study chapters:
- Hydrochemical field parameters like temperature and specific electrical conductance (included in all case studies)
- Major ions like potassium, sodium, chloride, sulphate, nitrate, ammonium (included in all case studies)
- Minor elements like boron (included in all case studies)
- Viruses (see Doncaster case study)
- Bacteria and phages (see Doncaster, Rastatt and Ljubljana case study)
- Pharmaceutical residues and iodated x-ray contrast media (see Rastatt case study)
- Chlorofluorocarbons and sulphur hexafluoride (see Doncaster and Mt Gambier case studies)
- Isotope compositions (see Ljubljana and Doncaster case studies)
- EDTA (see Rastatt case study)

These groundwater-quality sampling campaigns can follow different strategies. In the case of Doncaster, the vertical migration of contaminants was of major interest and a new construction method for multilevel piezometers was applied. The design of these piezometers allows the sample collection from defined depth intervals but requires only one drilling and consequently construction costs can be reduced. In Rastatt a pronounced horizontal groundwater flow leads to a strong dilution of the marker species in the aquifer and correspondingly it was necessary to sample rather close to the source. Drawing upon the complete coverage of CCTV monitoring of the sewer network, focus observation wells were specifically constructed close to leaky sewers. For layered and deeper aquifer systems like Ljubljana and where there are limited resources for the construction of new wells, vertical profiling techniques using packer systems are described. In the karstic setting of Mt Gambier, where stormwater is directly injected into the karst conduits without further treatment, it was of prime importance to monitor the stormwater quality. In this case new passive samplers were applied to monitor the episodic events.

Not for general application but for a deeper understanding of critical processes, two unique test sites were constructed in Rastatt and Ljubljana. In Rastatt the sewer test site was built to monitor a long-term time series of wastewater exfiltration from an operating sewer. The accumulated results constitute an important link between the laboratory studies described in the existing literature (Vollertsen and Hvitved-Jacobsen, 2003) and the extrapolation to real world conditions. Furthermore, double packer systems have been applied in the sewer to monitor exfiltration at a range of natural sewer defects. In Ljubljana, a special urban lysimeter was constructed beneath railroad tracks and a paved surface to monitor the infiltration of rainwater into the unsaturated zone under urban conditions.

The methods described were all chosen for their contextual suitability in assessing relative magnitude of sources of urban groundwater recharge and their effects on the quality and availability of water for public and private supply. They detail the distribution and persistence of standard sewage indicators and in some cases their seasonal fluctuations. For the AISUWRS model chain, they can provide key information for the calibration of the component models.

1.2.6 Socio-economic context and formal sustainability analysis

It is evident that any alternative urban water system is only feasible if it is planned and designed in tune with the society that pays for and benefits from the system. In order to explore these important boundary conditions several methodologies have been tried in recent years. This book demonstrates the use of a socio-economic model, which is fed by direct information from all relevant stakeholders. The stakeholders can be contacted via household surveys (example Rastatt), direct interviews of key institutions (performed in all cities) or in the case of high public awareness also with Internet questionnaires and specialised demonstration software (example Mt Gambier).

While these methods can be used to explore public awareness as well as the willingness to pay, a key element in the planning process remains the estimation of construction cost of alternative schemes. These are demonstrated in all case study cities using standard methods and regional pricing information. Although the costs of alternative measures are easy to calculate, it is much more difficult to quantify the benefit of the measures in monetary terms. A classical economic approach is the comparison between the costs for the treatment of the polluted groundwater versus the costs for the protection of the water resource. However, such comparisons neither take account of the precautionary principle nor are they supported by the legislative context in most countries. A comparison of different approaches to cost benefit analysis concerning groundwater is given in Rinaudo et al (2004).

Formal sustainability analysis is still relatively uncommon in Europe whereas regulatory standards for sustainability assessment have already been adopted in Australia. Most Australian authorities favour the so-called "triple bottom line" methodology, which addresses the three sectors of ecological, economic and social sustainability. The triple bottom line approach has been applied to a variable extent to the case studies provided in the respective chapters.

2

The Models

2.1 *Clare Diaper[1] and Victoria G. Mitchell [2]*

2.2 *Dhammika DeSilva[1], Stewart Burn[1], Magnus Moglia[1], Grace Tjandraatmadja[1], Scott Gould[1] and Paul Sadler[1]*

2.3 *Ray Correll[1], Peter Dillon[1], Ros Miller[1], Tony Miller[1], Jo Vanderzalm[1]*

2.4 *Ulf Mohrlok[3]*

2.5 *Peter Dillon[1], Majdi Mansour[4], Brian Morris[4], Christina Schrage[5], Joanne Vanderzalm[1], Goran Vizintin[6], Leif Wolf[5]*

2.6 *Stewart Burn[1], Stephen Cook[1], Stephen Meddings[1], Leif Wolf[5]*

[1] CSIRO, Melbourne, Australia
[2] Monash University, Melbourne, Australia
[3] Institute for Hydromechanics (IfH), University of Karlsruhe, Germany
[4] British Geological Survey (BGS), Wallingford, UK
[5] Department of Applied Geology (AGK), University of Karlsruhe, Germany
[6] Institute for Mining, Geotechnology and Environment (IRGO), Ljubljana, Slovenia

2.1 URBAN VOLUME AND QUALITY (UVQ)

2.1.1 Summary

Urban Volume Quality (UVQ) is a lumped, deterministic, conceptual hydrological model that simulates an integrated urban water system at a daily time step. UVQ estimates the contaminant loads and the volume of water flowing throughout the water system, from source to discharge point. The model has been developed to provide a means for rapidly assessing the impacts of conventional and non-conventional urban water supply, stormwater and wastewater development options on the total water cycle. The conceptual approach to simulating the urban water system permits significant flexibility in the manner in which water services are represented and provides the ability to mimic a wide range of conventional and emerging techniques for providing water supply, stormwater and wastewater services either to an existing urban area or to a site which is to be urbanised.

While there are several models devoted to urban water cycle modelling (Mitchell and Diaper, 2005), typical representations of the urban water cycle consider the man-made and natural systems as separate entities. Within these two systems, modelling approaches generally only concentrate on one aspect of the water cycle. UVQ integrates these networks into a single framework to provide a holistic view of the water cycle. UVQ uses simplified algorithms and conceptual routines to provide this holistic and integrated view.

Three nested spatial scales are used in UVQ to describe the components of an urban area, being the land block, the neighbourhood and the study area (Figure 2.1.1). The single allotment or plot (land block) represents a building and associated indoor and outdoor usages as well as paved and pervious areas such as paths, driveways and gardens. The proportion of these areas is specified by the user, allowing a range of building types such as flats, commercial premises and industry to be represented as well as detached dwellings (Figure 2.1.1). The neighbourhood comprises a number of identical blocks as well as roads and public open space. The study area represents the grouping of one or more neighbourhoods that may or may not have the same land use or water servicing approach. The sequence in which stormwater and wastewater flows from one neighbourhood to another can be specified by the user, providing the ability to represent how these streams actually flow through a catchment (see stormwater in the commercial neighbourhood in Figure 2.1.1).

UVQ requires configuration parameters for each spatial scale (such as occupancy, garden area, roof area and paved area in a residential land block and open space area and road area in a neighbourhood) before simulating the urban area. Rainfall data are supplied through a climate input file. As output, UVQ provides data on quantity and quality (quantified by contaminant levels) of wastewater generated by land blocks together with volumes of stormwater and irrigation water infiltrating through open spaces and gardens and of stormwater produced in the neighbourhoods.

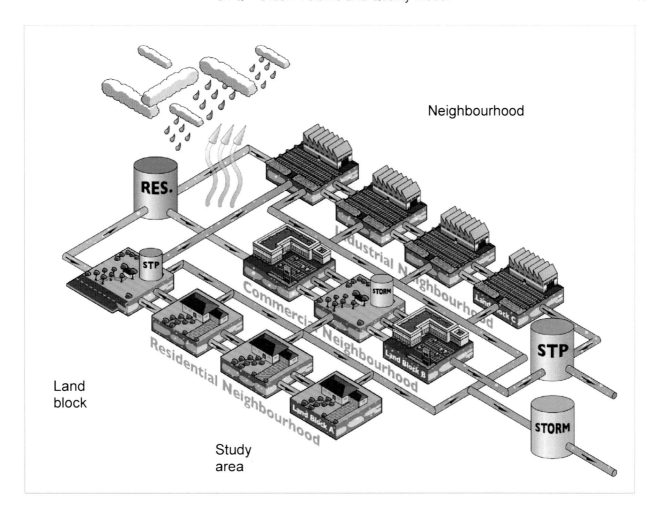

Figure 2.1.1. UVQ diagrammatic representation of the total urban water cycle in a study area.

2.1.2 Physical Principles

2.1.2.1 Governing Equations

Due to the complex interactions between water streams represented in UVQ there are many equations to describe the plethora of different water flow paths in a conventional supply system. The conventional flow paths represented by UVQ are shown in Figure 2.1.2, which also provides some of the abbreviations used in the equations. In addition, Figure 2.1.2 shows some of the alternative water management options that can be represented in UVQ, such as rainwater tanks, different scale wastewater and stormwater treatment options and aquifer storage and recovery. There are also many end-use options for these different management options, e.g. the raintank may supply garden irrigation or be used to supplement indoor water demand. However, for diagrammatic clarity these end use options for alternative sources of water are not shown.

The overall water balance for the study area can be described in terms of volumes for precipitation (and snow melt), evaporation, imported water, wastewater discharge and stormwater runoff. There are many other equations to describe processes that occur at smaller spatial scales (Diaper and Mitchell, 2005) but for clarity and demonstration only the study area input and output calculations are described here.

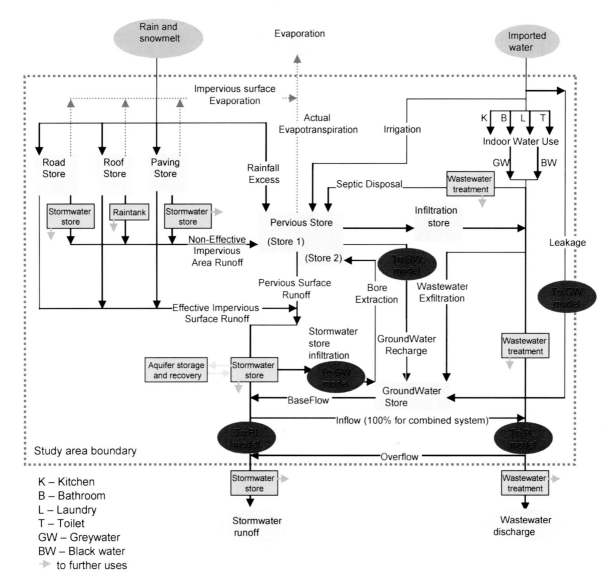

Figure 2.1.2. The UVQ flow diagram representation of the total urban water cycle.

Precipitation input to UVQ is direct from the recorded values in the climate file. Snow can accumulate on the pervious and impervious surfaces throughout the study area. The snowmelt rate is estimated according to a version of the degree-day method as follows:

$$SM = MRF \cdot (T-SMT) \tag{1}$$

Where SM is the amount of snowmelt, MRF is the user defined melt rate factor, T is the average air temperature defined in the climate input file and SMT is the user-defined snow melt threshold. A snowfall threshold temperature of 0°C is often used in hydrologic models that represent snow accumulation and ablation (Westerstrom, 1984). Melt rate factors vary depending on the condition of the snow and the local environmental conditions. The melt rate factor can range from approximately 2 to 6 mm/°C/day in sheltered forests to exposed fields, while a melt rate of 6 mm/°C/day has been reported for undisturbed clean suburban snow and 8.4 mm/°C/day for snow in a downtown park (Semadeni-Davies et al. 2000)

The evapotranspiration from a study area is calculated separately for impervious surfaces (E_{imp}) and the pervious surfaces (E_a) for each neighbourhood. The equation for E_{imp} for roof, paved and road surfaces in any one neighbourhood is given as:

$$E_{imp} = \max(E_p, RST) \cdot (\text{roof area/neighbourhood area}) + \max(E_p, PST)(\text{paved area/neighbourhood area}) + (E_p, RdST)(\text{road area/neighbourhood area}) \tag{2}$$

Where E_p is the daily potential evapotranspiration from the climate input file, RST is the roof surface storage level, PST the paved area surface storage level and RdST is the road surface storage level. In this equation E_{imp} will be calculated from the maximum value of either E_p or RST, thus calculated impervious area evapotranspiration will not exceed the maximum potential.

The approach taken to calculate evapotranspiration from the pervious stores (E_a) is based on work of Denmead and Shaw (1962), as used by Boughton (1966) to calculate actual evapotranspiration in a hydrological model. It assumes that the supply of water to a plant is a linear function of available water in the root zone (Federer, 1979). The maximum daily evapotranspiration is represented by the lesser of the daily maximum potential rate and the capacity of the vegetative cover to transpire, E_{pc}, assumed to be 7 mm/d within UVQ.

UVQ can represent the unsaturated zone of pervious areas (soil stores) in two different ways, employing either a partial area saturation overland flow (Boughton, 1990) or a two layer soil store saturation overflow conceptual representation, used in many hydrological models.

Therefore, the form of equation used to calculate E_a is dependent on the type of soil store to be modelled. The type of soil store used to represent a particular study area is user defined and will depend on known characteristics of the catchment to be modelled. Generally the two layer soil store approach is used for more permeable soils, such as sand or sandy loam, or where representation of the root zone and lower unsaturated zone is required. The partial area store approach is appropriate for less permeable soils, such as silty clay, or where the soil depths vary within the pervious areas. The final selection of soil store type should be dependent on the calibration of UVQ to observed runoff values whenever possible. Note that neither of the soil store representations accounts for soil infiltration excess, which can occur due to high intensity rainfall exceeding the capacity of the soil to infiltrate.

For any one neighbourhood using a partial area approach, E_a is given as:

$$E_a = A1 \min\{(PS1/PS1_c) \cdot E_{pc}, E_p\} + (100 - A1) \cdot \min\{(PS2/PS2_c) \cdot E_{pc}, E_p\} \tag{3}$$

Where A1 is the user defined percentage area of soil store 1 and $PS1/PS1_c$ and $PS2/PS2_c$ are the ratios of the storage level in the pervious soil stores 1 and 2 to user defined total capacity and E_{pc} and E_p are as above.

For a neighbourhood using the two-layer soil store approach, E_a is calculated separately for each of the two soil stores and is the sum of the upper and lower store actual evapotranspiration. Evaporation from the top store occurs in preference to the lower store first and is calculated as either:

$$\text{IF } (0.75 \cdot LS1_{max}) < LS1 < LS1_{max} \quad E_{a1} = \min(E_p, E_{pc}, LS1) \tag{4}$$

or

$$E_{a1} = \min\{(LS1/(0.75 \cdot LS1_{max})) E_{pc}, E_p\} \tag{5}$$

For the lower soil store the evaporation is calculated as either:

$$\text{IF } (0.75 \cdot LS2_{max}) < LS2 < LS2_{max} \quad E_{a2} = \min(E_p - E_{a1}, E_{pc}, LS2) \tag{6}$$

or

$$E_{a2} = \min\{(LS2/(0.75 \cdot LS2_{max})) \cdot E_{pc}, E_p - E_{a1}\} \tag{7}$$

Imported water to the study area (I) is calculated as the sum of the user defined indoor water uses (IWU) Garden irrigation (IR) and potable pipe leakage (LD) less the amount of water supplied by groundwater, rainwater tanks, greywater reuse, stormwater harvesting or wastewater recycling (ALT) as follows:

$$I = IWU + IR + LD - ALT \tag{8}$$

Two representations of the stormwater system can be modelled in UVQ, separate and combined drainage systems. The separate system in which all stormwater is channelled through a dedicated collection system is used in virtually all urban areas of Australia and some areas of the UK and USA, whereas the combined stormwater/sewage system is more predominant in Europe. For the separate stormwater system stormwater flow is calculated as:

$$SW = IRUN + SRUN + BF - ISI + OF - RW - SW \tag{9}$$

Where IRUN is calculated effective impervious surface runoff, SRUN is pervious surface runoff, BF is baseflow, ISI is the amount of water that flows into the wastewater system, OF is overflow of the wastewater system into the stormwater system due to cracks or breaks in dry weather or exceeding collection network capacity in wet weather, RW is the amount of rainwater used and SW is the amount of stormwater used.

In a combined stormwater/sewage system, stormwater flows are represented by the baseflow only and all surface stormwater infiltrates into the wastewater system (ISI = 100%{IRUN + SRUN}). The general equation for wastewater flows is:

$$WW = IWU + INF + ISI - EXF - OF - GrWR - WWR \qquad (10)$$

Where IWU is the indoor water usage, INF is the infiltration of water from the soil infiltration store into the wastewater network, ISI is the amount of stormwater that flows info the wastewater system (a percentage of the surface runoff), EXF is the amount of water leaking out or exfiltrating from the wastewater network, OF is as above, GrWR is the amount of greywater reuse and WWR is the amount of wastewater recycling.

Contaminant calculations are carried out once the daily water balance is complete. All contaminants are modelled conservatively and the equations describing them are all simple addition. The conservative nature of the modelling means care should be taken in the selection of contaminants for modelling and in the interpretation of results. Detail of the concentrations or loads of a number of input streams is required (Table 2.1.1). In addition, % removals of contaminants (%RemEff) of alternative water servicing approaches such as raintanks, on-site wastewater treatment systems, neighbourhood and study area scale stormwater and wastewater, are also required.

Table 2.1.1. Specified contaminant stream and their units.

Contaminant stream	Units	Contaminant stream	Units
Bathroom	mg/c/d	Roof runoff	mg/l
Kitchen	mg/c/d	Road runoff	mg/l
Toilet	mg/c/d	Fertiliser to POS	mg/l
Laundry	mg/c/d	Evaporation	mg/l
Imported water	mg/l	Ground water	mg/l
Rainfall	mg/l	Roof first flush	mg/l
Pavement runoff	mg/l		

All treatment processes are modelled as continuously stirred tank reactors (CSTRs) in which sludge accumulates and a certain volume is retained at the end of each daily timestep. An example of a treatment process modelled in UVQ, the on-site wastewater treatment system, is given in Figure 2.1.3 and a description of the sludge calculations for any given contaminant are given in Equations 11-13, where Equations 11 and 12 are load based calculations (Subscript L) and Equation 13 converts this load into a concentration (Subscript C). Generic calculations for any treatment process are also given. Treatment process calculations occur on a daily basis and the retained volume and contaminants from the previous day are the starting volume and contaminants for the current day. The retained volume and contaminants reported in results screens are for the final day only.

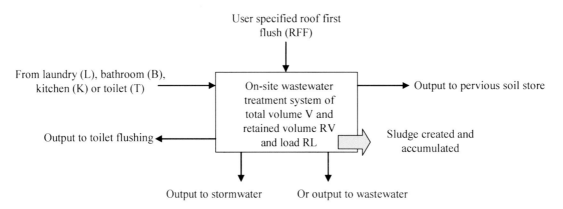

Figure 2.1.3. Representation of On-site wastewater treatment system.

$$OnWWTmix_L = RFF_L + L_L + B_L + K_L + T_L + RL_{n-1} \qquad \text{or } Mix_L = \Sigma\,(Inputs_L) + RL_{n-1} \qquad (11)$$

$$OnWWToutput_L = OnWWTmix_L \cdot (1 - \%RemEff) \qquad \text{or } Out_L = Mix_L(1-\%RemEff) \qquad (12)$$

$$OnWWToutput_C = OnWWToutput_L/(Output_V + RV_n) \qquad \text{or } Out_c = Out_L/\{\Sigma(Output_V) + RV_n\} \qquad (13)$$

Where the subscript n denotes the current day and $Output_V + RV_n$ are total volume of output streams and the current days retained volume calculated in the water balance. In addition to sludge calculations in treatment processes, sludge operations are used to calculate differences in user specified rainfall and runoff data and provide the user with a value for the assumed load to pavements, roofs and roads.

2.1.2.2 Model Structure

UVQ source code consists of two discrete modelling loops, the water balance and the contaminant balance. The water balance program loop calculates the flows through the urban water system on a daily basis. A call to the contaminant balance code at the end of this program loop utilizes the pre-calculated daily water flows to calculate the volumes and concentrations of contaminants through the system (see Figure 2.1.4). The contaminant balance was implemented by adding a code module to the code already written for the water balance (Mitchell, 2005; Mitchell and Maheepala, 1999). The water balance and contaminant balance operations occur sequentially for each daily time step. The water balance calculations are based on the concept of the urban volume and the fundamental unit of operation is depth in mm. The contaminant balance operations are based on the water volumes calculated in the water balance and user specified concentrations, loads and performance criteria (Figure 2.1.4).

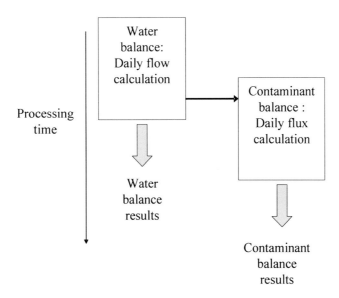

Figure 2.1.4. UVQ water balance and contaminant balance interaction.

Within both the water and contaminant balance-processing calculations occur in a specific order, primarily based on spatial scale. Generally, unit block scale operations are undertaken first, then neighbourhood scale operations and finally the study area calculations. This is illustrated in Figure 2.1.5.

2.1.2.3 Time Discretisation and Boundary Conditions

A daily computational time step is used for all processes represented in UVQ. Many of these processes, such as a rainfall-runoff or a toilet flush event, can take a matter of seconds to minutes and can occur many times in a given day. However, the computational time and input data requirements and modelling assumptions required to represent all of the urban water cycle processes at a time resolution finer than one day was considered to compromise the usefulness of the software as a what-if scenario analysis tool. For example, a time resolution of the order of 1 to 30 minutes would be required in order to represent peak flow rates in the water supply, but this time step would increase the simulation time by 48 to 1440 times, requiring a quantum increase in the complexity of the software. As UVQ was developed as a "conceptual hydrology" tool to examine the effects of alternative water management scenarios on water and contaminants flows in urban areas, the increased complexity and input data requirements of using less than daily time-steps was not considered to improve either the accuracy or the functionality of the model.

UVQ can be used to conduct daily simulations from a minimum of 1 to a current maximum of 80 years. In order to gain an understanding of the importance of seasonal and annual variations in climate, which impact the urban water budget, simulation periods of multiple years to decades are recommended.

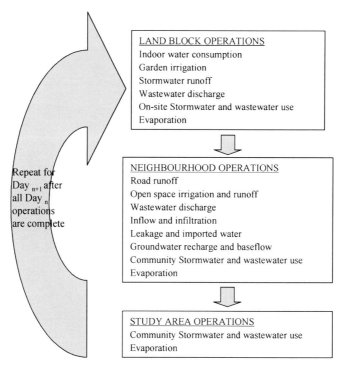

Figure 2.1.5. The structure of UVQ water and contaminant operations.

2.1.3 Known Limitations/Capabilities

A number of assumptions are used in UVQ for the representation of different processes:

2.1.3.1 Snow processes

- Precipitation falls either as all snow or all rain on any given day, depending on the average daily temperature and the user specified snowfall threshold temperature
- There is no variation in snowfall threshold temperature and melt rate factor due to variations in elevation within the study area. The effect of elevation variations is assumed to be minimal
- There is no variation in melt rate factor due to season, snow condition or snow density
- The melt rate factor represents the water depth equivalent amount of snow
- Snow automatically accumulates in garden and open space surfaces. The user can specify whether there is accumulation on paved, roof and road surfaces.
- Rainfall passes straight through the snow pack onto the surface below

2.1.3.2 Evaporation from surfaces

- The effect of wind turbulence due to increased surface roughness, sheltering by buildings, and other microclimate variations due to urbanisation, does not have a significant impact on the accuracy of the method used to calculate actual evapotranspiration from pervious areas and evaporation from impervious areas. There is little known about the actual difference between urban and non-urban evapotranspiration.
- Actual evapotranspiration of pervious areas varies depending on the soil moisture storage at the beginning of the day, and the evaporative demand estimated by potential evapotranspiration as supplied in the climate input file. This accords with the approach of Boughton (1966) describing a simplified Denmead and Shaw (1962) relationship
- The presence of a layer of snow covering a particular surface (garden, public open space, roof, road, paved) does not alter the calculation of actual evapotranspiration from these surface stores
- The maximum rate of evaporation from impervious surfaces is assumed as the potential evapotranspiration, which is supplied in the climate input file. No allowance is made for the effect of heating of impervious surfaces on the actual evaporation rate. Evaporation is removed from the impervious surface store at the end of the day (effectively after the rain event).
- The concentration of evaporated contaminants is assumed to be the same from all surfaces. Evaporation of all contaminants can be set to zero. Contaminants can evaporate from surface stores, all impervious surfaces and from subsurface stores of pervious surfaces.

2.1.3.3 Combined sewer systems

- Each neighbourhood can have either a separate or combined sewer system
- In a neighbourhood with a combined sewer system, all of the surface runoff generated from impervious surfaces in that neighbourhood (which has not been intercepted and utilised by rainwater tanks or stormwater stores) is directed into the wastewater system. The parameter percentage surface runoff as inflow should be set to 100%.
- Base flow from the groundwater store flows into the stormwater system, regardless of whether a separate or combined sewer system is selected in a neighbourhood.
- Stormwater flowing into a neighbourhood from an upstream neighbourhood stays in the stormwater system (this can be used to represent streams and creeks flowing through a neighbourhood).
- Overflows from a combined sewer system are directed into the neighbourhood's stormwater system

2.1.3.4 Groundwater store

- The groundwater store is assumed to be an unconfined aquifer.
- Groundwater recharge spreads uniformly over the entire groundwater store below a neighbourhood; transmissivity is assumed to be infinite. This assumption has little effect on model accuracy unless there is a large amount of water recharging at a fixed point within the modelled area. Any impact on base flow estimation is not significant and does not warrant more sophisticated modelling of the groundwater store.
- There is no deep seepage or lateral flow from the groundwater store. The groundwater store is an infinite source of water and the only discharge from the store is through base flow and/or extraction by a bore.

2.1.3.5 Impervious surfaces

- All roof, paved and road areas are 100% impervious.
- The maximum initial loss from an impervious surface and the effective impervious area are assumed to be constant throughout a rain event and for all seasons during a year.
- The runoff from unconnected impervious areas is assumed to spread evenly across the entire adjacent pervious area (therefore being added to both pervious stores in equal areal depths). Roof and paved area runoff spills on to the pervious area within the same land block. Any road runoff from unconnected areas (non-effective area) spills on to all pervious area within the neighbourhood. In actuality, the runoff would spill on to the edge of the adjacent pervious area and cause an increase in the moisture content of a small marginal area.
- If there is no pervious area adjacent to an impervious area, then the effective impervious area is 100%. All of the impervious surface must be directly connected to the stormwater system since there are no adjacent surfaces for the runoff to spill on to.

2.1.3.6 Contaminants from impervious surfaces

- Contaminant concentrations in the runoff from the garden and public open space are calculated separately from respective input loads
- Contaminant concentrations in the runoff from the pavement to the garden or stormwater are identical
- No user specified contaminant load is added to stormwater from impervious surfaces as the availability of this data is often limited. However, the model calculates the difference between rainfall and stormwater runoff EMCs (event mean concentrations) to provide users with an indicator of this load.

2.1.3.7 Pervious soil store

- All public open spaces are 100% pervious
- The input and output of water occurs in a set order each day. Precipitation is added to and actual evaporation is removed from the soil moisture stores simultaneously at the beginning of the day. The irrigation demand is calculated and is applied at the end of the day.
- Precipitation and irrigation wet the entire root zone to a constant depth. This assumes the moisture is instantaneously distributed throughout the root zone when, in reality, a wetting front forms and the soil is slow to reach a constant soil moisture content throughout.
- Surface ponding and overland flow do not occur until the soil moisture storage capacity of the store is exceeded. This may over-estimate the ability of precipitation and irrigation to wet the soil profile and underestimate runoff in intense rainfall events when infiltration capacity of the soil profile is exceeded.
- There is no lateral movement of moisture in the soil profile. Therefore, there is no transfer of moisture between the soil and groundwater stores in different neighbourhoods.
- All soil below impervious surfaces is regarded as dry.
- If there is no garden on the land block, there can be no leach field associated with a septic tank.

2.1.3.8 Partial area soil store

Assumptions specific to the partial area soil store approach are:
- There is no transfer of moisture between the two pervious stores.
- Any moisture in excess of the capacity of either of the two partial area soil stores overflows the store and is separated into surface runoff, groundwater recharge, and infiltration into the wastewater system according to user-defined calibration parameters.
- The septic tank system leach field drains into both soil stores.

2.1.3.9 Two layer soil store

Assumptions specific to the 2-layer soil store approach are:
- Any water entering the upper soil store, in excess of capacity, becomes runoff.
- Irrigation is applied to the upper soil store only.
- Drainage of the soil stores is modelled as a simple decay function
- The septic tank system leach field and the soakaway or spoon drain drain into the lower soil store.
- Infiltration is a constant proportion of the drainage from the lower soil store.

2.1.3.10 Irrigation

- The model assumes irrigation to be fully effective in recharging the soil moisture stores to the prescribed level with no wastage. In reality part of the water applied to a garden or open space will be wasted as some will evaporate before soaking into the soil, depending on the timing of irrigation and the method used, and some may run off.
- All outdoor water use is due to irrigation of either gardens or public open space.

2.1.3.11 Treatment processes

- All treatment processes are modelled as continuously stirred tank reactors (CSTRs) and contaminant removal is described as a percentage
- Sludge accumulates in the treatment process
- Treatment process calculations occur on a daily basis and the retained volume and contaminants from the previous day are the starting volume and contaminants for the current day. The retained volume and contaminants reported in results screens are for the final day only

2.1.3.12 Wastewater exfiltration and overflow processes

- Exfiltration from the wastewater network is a constant proportion of the generated wastewater flow and has the same contaminant concentration.
- Wastewater overflow comprises two components; dry weather overflow and wet weather overflow. Dry weather overflow is a constant proportion of generated wastewater flow up to capacity flow levels. Wet weather flow is all generated wastewater flow in excess of the system capacity.

2.1.3.13 Wetting and drying of pervious and impervious surfaces

- Only one wetting and drying cycle occurs within a day. In reality, there may be multiple wetting and drying cycles, due to multiple rain events occurring within the day
- Precipitation is spread evenly over the entire area with no variation due to wind turbulence and localised storms.

2.1.3.14 Water supply sources

- If there is more than one source selected to supply a particular demand (e.g. both rain tank and on-site wastewater treatment unit) then there is a set order in which these sources will be used to meet that demand. The rules used to determine the priorities for each demand are as follows:
(1) First use the lowest quality water source available, which meets the requirements of the demand.
(2) Supply indoor water demands before outdoor demands.
(3) Use the water sources within the land block before neighbourhood sources.
(4) Use neighbourhood scale water sources before study area scale water sources.
(5) Use all local sources of water before importing water (reticulated water).

If a particular potential source of water has not been selected by the user, then the next highest priority source is used instead.

2.1.4 Running the Model

The installation of UVQ if used as a stand-alone model is performed through the use of a setup application, by directly running the SETUP.exe file provided. During setup two sub-directories called INPUT and RESULTS will be generated in the folder selected for installation. The installation will also provide UVQ modelling software as an executable file and associated files required for graphical presentation of results and operating system needs. The software can be launched by running the UVQUI executable file.

Data is input to UVQ in two ways. The climate file, containing daily rainfall, potential evaporation and temperature data, needs to be in the form as described in the MODEL INPUT section. In this format, the climate file for the modelling period can be loaded directly into the modelling software. All other data are input through a series of interface screens; project information, physical characteristics, calibration variables, land block information, neighbourhood information, study area information, snow variables and water flow. Details of the data required for these input screens are given in the MODEL INPUT section. When the climate and other project data are saved, a UVQ project file is created.

The user interface allows the selection of contaminants to be modelled, with the choice of 12 set and 3 user-defined contaminants. The selection of the representation of the soil store is also made via the user interface, where a partial area or a two-layer approach can be chosen.

2.1.4.1 Model Input

The climate file used for modelling in UVQ contains the information as described in Table 2.1.2 in a Microsoft Excel comma-separated variable format. The column headers in this file aid in the description of the data and ensure information on rainfall and evaporation sources is always associated with the data.

Table 2.1.2. Climate input file format and configuration.

	Date	Rainfall	Evaporation	Temperature
Format	YYYYMMDD	mm	mm	$\pm °C$
Example	20060421	6.6	4.0	-8
Column in .csv file	Column A	Column B	Column C	Column D
Column header description	Start date	Rain gauge site and evaporation type and site	End date	SNOW
Column header example	19600101	Rain 40214 infilled unfactored pan evap 40214 infilled	19851231	SNOW

All other data input to UVQ is via the seven user interfaces screens. The information required in these screens, the units used and a description of each variable is given in Table 2.1.3 to 9. Table 2.1.3 to 5 describe data that are required to simulate a standard urban water system. Table 2.1.7 to 9 describe data required for simulating alternatives to the traditional urban water system, for example, rain tanks, on-site treatment systems, greywater use and neighbourhood stormwater or wastewater collection, treatment and use.

Table 2.1.3. Project information screen data input requirements.

Field Data Description	Format	Units	Description
Project Description	Text only	-	Brief description of the project. Max. 32 characters. This is usually the name of the area being modelled.
Study Area Size	Positive fraction	ha	The total area of the site to be modelled
Number of Neighbourhoods	Positive integer	1 – 75	Total number of neighbourhoods representing the study area. Maximum number is 75.
Soil Store Type	Partial area/two layer option	-	Selection of soil store type to be used in modelling calculations
Common Contaminants for Analysis in this Study Area, Neighbourhoods and Land Blocks*	User selection	-	This is a list of common contaminants in the urban water cycle. The contaminants to be represented within the study area are selected here.
Additional Optional User Defined Contaminants	User selection	-	Define any specific contaminants to model

*The elements B (Boron), K (Potassium), Na (Sodium), Cl (Chlorine), N (Nitrogen) and P (Phosphorous) are available for selection. In addition the following water quality parameters are also available; SS (Suspended solids), SO_4^{2-} (Sulphate), TOC (Total organic carbon) as are three microbial species and three user-defined contaminants.

Table 2.1.4. Physical characteristics screen data input requirements.

Field Data Description	Format	Units	Description
Total Area	Positive fraction	ha	The total area of a neighbourhood. The sum of the open space areas, road areas and the individual properties within the neighbourhood.
Road Area	Positive fraction	ha	The number of hectares of roads within a neighbourhood Note: road area is the sum of the roads and the footpaths.
Open Space Area	Positive fraction	ha	The number of hectares of open space such as parks and wildlife corridors within a neighbourhood
Percentage of Open Space Irrigated	0 – 100	%	The percentage of open space irrigated.
Imported Supply Leakage	0 – 100	%	The percentage of imported water leaking into the groundwater through broken and cracked pipes.
Wastewater as Exfiltration	0 – 1	Ratio	The ratio of wastewater exfiltrating (leaking) from pipes to total wastewater produced
Number of Land Blocks	Positive integer	-	Total number of identical land blocks within a neighbourhood
Block Area		m^2	Average size of land blocks in a neighbourhood
Average Occupancy	Positive fraction		Average number of people per land block. UVQ accepts whole and decimal numbers
Garden Area	Positive fraction	m^2	The average garden area per land block
Roof Area	Positive fraction	m^2	The average roof area including sheds and garages per land block
Paved Area	Positive fraction	m^2	The average paved area per land block
Percentage of Garden Irrigated	0 – 100	%	The percentage of the garden that is irrigated
Roof Runoff to Spoon drain or Soakaway	0 – 1	Ratio	The ratio of runoff draining to an on-site soakaway or spoon drain to the total roof runoff
Bathroom, Toilet, Kitchen and Laundry	Positive fraction	L/capita/day	The number of litres of water used in the bathroom, toilet, kitchen and laundry per person per day
Bathroom, Toilet, Kitchen and Laundry Contaminant Loads	Positive fraction	mg/capita/day	The contaminant load inputs from the bathroom, toilet, kitchen and laundry within a land block per person per day.
Contaminants Concentrations	Positive fraction	mg/l	The contaminant concentrations in imported water, rainfall, roof, pavement and road runoff, roof first flush, evaporation and groundwater
Contaminant Loads	Positive fraction	mg/ha/day or $mg/m^2/day$	Fertiliser loads to open space and garden

Table 2.1.5. Potable water and observed data calibration input requirements.

Potable water			
Garden and Open Space Trigger to Irrigate	0 – 1	Ratio	Represents the level of soil moisture that the garden or open space irrigator wishes to maintain. If the soil water storage level in the irrigated area drops below this trigger level then irrigation water is requested from the various sources available to it.
Observed data			
Observed Imported, Wastewater and Stormwater	Positive fraction	kL/y or ML/y	This volumetric parameter should be obtained from actual site data.
Wastewater/ Stormwater Observed Contaminants	Positive fraction	mg/L or kg/y	This flux parameter should be obtained from actual site data.

Table 2.1.6. Stormwater and wastewater data calibration input requirements

Field Data Description	Format	Units	Description
Stormwater			
Percentage Area of Soil Store 1 (%)	0 – 100	%	The proportion of the pervious area (garden and open space) in the neighbourhood that is covered by Soil Store 1 in a partial area soil store.
Capacity of Soil Store 1	Positive fraction	mm	The maximum depth of water partial area soil store 1 can store.
Capacity of Soil Store 2	Positive fraction	mm	The maximum depth of water partial area soil store 2 can store
Maximum Soil Storage Capacity	Positive fraction	mm	The maximum depth of water the upper and lower soil stores can hold in a two layer soil store,
Soil Store Field Capacity	Positive fraction	mm	The level to which the water in the upper and lower soil store in a two-layer soil store freely drains due to the action of gravity.
Maximum Daily Drainage Depth	Positive fraction	mm	The maximum depth of water which will drain from the upper and lower soil store in a two layer soil store in a day due to the action of gravity.
Roof Area Maximum Initial Loss	Positive fraction	mm	The amount of water it takes to wet the roof surface before runoff occurs.
Effective Roof Area	0 – 100	%	The proportion of roof area directly connected to the roof drainage system.
Paved Area Maximum Initial Loss	Positive fraction	mm	The equivalent depth of water it takes to wet the paved surface before runoff occurs.
Effective Paved Area	0 – 100	%	The proportion of paved area directly connected to the land block stormwater system.
Road Area Maximum Initial Loss	Positive fraction	mm	The equivalent depth of water it takes to wet the road surface before runoff occurs.
Effective Road Area	0 – 100	%	The proportion of road area directly connected to the neighbourhood stormwater system.
Base Flow Index (partial area stores only)	0 – 1	Ratio	The ratio of excess water recharging groundwater from partial area soil stores to total excess water from stores
Drainage Factor Ratio (two layer stores only)	0 – 1	Ratio	The ratio of the amount of drainage from the upper and lower two layer soil stores to the soil store water storage level over and above the field capacity. Controls upper and lower soil store drainage rate
Base Flow Recession Constant	0 – 1	Ratio	The ratio of baseflow to groundwater storage level. This controls the rate at which water leaves the groundwater store and contributes to the stormwater flowing out of the neighbourhood.
Contaminant Soil Store Removal	0 – 100	%	The percentage of the user-specified contaminants removed from the water as it drains through the soil stores.
Wastewater			
Infiltration Index	0 – 1	Ratio	The ratio of drainage or excess water routed to the temporary infiltration store to the total drainage or excess water from stores. Controls the flow from soil stores to the temporary infiltration store.
Infiltration Store Recession Constant	0 – 1	Ratio	The ratio of infiltration to wastewater pipes to the square root of the level in the temporary infiltration store. Controls rate of

Field Data Description	Format	Units	Description
			water flow into wastewater pipes from the temporary infiltration store.
Percentage Surface Runoff as Inflow	0 – 100	%	Proportion of surface runoff generated in the neighbourhood which flows into the wastewater pipe system rather than the stormwater system. Set to 100% for combined sewers
Dry Weather Overflow Rate	0 – 100	%	The proportion of wastewater, which overflows from the wastewater system to the stormwater system in dry weather due to pipe breaks or blockages.
Wet Weather Overflow Trigger	Positive fraction	kL	The maximum amount of wastewater the neighbourhood wastewater system can convey each day. Wastewater in excess of this capacity becomes overflow to stormwater.

Table 2.1.7. Land block Water Management screen data input requirements.

Field Data Description	Format	Units	Description
Rain Tank			
Storage Capacity	Positive fraction	kL	The maximum volume of water that an individual rainwater tank within each land block can hold. All of this volume is available for use.
Initial Storage Level	Positive fraction	kL	The amount of water that is already held in the rainwater tank on the first day of the simulation run.
First flush	Positive fraction	kL	The volume of roof runoff that is diverted away from the rainwater tank at the beginning of a rainfall-runoff event.
Supplies to: Bathroom, Laundry, Kitchen, Toilet, Garden	User selection	-	Land block water uses that the rainwater tank can supply
First flush to: Stormwater, Garden, On Site WW	User selection	-	Disposal options for the rain tank first flush. Standard practice in this regard varies for different countries
Storage Backup Trigger Level	0 – 1	Ratio	Ratio of tank water storage: tank capacity below which backup water is requested from stormwater or wastewater stores
Contaminant Removal Efficiency	0 – 100	%	The percentage of the user specified contaminants removed in the rain tank.
Greywater			
Sub Surface Greywater Irrigation Collected from: Bathroom, laundry or kitchen	User selection	-	Sources of greywater that are used for sub-surface garden irrigation. One or more can be selected.
On Site Wastewater Unit			
Storage Capacity	Positive fraction	kL	The maximum volume of treated wastewater that the onsite treatment unit can hold. All of this volume is available for use.
Exposed Surface	Positive fraction	m^2	The surface area of the on-site wastewater treatment unit, which is open to the elements rather than covered.
Initial Storage Level	Positive fraction	kL	The amount of wastewater that is already held in the on-site wastewater treatment unit store on the first day of the simulation run.
Treat the Following Sources; Bathroom, Laundry, Kitchen, Toilet	User selection	-	The wastewater sources which are directed to the on-site wastewater treatment unit. One or more can be selected.
Supplies to: Toilet, Garden	User selection	-	Land block water uses that the on site wastewater unit can supply
Excess Drain to Stormwater, Sewer, Leach field	User selection	-	Disposal options for excess treated wastewater. Selecting the leach field allows the user to represent the behaviour of a septic tank.
Contaminant Removal Efficiency	0 – 100	%	The percentage of the user-specified contaminants removed in the on site wastewater unit
Imported Water Supplies Garden Irrigation	User selection	-	

Table 2.1.8. Neighbourhood Water Management screen data input requirements.

Field Data Description	Format	Units	Description
Stormwater store			
Storage Capacity	Positive fraction	kL	The maximum volume of water that the stormwater store can hold. All of this volume is available for use.
Exposed Surface	Positive fraction	m^2	The surface area of the stormwater store which is open to the elements rather than covered.
First Flush	Positive fraction	kL	An initial volume of stormwater that is diverted away from the stormwater store.
Initial Storage Level	Positive fraction	kL	The amount of stormwater that is already held in the store on the first day of the simulation run.
Act as Infiltration Basin	User selection	-	When this option is selected, the stormwater store acts as an infiltration basin, with the floor of the store being pervious rather than impervious.
Contaminant Removal Efficiency	0 – 100	%	The percentage of the user-specified contaminants removed in the neighbourhood stormwater store.
Stormwater Sources: Road Runoff, Landblock Runoff, Upstream Neighbourhoods, Open Space Runoff	User selection	-	The stormwater sources which are directed to the neighbourhood stormwater store. One or more can be selected.
Stormwater Store Supplies to: Garden or Open Space Irrigation, Toilet	User selection	-	Land block and neighbourhood uses that the stormwater store can supply. One or more can be selected.
Aquifer Storage and Recovery			
Storage Capacity	Positive fraction	kL	The maximum volume of water held in the underground aquifer store. All of this volume is available for use.
Storage Level	Positive fraction	kL	The amount of water that is already held in the aquifer store on the first day of the simulation run.
Max Recharge Rate	Positive fraction	kL/d	The maximum volume of water that can be pumped into (injected into) the aquifer store each day.
Max Recovery Rate	Positive fraction	kL/d	The maximum volume of water that can be pumped from (recovered from) the aquifer store each day.
Wastewater store			
Storage Capacity	Positive fraction	kL	The maximum volume of water that the wastewater store can hold. All of this volume is available for use.
Exposed Surface	Positive fraction	m^2	The surface area of the wastewater store which is open to the elements rather than covered.
Initial Storage Level	Positive fraction	kL	The amount of wastewater that is already held in the store on the first day of the simulation run.
Collected Wastewater from Land Blocks in this Neighbourhood and/or from Upstream Neighbourhoods	User selection	-	Selection of land block and neighbourhood wastewater sources, which are directed to the neighbourhood wastewater store. One or more can be selected
Overflow to Stormwater/ Wastewater	User selection	-	Selection of overflow route from the wastewater store
Contaminant Removal Efficiency	0 – 100	%	The percentage of the user-specified contaminants removed in the neighbourhood wastewater store.
Wastewater Store Supplies to: Garden Irrigation, Open Space Irrigation, Toilet	User selection	-	Land block and neighbourhood uses that the wastewater store can supply. One or more can be selected.
Groundwater Store			
Initial Storage Level	Positive fraction	kL	The amount of water that is already held in the groundwater store on the first day of the simulation run.
Supplies to: Garden or Open Space Irrigation	User selection	-	Groundwater can be used to provide garden and/or open space irrigation within the neighbourhood

Table 2.1.9: Study Area Water Management screen data input requirements.

Field Data Description	Format	Units	Description
Wastewater store			
Storage Capacity	Positive fraction	kL	The maximum volume of water that the study area wastewater store can hold. All of this volume is available for use.
Exposed Surface	Positive fraction	m^2	The surface area of the study area wastewater store which is open to the elements rather than covered.
Initial Storage Level	Positive fraction	kL	The amount of wastewater, which is already held in the store on the first day of the simulation run.
Overflow to Stormwater /Sewer	User selection	-	Selection of the overflow route from the study area wastewater store
Supplies Selected Neighbourhoods: Garden or Open Space Irrigation, Toilet	User selection		Land block and neighbourhood uses that the wastewater store can supply. One or more can be selected.
Contaminant Removal Efficiency	0 – 100	%	The percentage of the user specified contaminants removed in the study area wastewater store.
Stormwater Store			
Storage Capacity	Positive fraction	kL	The maximum volume of water that the study area stormwater store can hold. All of this volume is available for use.
Exposed Surface	Positive fraction	m^2	The surface area of the study area stormwater store which is open to the elements rather than covered.
First Flush	Positive fraction	kL	The initial volume of stormwater that is diverted away from the study area stormwater store.
Initial Storage Level	Positive fraction	kL	The amount of stormwater that is already held in the store on the first day of the simulation run.
Supplies Selected Neighbourhoods: Garden or Open Space Irrigation, Toilet	User selection	-	Landblock and neighbourhood uses that the stormwater store can supply. One or more can be selected.
Contaminant Removal Efficiency	0 – 100	%	The percentage of the user specified contaminants removed in the study area stormwater store.

The final user input screen is used to route stormwater and wastewater flows between neighbourhoods.

2.1.4.2 Model Output

UVQ produces a number of output files in addition to results accessible through the user interface. Output files detailing results of modelled water flows are produced for daily, monthly and yearly time periods at the land block, neighbourhood and study area spatial scales. Contaminant balance output files are produced for the total modelled time period at land block, neighbourhood and study area spatial scales. Both water and contaminant balance output files are in comma-separated variable file format. Output files for input to other models in the AISUWRS modelling suite are also produced, and provide data on modelled water flows and contaminants to pipe networks and to groundwater. These AISUWRS output files are in TEXT format so they are compatible with the requirements of the DSS. The generation of all output files is by user selection. These output files are described in Table 2.1.10.

All other UVQ output is accessible through a series of results screens in the user interface and can be viewed in both tabular and time series graphical form (Table 2.1.11). These results screens provide the user with specific modelling results for the overall study area and for alternative water management options. There is also a capability for the user to define results to be presented in a variety of ways; histograms, pie charts and flow, load and concentration time series. This allows the user the flexibility to produce results in their required format. Descriptions of the results screen output files, their units and applicability are also given in Table 2.1.10.

Table 2.1.10. UVQ output file names and description.

Output file name	Flow Units	Contaminant Units	Description
AISUWRS			
UFMGardenToGW.txt	kL	g	Water flows and contaminant loads from the garden to the groundwater store for each neighbourhood
UFMPOStoGW.txt	kL	g	Water flows and contaminant loads from the open space to the groundwater store for each neighbourhood
UFMSWInfiltrationBasintoGW.txt	kL	g	Water flows and contaminant loads from the stormwater store when it acts as an infiltration basin for each neighbourhood
UFMTapToGW.txt	kL	g	Water flows and contaminant loads from potable pipe leakage to the groundwater store
PlmUVQSWinput.txt	m^3/land block connection	g	Water flows per connection and contaminant loads in the stormwater network for each neighbourhood
PlmUVQWWinput.txt	m^3/land block connection	g	Water flows per connection and contaminant loads in the wastewater network for each neighbourhood
Water balance			
StudyAreaBalance.csv	mm or m^3/d	-	Daily flows or depths relating to water flow paths at the study area scale
DailyNeighbourhood*.csv	mm or m^3/d	-	Daily flows or depths relating to water flow paths at the neighbourhood scale
DailyLandBlock*.csv	mm or m^3/d or m^3/household/d	-	Daily flows or depths relating to water flow paths at the land block scale
MthlyStudyArea.csv	mm/month or m^3/month	-	Monthly flows or depths relating to water flow paths at the study area scale
MthlyNBH*.csv	mm/month or m^3/month	-	Monthly flows or depths relating to water flow paths at the neighbourhood scale
YearStudyArea.csv	mm/year or m^3/year	-	Yearly flows or depths relating to water flow paths at the study area scale
YearNBH*.csv	mm/year or m^3/year	-	Yearly flows or depths relating to water flow paths at the neighbourhood scale
Contaminant balance			
Cont Bal – Neighbourhood*.csv	kL	mg	Water and contaminant balance for input and output paths at the land block and neighbourhood scale for the entire modelling time period
Cont Bal – Study area.csv	kL	mg	Water and contaminant balance for input and output paths at the study area scale for the entire modelling time period

*Where n is the neighbourhood number 1 – 75

Table 2.1.11. UVQ results screen descriptions.

Screen name	Flow Units	Contaminant units		Description
Water and contaminant balance	mm or kL/year	kg/y Tonne/y mg/L	or or	Neighbourhood and study area input and output flows and contaminants for a yearly time step
Climate statistics	mm	-		A summary of the climate file rainfall and evaporation data on a yearly basis
Land block water use	L/household/d or L/household/y	-		Daily average and annual indoor and outdoor water usage for total study area or individual neighbourhoods
Land block irrigation	mm or kL/year	kg/y Tonne/y mg/L	or or	Average annual irrigation supply and demand for land block gardens
Public open space and land block irrigation	mm or kL/year	kg/y Tonne/y mg/L	or or	Average annual irrigation supply and demand for land block gardens and public open spaces
Rain tank	kL/household	kg/y Tonne/y mg/L	or or	Input and output paths for rain tanks in individual neighbourhoods
Sub-surface greywater irrigation	kL/household	kg/y Tonne/y mg/L	or or	Supply demand flows and input contaminants for greywater use to garden irrigation
On-site wastewater	kL/household	kg/y Tonne/y mg/L	or or	Supply demand flows and contaminant balance for on site wastewater units
Neighbourhood stormwater	kL or kL/y	kg/y Tonne/y mg/L	or or	Supply demand flows and contaminant balance for neighbourhood stormwater store
Neighbourhood wastewater	kL or kL/y	kg/y Tonne/y mg/L	or or	Supply demand flows and contaminant balance for neighbourhood wastewater store
Aquifer storage and recovery	kL or kL/y	-		Supply demand flows for aquifer storage and recovery
Study area stormwater	kL	kg/y Tonne/y mg/L	or or	Supply demand flows and contaminant balance for study area wastewater store
Study area wastewater	kL	kg/y Tonne/y mg/L	or or	Supply demand flows and contaminant balance for study area stormwater store

2.1.4.3 Model Calibration

The calibration of UVQ is a complex task, with 19 calibration parameters influencing the model ouput stormwater, wastewater and potable water flows. Also, a further set of calibration parameters control the contaminants needed to detail the traditional water servicing approach (Table 2.1.5 and 2.1.6). There are twelve stormwater calibration parameters, five wastewater calibration parameters and two potable water calibration parameters. Additional calibration parameters are required for modelling alternative water management options, the number of which will depend on the complexity of the option selected by the user. Depending on the characteristics of the study area to be modelled and the availability of observed water usage, wastewater and stormwater flow and concentration information, different calibration parameters may be required for each neighbourhood modelled. The calibration of UVQ is a manual, trial-and-error process and whilst the calibration sounds an onerous task, an appreciation of the model assumptions, algorithms and output files and a detailed knowledge of the study area to be modelled can reduce the iterations required to achieve a good fit. The setup and calibration process in the AISUWRS case study cities is described in Rueedi & Cronin (2005), Souvent & Moon (2005), Cook et al (2005) and Klinger et al (2006), In addition, Aquacycle calibration data can be sourced for water flow calibration, as Aquacycle is the predecessor to UVQ and contains similar modelling algorithms and assumptions (Mitchell et al., 2001).

Due to the interactions between water streams represented in UVQ, it is recommended that a step wise approach to model calibration be undertaken:

(1) Obtain initial estimates of calibration parameters from study area knowledge or previous modelling experience (see literature sources for parameter values developed for specific case study sites in References and Further Reading).
(2) Adjust stormwater parameter values to improve the fit to observed stormwater flows. It is suggested that this initial stormwater calibration compares to total yearly flows rather than to shorter-term temporal variations in flow.
(3) Adjust wastewater parameter values to improve the fit to observed wastewater flows. As above, it is suggested above this initial wastewater calibration be compared to total yearly flows rather than shorter-term temporal variations in flow.
(4) Adjust potable water parameter values to improve the fit to observed potable water use. Note that if estimates of drinking water usage indoors have been used rather than observed values (from zonal metering) these may also need to be adjusted to obtain a good fit.
(5) Readjust stormwater parameter values to account for changes following adjustment of water and wastewater values. During this iteration use daily or monthly output files as detailed in Table 2.1.10 to examine the temporal variability of stormwater flows and adjust calibration parameters to achieve a good fit.
(6) Repeat steps 3, 4 and 5 comparing outputs to daily or monthly observed water and wastewater data, until the best fit to observed stormwater, wastewater and potable water flows is achieved.

Objective functions such as the coefficient of efficiency (Nash and Sutcliff, 1970) and graphical plots (hydrographs and X-Y plots comparing observed and simulated flows) are used to determine the 'goodness-of-fit' achieved by a particular parameter set. Appropriate objective functions were found to be i) SIM/REC, the sum of simulated flow (SIM) divided by the sum of recorded flow (REC) and ii) SDOF, the sum of squares of differences of simulated and recorded flows (Diskin and Simon, 1977) as well as daily or weekly coefficient of efficiency (Nash and Sutcliff, 1970) for use in the verification period.

2.1.5 LINKS WITH OTHER MODELS

UVQ provides data on water flows and contaminants loads to all other models in the AISUWRS modelling suite. Wastewater flows and contaminant loads per connection are provided to NEIMO (which drives SLeakI) in the PlmUVQWWinput.txt file (Table 2.1.10). In a combined sewer system this file will contain data describing the sum of stormwater and wastewater flows and contaminants, whereas in a separate stormwater and wastewater system, stormwater flow and contaminant data will appear in PlmUVQSWinput.txt. UVQ also provides data to POSI with individual files describing flow and contaminant data due to the consequences of garden irrigation (UFMGardenToGW.txt), public open space irrigation (UFMPOStoGW.txt), infiltration from stormwater ponds or storages (UFMSWInfiltrationBasintoGW.txt) and potable pipe leakage (UFMTapToGW.txt). The flow data in these files is also used by the chosen groundwater flow model (MODFLOW and FeFlow in the AISUWRS case) as recharge because all evaporation is assumed to occur in the soil stores represented in UVQ.

2.1.6 SUMMARY & CONCLUSIONS

The conceptual approach to simulating the urban water system in UVQ permits considerable flexibility in the manner in which water services are represented and provides the ability to mimic a wide range of conventional and emerging techniques for providing water supply, stormwater and wastewater services to either an existing urban area or a site which is to be urbanized. In addition, UVQ can provide insight into the potential consequences of implementing a number of non-structural changes to the system such as changing household occupancy, water usage behaviour, use of household chemical products or amount of fertilizer applied to gardens and open spaces (Mitchell et al., 2003; Mitchell and Diaper, 2005).

Following initial assessment of the applicability of UVQ to the AISUWRS project some modification was needed to ensure UVQ met the necessary requirements. Following initial upgrades further minor changes were required during the course of the project. UVQ has now been calibrated to each case study site in the AISUWRS study and has been successfully used to explore alternative water supply reticulation, wastewater collection and stormwater drainage scenarios detailed in the project (see Chapter 4).

2.2 NETWORK EXFILTRATION AND INFILTRATION MODEL (NEIMO)

2.2.1 INTRODUCTION

Leakage from non-pressure wastewater pipelines has been suggested as a potential source for contaminants in urban aquifers (e.g. Foster et al. 1993, Eiswirth & Hötzl 1997, Barrett et al. 1999, Dohmann et al. 1999, Ellis 2001) and although leakage is an unavoidable consequence of prevailing construction practices, emerging regulatory regimes in many countries require that leakage be addressed for sustainable use of water resources. This implies that construction and maintenance of non-pressure wastewater pipelines should be carried out to ensure minimal leakage occurs. The European Union standard EN 752-2 (1996) recognises these issues and imposes structural integrity requirements for urban sewer systems, which includes their leak-tightness (Keitz, 2002). A primary requirement for compliance with this standard and emerging legislation is the ability to demonstrate the amount of leakage in current and future states. However, significant difficulties still exist in quantifying leakage from sewers.

Sewer leakage encompasses infiltration and exfiltration. To overcome the difficulty of direct measurement, several infiltration assessment studies have been carried out to quantify infiltration by indirect methods such as flow balances using daily hydrograph analysis or by chemical tracers based on dilution of wastewater (De Benedittis and Bertrand-Krajewski, 2005). Karpf and Krebs (2004a) have proposed a partially deterministic method that applies groundwater level and wetted pipe surface area as inputs to quantify groundwater infiltration. Quantifying exfiltration is even more challenging. During exfiltration, organic matter, which flows through defects in the pipe structure, causes clogging of the surrounding soil. This clogging generates a colmation layer that reduces the conductivity of the soil and hence the flow rate through the defects. The fragility of the colmation layer means that measurement of exfiltration in operational sewers is difficult as any disturbance to the surrounding soil and the colmation layer will change the dynamics of the exfiltration.

Due to the difficulty in measuring these flows, a modelling approach that provides some predictive capacity is highly desirable. Research has thus been undertaken to consider the modelling of infiltration and exfiltration, and has resulted in the Network Exfiltration and Infiltration Model (NEIMO). NEIMO is a deterministic model designed to provide water authorities with quantitative and qualitative data on leakage from wastewater networks to address at least partially the information requirements of future legislation. The model has the flexibility for adjustment to suit specific networks taking into consideration defect size and type obtained from CCTV inspection records.

2.2.2 GOVERNING EQUATION

The flow of a fluid through a porous medium as formulated by Henry Darcy can be applied to model exfiltration from and infiltration into sewers. Darcy's Law relates the steady-state discharge rate through a porous medium to the hydraulic gradient (change in hydraulic head over a distance) and the hydraulic conductivity as,

$$Q = -k \cdot A \cdot \frac{(h_a - h_b)}{L} \qquad (1)$$

where,

Q = total flow between two points a and b (m³/s)
k = hydraulic conductivity of the soil (m/s)
A = area of flow (m²)
$h_a - h_b$ = drop in hydraulic head between the two points a and b (m)
L = Distance between the points a and b (m)

Equation 1 demonstrates that in a leaking sewer situation the total leakage is linearly dependant on the area of the damage and the pressure head. Additionally in applying this model to sewer exfiltration and infiltration, the soil layer immediately in contact with a sewer defect is considered to control the rate of movement of water into and out of the pipe and to be responsible for defining the hydraulic gradient.

The micro-environment outside the pipe defect differs between infiltration and exfiltration because exfiltrating wastewater clogs up the soil with organic matter, reducing the soil permeability, whilst with infiltration clogging is not an issue. Thus it is best to consider the two mechanisms separately.

2.2.2.1 Infiltration model

In the infiltration situation, the pipe is below the groundwater table as shown in Figure 2.2.1. In this scenario the pipe is ΔH_{Gw} m below the groundwater (GW) table elevation and is assumed to have a defect of area A m², located on the crown of the pipe to simplify the scenario.

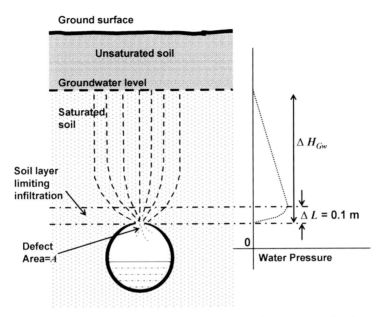

Figure 2.2.1. Schematic view of pipe below groundwater level (defect on pipe upper surface).

In saturated soil, the diffusion of water towards the defect is considered to be without resistance except in the zone near the defect. In this zone the water accelerates towards the defect and the permeability of the soil is assumed to control the rate of diffusion. The thickness of the soil layer that influences the diffusion rate for a typical range of defect sizes has been determined to be approximately 100 mm by modelling of flow and transport processes in porous media under saturated conditions (Wolf, 2005). Intuitively, the thickness should vary depending on the soil type (i.e. sand, clay), but in the absence of reliable modelling data and because pipes are generally laid in porous bedding, 100 mm is assumed for all soil types. On this basis, by applying the Darcy equation to the infiltration scenario, the infiltration volume $Q_{Infiltr.}$ is:

$$Q_{Infiltr.} = -k_{Infiltr.} \cdot A \cdot \left(\frac{\Delta H_{Gw}}{0.1} \right) \quad (2)$$

where $k_{Infiltr}$ is the specific conductivity of the soil (m/s). To apply this model to sewer pipes, the defect area A is estimated from direct measurements via CCTV inspection and the hydraulic head ΔH_{Gw} is determined from data sourced through an asset database and groundwater levels.

When the defect is on the pipe base (Figure 2.2.2), the water level within the pipe also needs to be considered to determine the hydraulic head. Then Equation (2) becomes:

$$Q_{Infiltr.} = -k_{Infiltr.} \cdot A \cdot \left(\frac{\Delta H_{Gw} - H_{Fill}}{0.1} \right) \quad (3)$$

where ΔH_{Gw} is the height of water table to bottom of pipe and H_{Fill} is the height of wastewater within the pipe. ΔH_{Gw} in this scenario is the sum of the distance from the GW table to the pipe crown (D_{BG}) and the pipe diameter (D_{Pipe}); that is:

$$\Delta H_{Gw} = D_{BG} + D_{Pipe}$$

Adjustments to ΔH_{Gw} and H_{Fill} are necessary when the defect is on the sidewall.

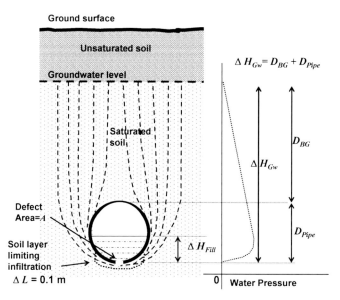

Figure 2.2.2. Pipe below groundwater table, with defect on pipe base.

2.2.2.2 Exfiltration model

As noted previously, during exfiltration the wastewater passing through a defect clogs up the soil adjacent to the defect with organic matter. This clogging layer, also known as the colmation layer, is an organic mat populated with protozoa, bacteria, algae, and other micro-organisms and their by-products (Dizer et al., 2004). By its very nature the colmation layer has a lower conductivity than the supporting soil and therefore has a limiting influence on the exfiltration rate. Thus in applying the Darcy equation for exfiltration, the properties of the colmation layer need to be considered for the specific conductivity and hydraulic gradient parameters (Rauch & Stegner, 1994 and Vollertsen & Hvitved-Jacobsen, 2003). Figure 2.2.3 illustrates the scenario for exfiltration.

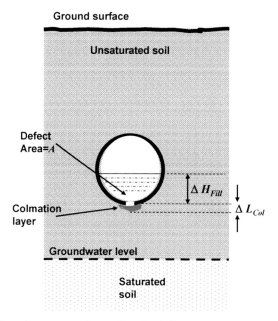

Figure 2.2.3. Schematic view of pipe above groundwater level – defect on pipe causes exfiltration.

In this case the pipe is above the level of the groundwater table. It has a defect of area A (m^2), located on the base of the pipe and a wastewater level to a fill height of ΔH_{Fill}. The colmation layer formed adjacent to the defect has a thickness of ΔL_{Col}. The hydraulic gradient is established across this layer with a pressure of ΔH_{Fill} on the pipe side and atmospheric pressure on the external side. Applying the Darcy equation to this scenario, the exfiltration volume $Q_{Exfiltrn.}$ is,

$$Q_{Exfiltr.} = k_{Exfiltr.} \cdot A \cdot \left(\frac{\Delta H_{Fill}}{\Delta L_{Col}} \right) \quad (4)$$

Where $k_{Exfiltrn.}$ is the specific conductivity of the colmation layer. To apply this model to sewer pipelines ΔH_{Fill} is calculated from the pipe geometry and fill level data. A method to accurately quantify the thickness of the colmation layer is currently not available. A thickness of 10-20 mm in sand bedding is assumed, for all soil types in the NEIMO model based on studies at Aalborg University (Vollertsen & Hvitved-Jacobsen, 2003). As knowledge on characterising the colmation layer improves, values specific for different soil types may be applied

2.2.3 VALIDATION OF MODELS

2.2.3.1 Validation of the infiltration model

An experimental setup with concrete, vitreous clay and PVC pipes was utilised for validation of the infiltration model. 100 mm diameter pipes of the three materials were installed at a 2% slope (to ensure flow) in a 0.6 m(W) x 0.8 m(L) x 1.0 m(H) box with the downstream end projecting out from the base of the side wall (Figure 2.2.4). 3mm holes, 15 in each pipe, were drilled through the upper pipe surface to simulate a crack for infiltration. However, in reality cracks will behave different to holes.

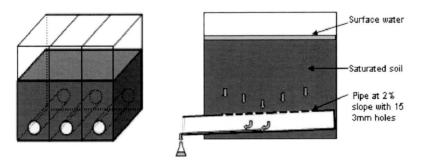

Figure 2.2.4. Experimental set-up for infiltration test.

The box was filled with uncompacted loamy soil to pipe depths of 0.65m and 0.85m, saturated with water and allowed to stabilise for 24 to 48 hours prior to measurement of infiltrating water. A hydraulic conductivity of 0.0002 m/s for loamy soil (Delleur, 1999) was applied to the model (Eq. 2) for the calculations. Comparison of the calculated against the measured infiltration (Table 2.2.1) showed a variation of 76% to 86%.

Table 2.2.1. Infiltration: measured versus predicted values.

	Predicted (ml/min)		Measured (ml/min)		% Variation	
	0.85 m	0.65 m	0.85 m	0.65 m	0.85 m	0.65 m
Concrete	1.04	0.81	4.34	3.78	76%	78%
Vitreous clay	1.07	0.83	7.41	3.03	86%	73%
PVC	1.08	0.88	7.91	5.16	86%	83%

Reducing the assumed 0.1m thickness of the limiting soil layer improved the match between calculated and measured values. However, it was also recognised that the degree of soil compaction influences infiltration and in this case the conductivity value used of 0.0002 m/s may be an underestimate of the actual conductivity as the soil was not compacted. Repeating the experiment after compaction of the soil around the pipe stopped infiltration, probably because compacted soil plugs sealed the 3 mm holes. Additional work is required, especially on determining the size and location of defects, and the degree of soil compaction, before adjustments to the assumed thickness value of 0.1 m can be justified.

2.2.3.2 Validation of the exfiltration model

The practical difficulty of measuring exfiltration in established sewer pipes has limited field validation of laboratory and pilot-scale studies (Gould et al., 2004). The estimates of colmation layer conductivity ($k_{Exfiltrn.}$) from a number of studies range from 3.1×10^{-8} m/s to 1.0×10^{-4} m/s (Table 2.2.2).

Table 2.2.2. Colmation layer conductivity from laboratory studies.

Soil type	Conductivity (m/s)	Reference
Sand	1.0×10^{-5} to 1.0×10^{-4}	Rauch & Stegner, 1994
Sand	3.1×10^{-8}	Dohmann et al., 1999
Sand	1.1×10^{-6} to 2.0×10^{-6}	Vollertsen & Hvitved-Jacobsen, 2003

The variation in conductivity of the colmation layer ($k_{Exfiltrn}$) values calculated in those studies is a reflection of the complexity of the factors controlling the exfiltration process and the biological nature of the colmation layer. The uncertainty of biofilm characteristics, their establishment in experimental set-ups and the extent to which they are breached and reformed after strong flow events is discussed by Vollertsen & Hvitved-Jacobsen (2003). To overcome some of the implied uncertainty, in-situ measurements were carried out on several Melbourne sewers. In each case a section of sewer was isolated between manholes, filled with wastewater through the upstream manhole (Figure 2.2.5), and the exfiltration monitored over several hours using the reduction in water level in the manhole. Prior to the tests, the defects in the pipeline sections were measured and quantified through CCTV inspection. Analysis of the exfiltration data using Equation 4 produced $k_{Exfiltrn}$ values ranging from 5.9×10^{-5} to 7.1×10^{-4} m/s (Table 2.2.3) based on a colmation layer thickness of 15 mm.

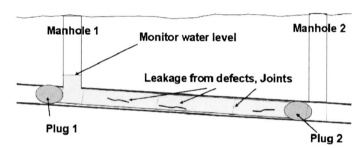

Figure 2.2.5. Schematic of Melbourne sewer exfiltration measurements

Table 2.2.3. Conductivity of colmation layer ($k_{Exfiltrn}$) from exfiltration measurements in Melbourne.

Pipe Location	Material	Diameter (mm)	Length (m)	Construction Date	Soil Type	$k_{Exfiltrn}$ m/s
Test Site 1	Vitreous clay	150	41	1901	Clay	7.1×10^{-3}
Test Site 2	Vitreous clay	150	95	1907	Clay	3.4×10^{-4}
Test Site 3	Concrete	150	25	1935	Sand	8.74×10^{-3}
Test Site 4	Concrete	225	31	1933	Sand	5.95×10^{-5}

While these values are within or marginally higher than the limits reported in the literature, the experimental procedure has several weaknesses. As the manholes were filled to a depth over 1m during the tests, the water in the pipe was at a significantly higher-pressure head than normal and this could have damaged the colmation layer. Furthermore, the pipe was always full, allowing exfiltration through defects in upper sections of the pipe where a colmation layer would typically be very thin or non existent. The occurrence of both of these factors may explain the variation in $k_{Exfiltrn}$ and the high exfiltration estimates. On the positive side, the pipe surroundings were undisturbed and, the colmation layer when present would have been fully established. Therefore the measurements would have reflected the natural sealing properties of the soil-surround, in contrast to those in laboratory test conditions in previous studies.

An additional validation exercise was performed using data from an exfiltration study conducted by the University of Karlsruhe in Rastatt, Germany (Wolf et.al., 2005). A 2000 mm^2 defect (10 mm wide, 200 mm long) was cut into the base of a 500 mm diameter sewer pipe (Figure 2.2.6), and the wastewater flow and exfiltration monitored over a 12-month period beginning July 2004. Although the bedding was disturbed during the defect cutting operation, only measurements in the final months (March to May 2005) were utilised for validation, allowing time for the re-established colmation layer to mature.

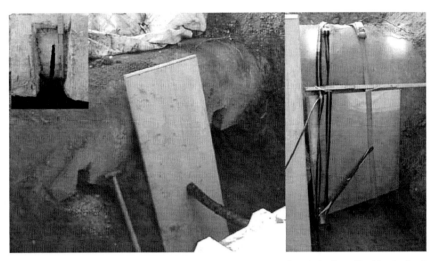

Figure 2.2.6. Excavated sewer during set-up of Kehler Strasse test site. Top left: Detail of leak design. Right: Collecting tank underneath leak (Held et al 2006).

After the exfiltration rate stabilised, the measured exfiltration was compared with predictions from the model. The measured dry weather exfiltration fluctuated between 0.04 L/hr and 0.06 L/hr over March and May 2005. However, some uncertainty was associated with measurement of exfiltration volumes and wastewater flow rates in the experimental set-up. The model predicted exfiltration of 0.05 L/hr based on a flow rate of 300 m³/day and assumed specific conductivity ($k_{Exfiltrn.}$) of 3.5×10^{-5} m/s and thickness of 15 mm for the colmation layer. Whilst the measured and predicted values are in close agreement, the results should be regarded in the context of uncertainty in the field data and the assumptions for the colmation layer parameters. Further work is required to develop a method of characterising the colmation layer to enable those parameters to be measured accurately.

2.2.4 APPLYING THE MODELS TO A SEWER NETWORK

2.2.4.1 Applying the Infiltration Model to a Network

To simplify application of the infiltration model to a gravity sewer network, Equation (3) was used in all situations irrespective of defect positions. The data requirements for Equation (3) are,

- Distance from GW table to pipe bottom (ΔH_{Gw}, m)
- Height of water within the pipe (H_{Fill}, m)
- Defect size (A, m²)
- Specific conductivity of the soil ($k_{Infiltr.}$, m/s)

As a further simplification, wastewater fill level within the pipe (H_{Fill}) was discounted in comparison to the distance from GW table to pipe bottom (ΔH_{Gw}) (i.e. ($\Delta H_{Gw} >>> H_{Fill}$). Therefore Equation (3) reduces to,

$$Q_{Infiltr.} = -k_{Infiltr.} \cdot A \cdot \left(\frac{\Delta H_{Gw}}{0.1}\right) \qquad (5)$$

Equation (5) is applied to every asset in the network that is below the groundwater table. ΔH_{Gw} is the sum of depth of pipe below GW table and pipe diameter. This data may not be explicitly available but can be determined by combining the asset depth from the asset database and the groundwater depth from hydrogeological contour maps for the study area.

Depth of pipe below GW = (Depth of asset from Surface) – (Depth of GW table from Surface) (6)

Given the uncertainty in both these values for any asset, and also that the asset will probably be at a gradient, assets that are partially below the GW table are considered to be totally below GW. The determination of defect size is discussed in the following section on the exfiltration model.

2.2.4.2 Applying the Exfiltration Model to a Network

Applying the model in Equation (4) to a network that has a large number of pipes and a fluctuating wastewater flow over a diurnal period requires a procedure that determines fill level at hourly intervals, calculates exfiltration and integrates the total over 24 hours for each pipe in the network. The hourly wastewater fill level in each pipe was determined on the assumption that the network has a tree structure and the water flow capacity within the pipe follows a diurnal pattern which is a reflection of morning and evening peaks with peak smoothing as the pipe size increases. The defect sizes for each pipe are supplied either from CCTV data or from a generic defects file. The method of compiling the defects data from CCTV records is described in the input data section (Sect. 2.2.7.3.5, cctv_data.csv).

The likelihood of defects being present at any position on the pipe circumference, and the need to consider the hydraulic head at all of these positions, required development of a mathematical treatment which considered the wetted perimeter calculated from the flow volume. The following sections describe the mathematical basis of exfiltration calculations from flow volumes, wet perimeter determinations, circumferential distribution of defects, calculation of wet perimeter angle and integration of exfiltration over the full pipe circumference.

2.2.4.2.1 Determination of Flow Volumes

The network model assumes the pipes in the network follow a tree structure from the first upstream pipes to a single final pipe leading out of the study area.

The inflow to a pipe asset is the sum of input volumes from customer connections and input volumes from any upstream pipes (Figure 2.2.7). The outflow volume from a pipe to a downstream pipe is the inflow volume less the exfiltration volume (or added infiltration). Initial flow into first pipes is the sum of the flow from the property connections to that pipe. The 24-hour flow volumes from property connections are supplied by the Urban Volume Quality (UVQ) model (see section 2.1). To take into account the diurnal flow variation, the 24 hour flow volume is converted to a series of hourly flow volumes by applying flow profiles. The flow profiles were developed to reflect the reduction in morning and evening peaks typically observed as the flow passes from the smaller collector pipes (upstream) to larger mains sewers. The hourly flows volumes are utilised in subsequent water level calculations.

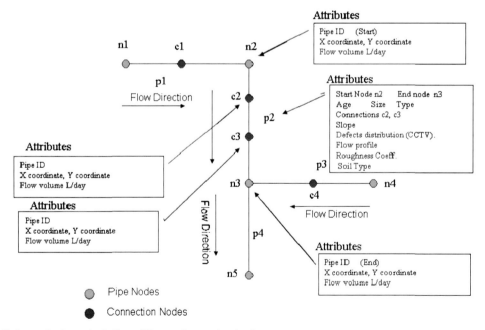

Figure 2.2.7. Schematic for calculation of flow volumes in pipelines.

2.2.4.2.2 Determination of water levels

The water level of a pipe affects its potential for leakage. This section describes how the water level of a pipe can be calculated. The water level, or pipeline fill level, depends on a combination of variables: the water inflow volume per second, the diameter of the pipe, and the velocity of the water, which is in turn affected by the slope of the pipe. Figure 2.2.8 shows a pipe of radius r with the water level forming an angle 2θ with the pipe centre. The arc described by angle 2θ is the wet perimeter (surface wetted); θ is referred to as the wet perimeter angle.

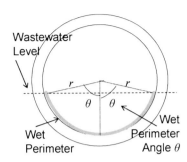

Figure 2.2.8. Wet perimeter of a pipe.

Calculating the wet perimeter angle, θ, is achieved by considering geometrical relationships in conjunction with Manning's equation. Manning's equation, which is an empirical relationship, states the following:

$$V = \frac{(W/P)^{2/3} \cdot S^{1/2}}{n} \qquad (7)$$

where, V is the velocity of the water, W is the wet area of the pipe, P is the wet perimeter of the pipe, S is the slope of the pipe, and n is Manning's number with values ranging from 0.009-0.011 for new PVC pipe and 0.012-0.013 for VC and concrete pipes being suggested in literature. Whilst a value of 0.013 is applied in NEIMO across all pipe types, selection of values specific to pipe material could be introduced in future improvements.

The relationship between inflow volumes, wet area and velocity is:

$$W = \frac{Q}{V} \qquad (8)$$

where Q is the inflow volume. From (7) and (8), the following can be deduced:

$$W = P^{2/5} \cdot \left(nQS^{-1/2}\right)^{3/5} \qquad (9)$$

Basic trigonometric manipulations give the relationship between wet area and θ:

$$W = (\theta - \sin\theta) \cdot \frac{r^2}{2} \qquad (10)$$

The relationship between θ and wet perimeter is:

$$P = \theta \cdot r \qquad (11)$$

From (9) and (11):

$$W = [\theta \cdot r]^{0.4} \cdot f \qquad (12)$$

where $f = \left[n \cdot Q \cdot S^{-0.5}\right]^{0.6}$

From (10) and (12):

$$(\theta - \sin\theta)\frac{r^2}{2} = [\theta \cdot r]^{0.4} \cdot f \qquad (13)$$

Rearranging the terms and factors in (13) gives:

$$g(\theta) = (\theta - \sin\theta) \cdot \theta^{-0.4} - 2 \cdot f \cdot r^{-1.6} = 0 \qquad (14)$$

where $g(\theta)$ is a non-linear equation that has to be solved in order to find θ. To do this Newton-Rhapson's method is used, which requires the derivative of the function $g(\theta)$. $g(\theta)$ is derived as follows:

$$\frac{dg(\theta)}{d\theta} = g'(\theta) = 0.6 \cdot \theta^{-0.4} - \theta^{-0.4} \cdot \cos(\theta) + 0.4 \cdot \theta^{-1.4} \cdot \sin(\theta) \qquad (15)$$

Newton Rhapson's method works iteratively towards a solution and each update is calculated as follows:

$$\theta_{i+1} = \theta_i - \frac{g(\theta_i)}{g'(\theta_i)} \qquad (16)$$

The Newton-Rhapson method will in most circumstances converge towards a single value. It should be noted, however, that $g'(\theta)$, which is found in the denominator in Eq. (16), can potentially take values equal or close to zero. This means division with zero in the update formula will cause the algorithm to fail.

Newton Rhapson's method requires an initial guess of θ (θ_0) in order to solve Eq. (14). θ can take values between 0, which represents an empty pipe, and π, which represents a full pipe. A reasonable starting guess is $\theta_0 = \pi/2$, which represents a half full pipe. Once θ is determined, it is used in leakage calculations as described in the following section. It should also be noted that when the water inflow exceeds the pipe's capacity, the above algorithm does not converge. An overflow test is therefore applied prior to the above calculations, and if such a test identifies a pipe to be overflowing, θ is simply set to π, meaning that the pipe is full.

Equations directly relating diameter, flow depth and wet perimeter have been developed by Swamee (2001). These equations may be applied in future development of the algorithms.

2.2.4.2.3 Calculating the exfiltration from wet radius/perimeter

The exfiltration calculation requires the defect size in a pipe, and its position in relation to the water level, to be quantified. To simplify calculations it is assumed that all crack defects in a pipe are aggregated to one location. To facilitate circumferential location of the defect, the pipe circumference is divided into 4 clockwise sectors as shown in Figure 2.2.9 (i.e. Sector A from 4.30 to 7.30, Sector B is from 3.30 to 4.30 and 7.30 to 8.30, etc). The total area of crack defects in a pipe is apportioned to the 4 sectors, as ratios of D:C:B:A. A pipe with a crack defect on the pipe crown will have the ratios as 1.0:0:0:0. If the defect runs from middle of the pipe to the base the ratios will be 0:0.3:0.3:0.4.

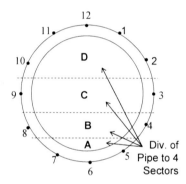

Figure 2.2.9. Division of pipe to 4 sectors.

2.2.4.2.4 Defect size and distribution

An estimate of total defect area in an asset is either given as an input from a CCTV inspection report (see Sect. 2.2.7.3.5, cctv_data.csv), or in the absence of CCTV, utilising data extracted from a generic defects file. The generic file has crack size and joint displacement sizes, which were based in the case of the crack sizes on CCTV records of several 50-year-old mortar-jointed vitreous clay pipelines, installed in sandy soil conditions (Sadler and De Silva, 2005). The values of joint displacement sizes were based on discussions with research groups active in sewer exfiltration studies, because it was acknowledged that experimental data on joint displacements is not available for pipes that have been in service for extended periods. Although almost all joints exhibit at least a slight degree of leakage, as observed through CCTV in infiltrating pipes, a practical method of measurement was not available especially at socket joints. The leakage occurs when displacement is at the overlapping surface of the socket, and CCTV cannot examine this for damage assessment. Removal of the pipeline for inspection disturbs the joint making the measurement redundant. To permit development of the generic data it was assumed a value less than 0.5 mm separation at the over lapping surface was insufficient to cause a leak and greater than 5 mm exceeds the designed interference fit. On that basis a 2 mm displacement (overlap separation) was selected as a first approximation for 50-year-old mortar joints to illustrate the model (Vollertsen, 2005).

Assuming a time dependant progression in crack defect sizes and joint displacements, this data, supplemented by expert opinion was utilised to compile a data base for pipe and joint defect sizes in concrete, vitreous clay, PVC and PE pipelines with mortar and rubber-ring jointing for a 200 year life span. The basis of the extrapolations is detailed in the NEIMO Manual (AISUWRS, 2006). For instance the joint displacement in rubber ring joints was assumed to be less than in mortar joints. The curves developed for vitreous clay pipe in sandy soil are shown in Figure 2.2.10. Further analysis of CCTV records of vitreous clay, concrete and PVC pipes of other ages and installed soil types is required to justify the validity of the assumptions. Therefore the generic data should only be treated as representative of defect sizes. Intuitively, crack defects in sewer pipes are site specific, and validation of the data in a universally generic basis may not be practical. A realistic estimate of leakage requires real data from CCTV inspection.

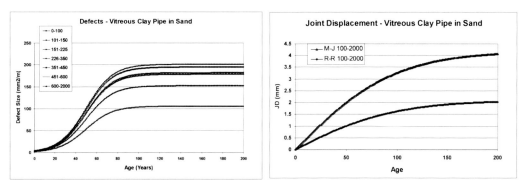

Figure 2.2.10. Generic curves for vitreous clay pipe in sand. Curves are provided for other soils and pipe types.

The generic defect data are provided for the model through a text file that has the following rows for each combination of material type, diameter range, soil and age:

Material Type, Diameter, Soil Type, Age, A, , pA, pB, pC, pD, jD

where,
- *A* is crack defect area per m length of pipe
- *pA* is proportion of pipe crack defects in section A
- *pB* is proportion of pipe crack defects in section B
- *pC* is proportion of pipe crack defects in section C
- *pD* is proportion of pipe crack defects in section D
- *jD* is joint displacement per joint, m

The defect areas are split between cracks and joint displacements. CCTV data, if available, is provided through a file with the same structure. The total defect area per pipe asset is scaled up based on the pipe's length. The total joint displacement of a pipe asset is scaled up based on the number of joints on the asset, based on the pipe material.

- $A_{tot} = A \cdot pipe\text{-}length$ (pipe length in m)
- $jD_{tot} = jD \; number\text{-}of\text{-}joints\text{-}per\text{-}km$

The calculations assume the crack defect areas are distributed over the pipe circumference as per the proportions (pA, pB, pC, pD,) in the generic defect curves file (or from the CCTV data input) and the joint displacement widths are constant around the circumference of the pipe. For each pipe, crack defects are calculated for each section *A*, *B*, *C* and *D* by calculating the,

(1) Length of the wet perimeter arcs (γ) of the pipe in each section:, [$\gamma = 2 \cdot \{pipe\;radius\;r\} \cdot \{wet\;perimeter\;angle\;\theta\}$]
(2) Equivalent Radial Width ($\Delta w_{defects}$) in each section: [$\Delta w_{defects} = A_{tot} / \gamma$]
(3) Equivalent Joint Displacement (Δw_{joints}) in each section: [$\Delta w_{joints} = jD_{tot} / \gamma$]
(4) Equivalent Radial Widths $\Delta w_{defects}$ and Δw_{joints} are the used in the leakage function.

2.2.4.2.5 Calculating leakage

Once the wet perimeter angle θ has been calculated, it is possible to calculate leakage volume for each section of the pipe. The leakage for each section is given by the leakage function, through integration of the leakage equation, separately for each of the pipe circumference's sections:

Leakage from section *A* (Defects or Joints) = $Q^A = 2 \cdot \dfrac{K}{\delta L} \cdot \Delta w^A \cdot r^2 \cdot lambda(\alpha_{A0}, \alpha_{A1})$

Leakage from section *B* (Defects or Joints) = $Q^B = 2 \cdot \dfrac{K}{\delta L} \cdot \Delta w^B \cdot r^2 \cdot lambda(\alpha_{B0}, \alpha_{B1})$

Leakage from section *C* (Defects or Joints) = $Q^C = 2 \cdot \dfrac{K}{\delta L} \cdot \Delta w^C \cdot r^2 \cdot lambda(\alpha_{C0}, \alpha_{C1})$

Leakage from section *D* (Defects or Joints) = $Q^D = 2 \cdot \dfrac{K}{\delta L} \cdot \Delta w^D \cdot r^2 \cdot lambda(\alpha_{D0}, \alpha_{D1})$

The angles α_0 and α_1 are determined as follows:

For sector A: $\alpha_0 = 0$, is the starting angle of the section, and $\alpha_1 = \min(\frac{2 \cdot \pi \cdot 1.5}{24}, \theta)$, is the end angle of either the section or the wet angle

For sector B: $\alpha_0 = \frac{2 \cdot \pi \cdot 1.5}{24}$, is the starting angle of the section, and $\alpha_1 = \min(\frac{2 \cdot \pi \cdot 2.5}{24}, \theta)$, is the end angle of either the section or the wet angle

For sector C: $\alpha_0 = \frac{2 \cdot \pi \cdot 2.5}{24}$, is the starting angle of the section, and $\alpha_1 = \min(\frac{2 \cdot \pi \cdot 3.5}{24}, \theta)$, is the end angle of either the section or the wet angle

For sector D: $\alpha_0 = \frac{2 \cdot \pi \cdot 3.5}{24}$, is the starting angle of the section, and $\alpha_1 = \min(\pi, \theta)$, is the end angle of either the section or the wet angle.

Having calculated Q for defects and joints in all 4 sectors, the total leakage is the sum of the components.

$$\text{Total leakage} = Q^A_{defects} + Q^B_{defects} + Q^C_{defects} + Q^D_{defects} + Q^A_{joints} + Q^B_{joints} + Q^C_{joints} + Q^D_{joints} \tag{17}$$

2.2.5 KNOWN LIMITATIONS

2.2.5.1 Model parameters

In the infiltration and exfiltration models assumptions have been made for several key parameters. Specifically, the thickness of the controlling layer in the infiltration model and the values of the colmation layer parameters in the exfiltration model. The assumptions are necessary as experimental data is not available nor methods developed to measure those values. However the models have been used in NEIMO with those assumptions to progress the development of a tool that can be applied across a network. As more information is available in the future, modifications can be structured into the tool. For example rather than a generic 100 mm for the controlling layer in the infiltration model, a thickness specific to each soil type may be applied. Likewise, in the exfiltration model the thickness and conductivity of the colmation layer may be adjusted depending on soil type and flow conditions.

2.2.5.2 Generic defects

Common to infiltration and exfiltration models is the reliance on the generic defects file for defect size quantification when actual defects data is not available. As noted previously, for a specific model study CCTV inspection reports must be available for defect size quantification. Due to the assumptions applied in compiling crack sizes and joint displacement values in generic data, application of generic data should be treated as a demonstration of the methodology. As large defects caused by specific local conditions would not be captured by generic defects, the leakage estimates based on generic data would at best be equivalent to background leakage. Intuitively, joint displacement is age dependant whilst cracks and fissures are site specific. It might therefore be possible to describe mathematically a relationship between joint displacement and pipe age, thereby allowing meaningful generic input data to be specified for the model. This would be advantageous, as it would allow leakage estimates to be made without CCTV data, which is expensive to collect. Further work is however necessary before these concepts can be applied.

2.2.5.3 Depth below groundwater

The elevation of the water table relative to the water level within the pipe determines if the active mechanism is infiltration or exfiltration. However, there is often uncertainty in water table elevation at the individual asset level because of the coarse resolution of groundwater level data from maps, making it impossible to determine the exact level with the required degree of accuracy. Therefore, pipes partially below the groundwater level are treated the same as those totally below with infiltration the prevailing mechanism. In infiltration calculations, wastewater fill level within the pipe (H_{Fill},) is discounted in comparison to the distance from the water table to pipe bottom (ΔH_{Gw}) (Eq. 6). The facility to account for temporal changes in groundwater levels is not available in this model.

2.2.6 RUNNING THE MODEL

The exfiltration and infiltration calculations across a network are implemented through a computer program coded in Python (Python, 2005). The Python script is executed only through the Python program. It is not possible to run the scripts as a stand-alone executable file in the batch mode. In the AISUWRS project the Decision Support System (DSS) interface (see Chapter 3) is designed to execute the Python script through the Python program without any user input. The DSS first executes the UVQ program (see Section 2.1) to generate a file that has the wastewater flow data (uvq_nbhoods.csv) before proceeding with NEIMO execution. The progress of the calculations through the different stages is shown on a display box. The time taken for execution of the combined UVQ and NEIMO models, depending on the size of the network, is typically 2 to 10 minutes on a fast computer.

The DSS is supplied as a compressed file consisting of a self-extracting-auto-installation package. The installation creates a folder structure and installs the Python script and associated NEIMO data files into appropriate folders. With this file structure in place, it is possible to run the Python script outside the DSS, but within the Windows environment, using PythonWin software.

2.2.7 MODEL INPUT

2.2.7.1 Data flow chart

For every pipe asset in a network, NEIMO applies the data on pipe size, slope, defect size, defect distribution and inflow volume to calculate exfiltration (or infiltration). The volume exfiltrated is subtracted from the inflow volume (or infiltrated volume added) and passed on to the next downstream pipe as input. The chart in Figure 2.2.11 details the integration of data from different sources as programmed in NEIMO. Data from 9 input files are required by NEIMO for the calculations.

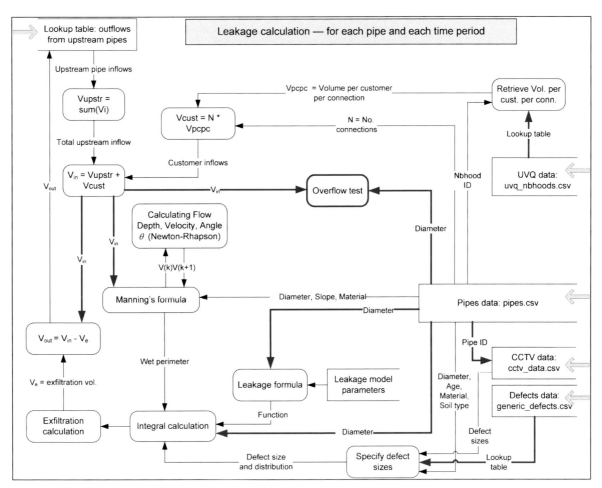

Figure 2.2.11. Data flow diagram for NEIMO.

2.2.7.2 Preparation of input files

Of the nine files, the user needs to prepare six (pipes.csv, connections.csv, successor.csv, cctv_data.csv, dbg.csv and grw_contaminants.csv). The other three files (uvq_nbhds.csv, genric_curves.csv, flow_fractions.csv) are either created by a host program or provided as default files. The structure of the files and the best sources for obtaining the data are given in the following subsections. Brief descriptions of the files and input protocols are given in Table 2.2.4 and 2.2.5.

Table 2.2.4. NEIMO Input files.

Name	Function	Format
pipes.csv	Information on assets including material, size, length and age.	Alphanumeric and text
connections.csv	Customer/property connections number in each asset	Alphanumeric
successor.csv	Provides upstream/downstream relationship between pipe assets	Alphanumeric
cctv_data.csv	Represents the defects shown in a CCTV inspection report.	Alphanumeric
dbg.csv	depth below groundwater for assets below GW level	Alphanumeric
grw_contaminants.csv	Provides the contaminant loads in infiltrating GW	Alphanumeric
uvq_nbhds.csv	Generated by UVQ; it provides the wastewater flow per property connection in a neighbourhood	Alphanumeric and text
genric_curves.csv	Generic defect sizes, distributions, and joint widths for pipes of all material types, ages, sizes (DN<100 to 2000 mm) and soils.	Alphanumeric
flow_fractions.csv	Diurnal variation in flow for different pipe sizes	Alphanumeric

Table 2.2.5. Protocols for NEIMO input data files.

Subject	Protocol
Numerical fields format in input data file	All numerical fields are assumed to use the dot as decimal delimiter. The scientific notation is allowed. The fields containing a year (installation or inspection) are integral, and thus must be composed exclusively of numerical characters.
Alphanumerical fields in input data file	Alphanumerical fields are allowed to be composed of any ASCII printable character, except the one used as field delimiter (see below). They must be no more than 49 characters wide.
File delimiter	NEIMO assumes all files it has to deal with have all their record fields delimited by the comma ",". NEIMO does not recognise the semicolon ";" as a valid delimiter.
Input data file headlines and input data files rows	The structures of the files are as shown in the subsections.

2.2.7.3 Input data file structures

2.2.7.3.1 pipes.csv

This file provides basic information about the pipe assets in the network and is generated from data in the utility asset database. It has 6 fields with the following structure;

ID - the unique number of the asset.
PTYPE - gravity flow pipe RETIC, or a pressure pipe MAIN). Only gravity pipes are processed by NEIMO.
DN - Nominal diameter of the pipe in mm.
LENGTH - length of the pipe in m.
GRADE - grade or slope of the pipe, length in m for 1m drop. If uncertain 100m is used as default.
MAT - pipe material, coded to include type of joint material:

HDPE	High Density Polyethylene
PVCRR	PVC with rubber ring joints
PVCSC	PVC with solvent cement joints
RCMJ	Concrete with mortar joints
RCRR	Concrete with rubber ring joints
VCMJ	Vitreous Clay with mortar joints
VCRR	Vitreous Clay with rubber ring joints

If material type is not known, use Unknown. Any wrong/misspelt names default to vitreous clay with mortar joint.
GROUND - soil type where pipe is installed; (Sand, Clay or Unknown).
CDATE - year pipe was installed (yyyy).
NBH_ID - the neighbourhood ID for the asset; every asset is within a particular neighbourhood; if any asset is across 2 neighbourhoods, choose one.

2.2.7.3.2 connections.csv

This file has two fields; the pipe ID and the number of customer/property connections for each asset. The information on property connections required for the connections.csv file is best sourced from the utility asset database.
ID - the unique number of the asset.
No - the number of properties that connect to this asset.

2.2.7.3.3 successor.csv

This file has two fields, UPSTREAM and DOWNSTREAM, which specifies the upstream/downstream relationship between pipe assets. The network operator may be able to generate this file from their GIS database. It is sometimes referred to as the 'connectivity' file. Figure 2.2.12 shows the entry at a joint with 2 entries for the DOWNSTREAM pipe (ID 789).

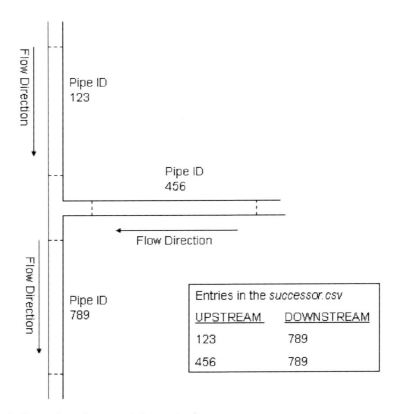

Figure 2.2.12. Schematic illustration of a correct data entry in successor.csv.

2.2.7.3.4 uvq_nbhoods.csv

This file is generated by UVQ; it provides the wastewater flow per property connection in a neighbourhood on a given date. During rain events a specified small percentage of rainwater is transferred to the wastewater flow. With combined sewer systems, 100% of the rainwater is transferred, so that rainwater is channelled via the property connections to the sewer. The contaminant load in wastewater is included for every entry. Fields in the file are NBH_ID, Date, Wastewater_generated_m^3/connection/day and the list of contaminant load/concentrations in the wastewater or combined flow, i.e. Boron_g/day, Potassium_g/day, Sodium_g/day, Chlorine_ g/day, Nitrogen_ g/day, Phosphorus_ g/day, Suspended_Solids_g/day, SO4(2-)_g/day, Total_Organic_Carbon_g/day, Faecal_Streptococci_g/day, Endocrine_Disruptors_g/day, Virus_g/day, User_Defined_Contaminant_1_ g/day, User_Defined_Contaminant_2_ g/day, etc.

2.2.7.3.5 cctv_data.csv

The defects shown in a CCTV inspection report are represented in this file. A single line entry is added for each asset with CCTV information. This line has seven fields (asset ID, *DA, A, B, C, D* and *JA*). *DA* (mm^2) is the total area of all the defects in the asset shown on the CCTV report. *A, B, C* and *D* are the proportional distribution of the defects around the circumference, as in Figure 2.2.9. *JA* is the average width (mm) of displaced joints shown in the report. The data in this file for a particular asset overrides the defect area and joint width information extracted from generic curves.

The values of *DA* and *JA* are estimated from the codes in CCTV inspection reports. The EN 13508-2.2 codes for cracks and joint displacements are BAB and BAJ respectively and their presence on the report is indicative of the presence those defects in an asset. Secondary characterisation codes for cracks give qualitative descriptions of the extent of the crack (surface crack, crack without separation, visibly open), their orientation (longitudinal, circumferential, complex) and width (mm) as a quantification. The distance index on the report provides the length of longitudinal cracks. Using this data the total area of the cracks on the asset can be calculated (*DA*). However the secondary data is rarely offered in reports and assumptions may be necessary The proportional distribution of defects around the circumference (*D:C:B:A*) is determined from the location information in the report. Secondary codes and quantification data on displaced joints are used to determine the total joint displacement width. This value is divided by the number of joints to calculate average width (*JA*). Note that only joints displaced significantly will be observable by CCTV.

2.2.7.3.6 generic_curves.csv

This file gives the defect sizes, distribution, and the joint width for pipes of all sizes (DN<100 mm to 2000 mm), installed soil types (clay or sand) and material type (VC, RC, PVC, HDPE, with mortar joint, rubber-ring joint). This file is provided with NEIMO and is not specific to a network. By cross referencing this file with data in pipes.csv, NEIMO generates the file generic_defects.csv which is specific to the network. Note that after the generic_defects.csv has been written once, it is accessed in subsequent runs, improving the speed of calculations.

2.2.7.3.7 flow_fractions.csv

In NEIMO, the input volume per customer connection varies over the course of the day on an hourly basis, and this file gives the diurnal variation in flow for different pipe sizes (on a diameter basis). This file is provided with NEIMO and is not specific to a network. The format of the file is that it has a header row, followed by a number of rows that each represents the diurnal customer inflow variations for a certain diameter range. The columns are in order: minimum of diameter range, maximum of diameter range, followed by 24 fields representing each hour of the day.

2.2.7.3.8 dbg.csv

This file gives the depth below groundwater for every asset in the pipes.csv file. After a first header row, in each subsequent row, the first column has the pipe ID, while the second row has the depth below groundwater. The limitations in determining depth are discussed in Sec. 2.2.5.2.

2.2.7.3.9 grw_contaminants.csv

This file duplicates the NBH_ID and DATE fields in the uvq_nbhoods.csv to give the contaminants in infiltrating groundwater. This file is accessed only if an asset is below the water table, but the file has to be present even if the data fields are not populated.

2.2.8 MODEL OUTPUT

The exfiltration and infiltration models operate on a daily time step. The output from the models for a study area is a single data file with the daily exfiltration volume and infiltration volume (m^3/d) for each asset in the network and contaminant load of the exfiltrated water (grams/d). A summary of leakage statistics for the entire network is also generated. Table 2.2.6 shows a typical output file from a network.

Table 2.2.6. NEIMO Output data file for a sewer network (where all assets are above groundwater level)

Pipe ID	Infilt. Vol in date 19960119	Exfilt. Vol. in date 19960119	B^+_grm	K^+_grm	Na^+_grm	Cl^-_grm	N_2_grm	Phsp_grm	Susp_Slds_grm	SO_4^{2-}_grm
1	0	.000026	.00000274	.00041	.00450	.00717	.00139	.000218	.00531	.00256
2	0	.0000109	.0000011	.000172	.00189	.00301	.000582	.0000916	.00222	.00107
3	0	.0000761	.00000799	.00119	.0131	.0209	.00404	.000636	.0155	.00746
4	0	.00517	.000544	.0814	.894	1.422562	.275	.0433	1.052125	.507
5	0	.00298	.000313	.0469	.515	.819	.159	.0249	.606	.292
6	0	.000313	.0000329	.00492	.0540	.0860	.0167	.00262	.0636	.0307
7	0	.000856	.0000899	.0134	.148	.235	.0455	.007163	.173	.0839
8	0	.0000033	.000000352	.0000528	.000579	.000922	.000178	.0000281	.000682	.000329
9	0	.0216	.00245	.369	4.054285	6.451002	1.248367	.196	4.770540	2.300961
10	0	.0000026	.000000278	.0000417	.000457	.000728	.000141	.0000222	.000539	.000259

The output data serves as input for the SLeakI model, a linked model within the AISUWRS project that attenuates the contaminants in the exfiltrated water as it traverses the unsaturated soil, or can be used directly as a pipe leakage component of recharge in a steady-state groundwater model.

2.2.9 MODEL CALIBRATION AND CODE VERIFICATION

While the underlying equations were verified by measurements on individual assets and leak sites as discussed previously, a method for validation of the model over a full network has not been identified because a reliable method of measuring exfiltration over a full network is not available.

The best possible validation would be to run the model on a network and assess the exfiltration output against values obtained from experimental studies. Towards this goal, the network model was applied to the sewer system of Mount Gambier, a medium sized city in South Australia with a population of 23,000 with about 9700 domestic households, 150 small to medium industrial/commercial businesses, 10 large industries, 11 schools and 1 large hospital. The sewer system servicing the city is made up of 177 km gravity sewers and 68 km of pressure mains, consisting of vitreous clay, concrete and PVC pipes.

The network specific input data files (pipes.csv, connections.csv and successor.csv) were compiled from information in the asset database maintained by the Mt. Gambier wastewater authority. In the absence of reliable CCTV inspection records, generic defect data was applied to determine defect sizes for assets in the network assuming clay as the prevalent soil type in the area. The wastewater flow volume file (uvq_nbhoods.csv) was generated by running UVQ with a climate file for Mt. Gambier.

The output from the model (Table 2.2.7) indicates an exfiltration of 1.2 % from a total inflow volume of 15,092 m^3. In a network of 245 km this averages to a leakage rate of 0.1 l/sec/km which compares satisfactorily with rates between 0.01 and 0.3 l/sec/km reported in the literature (Blackwood et al., 2005). A dry weather exfiltration rate of 1.3 % to 3.8 % has been estimated for a study in Dresden (Karpf and Krebs, 2004b) on the basis of a single exfiltration specific leakage factor representative of all pipes determined in a previous study.

Although generic data was utilised in this exercise, for a specific case like Mt. Gambier, real defect data is indispensable as an assessment of uncertainty would mask the order of magnitude of the estimate. Additional work, especially on determining validity of the assumptions which form the basis of the generic defects data, is required for greater confidence in estimates of leakage calculated solely based on generic defects.

NEIMO has been applied to 3 other case study cities with differing outputs for exfiltration rates. Except for the city of Rastatt, Germany, none of the cities had reliable knowledge of the condition of the pipes with respect to defects or the extent of exfiltration from the network. The Rastatt CCTV data was also not in a suitable format for direct processing through NEIMO. Therefore those results could not be used for definitive validation of the model.

Table 2.2.7. Model output for sewer network in Mount Gambier

Property	Output
Total number of connections	9387
Approximate total inflow volume per day:	15092.61 m^3
Average input volume per connection per day:	1.608 m^3
Total exfiltration volume:	182.33 m^3
Total infiltration volume:	0.08 m^3
Exfiltration proportion:	1.21 percent
Exfiltration through cracks:	4.45 percent
Exfiltration through joints	95.55 percent

2.2.10 LINKS WITH OTHER MODELS

NEIMO is intimately linked with the UVQ model in that an output from UVQ (uvq_nbhds.csv) is utilised as an input. The uvq_nbhds file provides wastewater flow volumes and contaminant loads in wastewater for NEIMO calculations. Furthermore the classification of an area to neighbourhoods in UVQ is carried on to NEIMO through the identification of pipes by neighbourhoods in the pipes.csv file.

The output from NEIMO serves as input for SLeakI, the unsaturated flow model that traces the passage of water exfiltrating from sewer pipes through the unsaturated soil zone to the saturated aquifer.

2.2.11 CONCLUSIONS

Development of a combined Network Exfiltration-Infiltration Model (NEIMO) based on individual exfiltration and infiltration models has been described in this chapter. The individual models are based on the classical Darcy's Law for fluid flow through porous media.

In adapting the Darcy equation for infiltration, the limiting hydraulic gradient is assumed to occur across the 100 mm of soil immediately surrounding the pipe. This has been based on available information. For exfiltration, the limiting layer is the soil clogged up by exfiltrating organic matter and biofilm (colmation layer), located immediately outside pipe defects. The thickness and conductivity of the colmation layer has been sourced from the literature.

The two models were applied to assets in a sewer network to determine the infiltration and exfiltration over a larger area. This required, amongst other attributes, the volume of wastewater flow through the network and, where sewer defects had not been quantified by in-situ inspections, a method of determining defect distributions. The water volume was established by utilising data imported from an urban water balance (UVQ) model, together with asset attributes. A generic defect file was developed which allows estimation of defect sizes based on age, size and material type of pipes. A mathematical model was developed to utilise these inputs as well as environmental attributes such as soil type and depth of water table to estimate the daily exfiltration and infiltration from each asset in the network. A significant uncertainty is associated with the estimated values of exfiltration and infiltration due to the uncertainty in the permeability of the colmation layer, its typical thicknesses, defect sizes, soil types, permeability of the soil and flow volumes.

Validation of the model will therefore always be challenging, especially as direct measurement of exfiltration is an inexact science. The model applied to the Mount Gambier sewer network predicted a 1.2 % exfiltration rate, which compared satisfactorily with pilot scale studies reported in the literature for individual leaks. There are no reports of exfiltration quantification by direct measurement over large sections of a network except for the estimation of 1.3 % to 3.8 % for Dresden (Karpf and Krebs, 2004b). When NEIMO was applied to a single leak at Rastatt, it predicted an exfiltration rate in close agreement the measured value, albeit with assumed values for colmation layer parameters. Of the case study cities, infiltration was suspected in some sections of Rastatt and deduced for a small part of the network in Doncaster. Reliable infiltration data are currently not available for those areas to be used in validation exercises.

In its present form NEIMO is a tool that can provide exfiltration and infiltration data on a network scale using input data generally readily available from water authorities or accessible from public institutions. The assumptions behind some of the input data contribute to uncertainty in the output values. Typically, with the generic defect data and assumed input parameters the uncertainty in leakage estimates would be high. Improvements are possible with more analysis of CCTV records to investigate the validity of the generic data. At best this data in its present form only permit estimates of background leakage. Real defect data is necessary for rational estimations, as large defects in sewer pipes are site specific and will only be identified by in-situ examination such as CCTV inspection. The values assumed for model parameters also contribute to the uncertainty. As additional research is conducted on groundwater movement through different soil types and on the properties of colmation layers, more experimental data will become available, allowing the flexibility to adjust the model parameters to suit specific conditions, further improving the accuracy of estimations.

2.3 CONTAMINANT TRANSPORT THROUGH THE UNSATURATED ZONE: THE SLEAKI AND POSI MODELS

2.3.1 Introduction

This section describes the modelling of contaminant transport through the unsaturated zone. As well as requiring an understanding of the movement of water through the unsaturated zone, this modelling had the additional constraint for its practical application in the AISUWRS model chain in that its execution needed to be sufficiently fast in the processing of thousands of scenarios. While the principles used in this section are the same as those used in the previous one, some different approaches were required to allow the rapid assessment of contaminants through the unsaturated zone and the magnitude of likely attenuation/magnification mechanisms which could mitigate the effect on the underlying groundwater resource. These contaminants originate either from infiltration through pervious surfaces or from point contamination sources such as leaks in pipes. In reality the soil is non-uniform - in the AISUWRS study it has been assumed that the soil is comprised of uniform layers, and in the cases studied this has been limited to two layers.

Two approaches have been developed in the course of this AISUWRS project to model the transport of contaminants in the unsaturated zone. The first assumes that infiltration occurs over a sufficiently large area such that the edge effects can be neglected, so that a one-dimensional model can be used. The second case considers point source leaks from pipes (typically sewers). Further details of these models can be found in Miller et al. (2006).

2.3.2 Simplifying approximations

The movement of water and contaminants through soil is very complex, and depends on many factors including soil and contaminant properties. To make progress with mathematical modelling, a number of simplifying approximations are necessary:
- That the unsaturated zone above the water table consists of two 'soil' layers. This approximation gives more flexibility than a single layer, and the theory developed could then be generalised to consider a number of layers.
- That the soil (or rock matrix) is uniform within a layer. This approximation is open to question because paths of preferential flow do exist. There is no currently accepted method of dealing with this challenge without a great deal more site-specific information than is, in practice, available.
- That the soil in each layer is one of a series that has been characterised by Van Genuchten (1980).

In addition, following the precautionary principle, generally conservative approximations have been made that would tend to overestimate contaminant loads reaching the groundwater.

2.3.3 Movement of water through the unsaturated zone

The first approach to modelling the transport of contaminants in the unsaturated zone assumes that infiltration occurs over a sufficiently large area such that the edge effects can be neglected. This results in a one-dimensional flow pattern (this problem is described below, and in further detail by Mohrlok and Bücker-Gittel (2005)). The approach uses established theory to derive a residence time for water in the unsaturated zone. It deals only with solute transport beneath a plane that is below the root zone, thus avoiding the effects of vegetation on water and nutrient removal. Hence uptake and recycling of nutrients within the root zone are not considered. Realistic estimation of residence time in the unsaturated zone is a key requirement if attenuation processes are to be modelled in a representative way, and this is further considered in Ch 2.4, where the UL_FLOW approach is also described.

The second approach considers leaks from buried pipes. This creates modelling difficulties because the movement is not only vertical but also horizontal. Under some conditions, water can also rise above the leaks due to capillary forces. New theory was developed by Robinson (2005) to address this problem.

Outlines of each of these approaches are given below. An operational constraint was computation run-time - the required algorithm needed to be sufficiently fast for many open areas or many leaks to be considered in a short period of time.

2.3.3.1 First approach: Infiltration in open spaces

The soil is treated as a layered medium with two homogeneous layers as illustrated conceptually in Figure 2.3.1. This model provides an acceptable approximation to infiltration over large open areas. In particular, it gives an estimate of the typical time for water to travel from the surface to the water table.

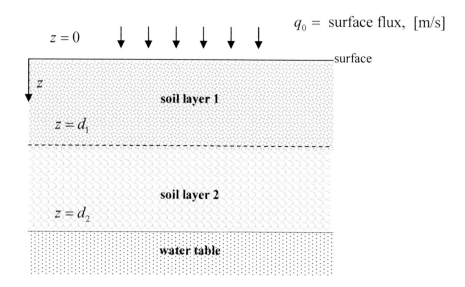

Figure 2.3.1. One-dimensional infiltration in large open spaces.

Darcy's law relates the volumetric flux of moisture to the pore water pressure by:

$$\theta u = -k \frac{\partial}{\partial z}(p - \rho g z) = -K \frac{\partial}{\partial z}(h - z) \qquad (1)$$

where θ is the volumetric water content (m^3/m^3), u is the average vertical velocity of water movement, ρ is the density of water [kg/m^3], g is the acceleration due to gravity [m/s^2], k is the permeability [m^2/(Pa s)], z is the depth [m], $K = k\rho g$ is the hydraulic conductivity [m/s] and h is the head of water [m]. The boundary and interface conditions on h are $h = 0$ (saturation) at the water table $z = d_2$, and the requirement that h be continuous at the soil layer interface.

Consider a soil consisting of two layers and assume that the conductivity of each layer can be described as a relationship of the form

$$K = K(h) = K_0 \exp(2\beta h) \qquad (2)$$

where $\beta > 0$ [1/m] and $K_0 > 0$ [m/s] are parameters (see Robinson, 2005, Equation 5). Note that K_0 is the conductivity at saturation. Suppose the soil moisture depends on the head h through a relation of the form $\theta = \theta(h)$. Then the result (2) can be used to calculate the moisture content θ as a function of the depth z, $\theta = \theta(h(z))$. It follows in turn from (1) that the average velocity profile with depth is given by

$$u(z) = \frac{q_0}{\theta(h(z))} \qquad (3)$$

A convenient parameterised form for the soil moisture capacity which, comes from Equation 93 of Robinson (2005) is:

$$\theta = \theta_r + \frac{\theta_s - \theta_r}{(1 + (\alpha|h|)^n)^{1-1/n}}$$

where θ_s is the saturation moisture content, θ_r is the limiting residual moisture content at very large suction, and α and n are soil property parameters.

To calculate a transit time t_0 for a typical fluid particle to travel from the surface $z = 0$ to the water table $z = d_2$, the average velocity profile (3) is employed to obtain:

$$t_0 = \int_0^{d_2} \frac{dz}{u} = \frac{1}{q_0} \int_0^{d_2} \theta(h(z)) dz = \frac{1}{q_0} \left[\int_0^{d1} \theta h(z) dz + \int_{d1}^{d2} \theta h(z) dz \right]$$

Although this integral cannot be calculated in a closed form, it can be readily evaluated by numerical methods such as two point Gaussian quadrature applied to both soil layers.

This approach is implemented in a software routine called POSI (Public Open Space Index).

2.3.3.2 Second approach: Infiltration from pipe leaks

2.3.3.2.1 Colmation layer

Leaks from pipes, especially sewer pipes, often form a colmation layer (or 'schmutzdecke'). Unfortunately, despite its importance, little has been published on the behaviour of water and contaminants in the colmation layer. Two studies on this topic are by Rauch and Stegner (1994) and Vollertsen and Hvitved-Jacobsen (2003). Recently new data on exfiltration dynamics has been generated within the AISUWRS project (Wolf et al 2005a).

The colmation layer has several very important effects. Firstly, it acts to limit the flow from a leak so that the rate of flow declines asymptotically to a value, which is generally less than the initial rate of leakage. Secondly, it is an area of intense biological activity with a high rate of organic carbon consumption. The high rate of biological activity, together with the high moisture content, generates anaerobic conditions. Associated with the anaerobic environment there is a large de-activation rate of pathogens, as described later. The colmation layer is therefore very important in the understanding of the movement of contaminants from a leaking pipe.

The effects of the colmation layer on various contaminants are considered below. Destruction of this layer by processes such as pressurisation of a sewer during very high flow rates could have a large effect even for a small fraction of the time, as can occur when storm water is also carried. Further work is required in this field.

2.3.3.2.2 Soil Layers

Several of the principles and assumptions required in the one-dimensional model described above are also required for this model. However, various other factors need to be considered. The first major difference is that largely capillary forces will influence movement of water in the X-Y plane. Once flow reaches a steady state, this horizontal force will be matched by the hydraulic resistance, which is in turn a function of the soil permeability and the rate of lateral movement.

Gravitational effects influence movement of water in the vertical direction in a similar manner to that described above for the one-dimensional model. However, unlike the one-dimensional model, there can be some small flow against the gravitational gradient as the water initially spreads horizontally. The flow from a leak is approximately radially symmetric (see Figure 2.3.2). The system in the neighbourhood of a single leak is therefore conveniently described using polar coordinates.

A further difference is that the path length (and hence the residence time) varies depending on whether the flow is in a vertical straight line (fastest) or by a more circuitous route. The worst case is where there is minimum residence time, so for the conservative approach this was the only case considered. Integration in the three-dimensional case is more complex than in the one-dimensional case. However, a fast Gaussian quadrature was found that gave an accurate evaluation of the integral and hence provides an estimation of the minimum residence time.

This approach is implemented in a software routine called SLeakI (Sewer Leak Index).

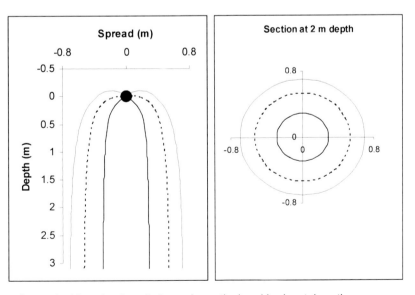

Figure 2.3.2. Flow lines from a leaking pipe in soil shown in vertical and horizontal sections.

2.3.4 Attenuation/magnification of potential groundwater contaminants

A range of contaminants from urban discharges can affect groundwater. Each contaminant behaves in a different manner, but they can be grouped into categories. Three such categories are described below, but other classes could be included. For example, organics would behave quite differently since partitioning can occur between the soil water and soil organic matter. A description of this topic can be found in Kookana *et al.* (2005).

2.3.4.1 Inorganic contaminants which are assumed to behave conservatively

Several of the commonest ionic species, such as chloride and other halides, sodium, potassium, calcium and magnesium, can be considered to behave conservatively as a first approximation. Under the steady state assumption, the entire contaminant load that enters the unsaturated zone would then also reach the water table. There will be some minor changes in load however. For instance, rock-water interaction or the breakdown of organic matter will release small quantities of these contaminants, but the effects of sorption, hydrolysis, complexation or volatilisation processes are typically small enough to be neglected.

The one exception is cation exchange, which could have a much larger effect. High concentrations of sodium can displace exchangeable potassium or calcium from clays or organic matter, depending on the relative concentration difference between the exchangeable surfaces on the organic matter and the clay and of the solute entering the unsaturated zone. Without detailed site-specific information on these concentrations, accurate predictions cannot be made. The simple case of the gains matching the losses is assumed. This approximation is consistent with the simplifying approximation of the system being in a steady state.

2.3.4.2 Inorganic contaminants which are not assumed to behave conservatively

These contaminants are typically polyatomic ions. Four examples are described below.

2.3.4.2.1 Nitrogen

Nitrate is one of the most common groundwater contaminants. Nitrogen (N) leached from open spaces such as parks, playing fields and gardens could be in a variety of forms, but under aerobic conditions will mostly be in the form of nitrate (Correll, unpublished data, Dillon 1988). Organic matter present in sewers is another source of N. Even though there is little or no nitrate-N in wastewater in sewer pipes, due to its anoxic state it does contain significant quantities of N in the form of ammonium or urea (and as organic N and possibly as nitrite).

Leaked material from sewers will initially pass through the colmation layer, which is a zone of high microbial activity. Rauch and Stegner (1994) recorded an almost halving in COD whilst Fuchs *et al.* (2004) observed a 30 to 40% reduction. C:N ratios between 5 and 15 are common in sewage (Koske 2005; Jenkins 2005), implying that commensurately large amounts of organic N are being converted to inorganic forms (mainly into the form of ammonium).

If nitrates were present in the source water, some could be lost by ammonification and the denitrification in the colmation layer (Figure 2.3.3). However, inorganic nitrogen in sewage effluent is more likely to be in a reduced state. Volatilisation of ammonia is another loss mechanism, but lack of exchange with the atmosphere is likely to inhibit gaseous losses. Finally, some inorganic nitrogen could be taken up by biomass in and near the colmation layer or by organic components sorbed to the soil matrix. Over time, the net effects of these processes are likely to decline as nutrients are recycled. Hence it is reasonable for a conservative risk estimate to adopt the assumption that all mineral N is retained in the recharge system.

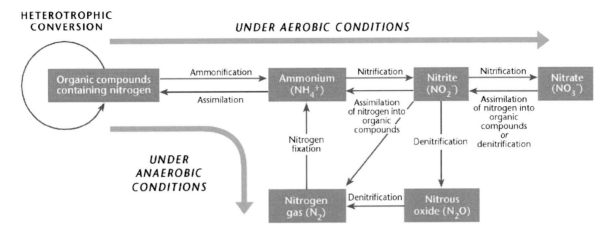

Figure 2.3.3: Transformations of nitrogen from waste water during groundwater recharge (from Lawrence et al. 1997).

Beyond the colmation layer, the soil/unsaturated zone will generally be aerobic in unconfined aquifer conditions, and mineral N will be in the form of nitrate. In the soil, organic nitrogen present in the effluent will ultimately nitrify and convert to nitrate via nitrite if the aerobic residence time in the unsaturated zone is sufficient. Hence, the worst-case scenario for nitrate contamination of groundwater is that all nitrogen lost through pipe leakage will ultimately be converted to nitrate in water that is recharging the aquifer.

The working approximation used in this study is that all of the N load, either applied to open space or lost through sewer leaks, will eventually arrive at the water table in the form of nitrate.

2.3.4.2.2 Boron

Boron has the potential to be an important contaminant when compared to the World Health Organisation guideline value of 0.3 mg/L. This remains the case despite the recent Australian Drinking Water Guideline (NHMRC and NRMMC, 2005), which has increased the boron guideline value from 0.3 mg/L to 4 mg/L. The effect in regulatory terms of increasing the contamination concentration threshold in Australia has been to reduce boron contamination hazard for drinking water sources. Effluent from South Australia's major wastewater treatment plant at Bolivar, for instance, has boron concentrations in the range 0.3 – 0.6 mg/L which would no longer be considered a constraint if the water was earmarked for recharge.

Boron is adsorbed on to soil particles. The degree of adsorption depends both on the soil conditions (pH, salinity) and type (contents of organic matter, iron and aluminium oxide, iron- and aluminium-hydroxy and clay). Boron adsorption can be reversible or irreversible, depending on the soil type and condition.

Munster (2003) considers that boron is suited for use as a conservative tracer because of its high aqueous solubility, its presence in nearly all wastewater, and the lack of effects by evaporation, volatilisation, oxidation-reduction reactions, mineral precipitation or dissolution in all but extremely saline waters. Boron is also suggested as a marker for sewage by Amick and Burgess (2003). DeSimone et al. (1997) found similar concentrations of boron both in contaminant plumes and in sewage, but at lower concentrations at the wetting front. This would suggest that in a steady state system, borate would be conserved. It is noted that the pH in DeSimone's trial was as low as 5.

For application to the Australian case study in Mount Gambier, boron as an indicator was also attractive because recent data from Vogel (2005) indicated that there was no adsorption of boron on core samples taken from the Gambier Limestone unconfined aquifer.

Other workers have noted that B is more likely to be found where the city retains, or has a history of, metal fabrication and processing (Morris, priv. comm). This may limit the usefulness of B as a conservative tracer.

Hasset et al. (2003) reported increased boron removal over time in leaching conditions, which was attributed to the formation of ettringite group minerals (sturmanite, charlesite) in the soil. The decrease in concentration was from 5 mg/L to 0.5 mg/L over about 200 days.

Wolf et al (2005b) reported boron concentrations of 0.2-0.3 mg/l in the soil water 30 cm beneath a sewer leak, indicating minor sorption compared to the typical range of 0.5-1 mg/l in the sewage.

According to Smallwood (1998), adsorption–desorption reactions are expected to be the only significant mechanism influencing the fate of boron in water. The extent of boron adsorption depends on the pH of the water and the concentration of boron in solution. The greatest adsorption is generally observed at pH 7.5 – 9.0. As mentioned above, Vogel (2005) did not detect any boron adsorption to carbonate material taken from Mount Gambier in a short term experiment. This carbonate material was low in clay (no clay minerals were quantified) and organic matter (<0.1%).

Boron is therefore considered to be conserved in the subsurface when the pH is less than 7.5, above which adsorption is expected, increasing with rise in pH and residence time.

For each layer, a reduction factor R of the form $R = \exp(-\lambda)$ is used, where
$\lambda = 0$ if pH $< a$, or
$\lambda = b$ (pH $- a$) if pH $> a$.

The current data are consistent with $a = 7.5$ and $b = 1.0$, but further data are required to give reliable estimates. It is noted that eventually the soil would saturate with borate, and under such conditions a steady state would exist in the soil as well as in the colmation layer, i.e. the contaminant would be transported unattenuated. This possibility has not been explored and is not considered by either SLeakI or POSI. A further possibility is that under very acid conditions (pH<3) boron could be removed from the profile, perhaps as a halide.

2.3.4.2.3 Sulphate

Although there is little information about the loss of sulphate in the colmation layer, factors that might affect loss rates include:
- An available microbial substrate – a useful surrogate would be DOC. Generally low dissolved oxygen levels, at least below 1 ppm (Cytoculture, 2004) and a correspondingly low redox potential would be required to make sulphate (which is difficult to reduce) a suitable electron acceptor for organic matter degradation (Fig 2.3.4),
- Temperature (Erichsen et al. 1999);
- Residence time (initially assumed constant).

Given the lack of quantitative data, the worst case that all the sulphate is leached to the subsurface is assumed.

2.3.4.2.4 Phosphate

Inorganic phosphorus (P) in wastewater will be in equilibrium between three states:

$$PO_4^{3-} \leftrightarrow HPO_4^{2-} \leftrightarrow H_2PO_4^-$$

with the relative amount in each state being largely affected by pH. At pH 5, approximately half the phosphate will be present as $H_2PO_4^-$. At high pH, much of the phosphate will be as PO_4^{3-}. The solubility of phosphate depends very much on the species of phosphate, and PO_4^{3-}. It will be strongly bound in the presence of iron oxides or precipitated at high pH. There are few references concerning the fate of phosphate from sewer leaks. There is a suggestion by Amick and Burgess (2003) that phosphate is a useful marker of sewage, but phosphate could be taken up by the organic matter in the soil humic layer at shallow depths and further down by sorption e.g. to calcite. Absence of phosphate would therefore not be an indicator of lack of contamination from sewage.

In the presence of high concentrations of Ca^{2+} or Mg^{2+}, phosphate will be insoluble and therefore precipitate. This suggests a relationship as outlined below.

For each layer, a reduction factor R of the form $R = \exp(-\lambda)$ is used, where
$\lambda = 0$ if pH < 5, or
b (pH $- a$) if pH $> a$.

The value of a was set at 5.0, corresponding to equal proportions of $H_2PO_4^-$ and HPO_4^{2-}. By pH 10, the main form would be PO_4^{3-} so most of the phosphate would be out of solution. Credible results are achieved by setting b to 10. These data are consistent with results from lysimeter studies in an area close to Mount Gambier, where a residence time of a minimum of 100 days is estimated in a soil of pH 8.5. In these cases, no P was detected in the drainage water. Further data are required for calibrating this part of the model.

Some of the leached phosphorus could be in an organic form. As discussed for the case of nitrogen, organic matter is considered as being hydrolysed releasing inorganic nitrogen. Similarly, it is assumed that organic P will also be released as inorganic P. There could be sorption of organic P by the colmation layer, but over time the net effect of this process would reduce. A conservative working assumption was therefore to assume that all P in the effluent is in the form of phosphate.

2.3.4.3 Pathogenic microbiota

The biological contaminants of concern are typically pathogens. These include a broad spectrum of organisms ranging in size from viruses through bacteria to protozoa.

An active colmation layer is reported to remove 99% of viruses (Dizer et al. 2003). This is a high level of reduction, and was consistent across the two studies. A 2-log reduction rate was therefore used in this study. However, at low temperature there would be little biological activity, so a correction for temperature was incorporated. Relevant temperature data are unlikely to be available so surrogate data have to be used. The temperature of the colmation layer was taken as being that of the effluent, while the temperature of the soil layers was approximated by the annual average air temperature.

It was assumed that higher removal rates would occur at temperatures greater than 20°C, although few data are available to quantify this assumption. Gordon and Toze (2003) showed that there was rapid attenuation at 28°C, especially in the presence of microbial activity.

A similar colmation layer removal result for bacteria was found by Vollerstsen and Hvitved-Jacobsen (2003). As in the case of viruses, it would be expected that higher removal rates of bacteria would occur at temperatures greater than 20°C. There is assumed to be microbial attenuation with a given half-life, varying with temperature. The rate of degradation is expressed as time for a 1-log reduction (or t_{90}). The log reduction is given by

Log reduction = (Residence time) (temperature correction) / (time for 1-log reduction)

where the temperature correction is $2^{(t-t_0)/10}$, t is the actual temperature and t_0 (typically 20°) is the temperature at which the experimental data are obtained.

The 1-log reduction times vary between microbe types. Two simplifying approximations that were necessary were:
(1) that residence time is assumed to be the same as for the water.
(2) that no new pathogen particles are assumed to form in the soil.

In summary, the models described above assume that chloride, sodium, potassium and sulphate behave conservatively, that all nitrogen converts to nitrate, that boron and phosphate sorption are pH-dependent and that pathogens are strongly attenuated in the colmation layer, beyond which they are inactivated at temperature-dependent rates which are related to experimental data.

2.3.5 Data requirements

The data required to run the packages SLeakI and POSI are presented in three files. The first file contains data that describes the local environmental features such as soil characteristics and temperature (neighbourhood data). A second file provides details of the effluent and the associated contaminants (contaminant data). These two files are described below.

A third file that is provided with the package supplies data concerning 1 log removal (t_{90}) times of pathogens. These data are the best available to the authors, but local data should be used when available. The package employs a number of constants, for which credible values are used. Although organic compounds were not contaminants of primary concern for this study, provision has been made for including half lives and partition constants (K_{oc}s) of organic compounds so that the packages POSI and SLeakI can provide estimates of the loads of these compounds entering the aquifer.

2.3.5.1 Neighbourhood data

The format is as given in Table 2.3.1

Table 2.3.1. Environmental parameters required to describe a neighbourhood.

Column number	Identifier	Unit
1	Neighbourhood identifier	Label
2	Thickness of soil layer	Metres
3	Texture of soil layer	See list
4	pH of soil layer	pH unit
5	Temperature of soil layer	Degrees C
6	Thickness of sub layer	Metres
7	Texture of sub layer	See list
8	pH of sub layer	pH unit
9	Temperature of sub layer	Degrees C

Notes:
a. Typically data in this table will be the same for all leaks in a neighbourhood.
b. Soil temperature may not be available so the annual average air temperature can be used as a surrogate. Annual average air temperature will be the average of the annual daily maximum and the annual daily minimum.

2.3.5.2 Contaminant data

The format for the effluent data is as shown Table 2.3.2. A similar file is used for both SLeakI and POSI, with slightly different input fields.

Table 2.3.2. Parameters required to describe effluent as input to SLeakI and POSI.

Column	SLeakI input	SLeakI units	POSI input	POSI units
1	Neighbourhood identifier	name	Neighbourhood identifier	name
2	Asset identifier	name	Open space identifier	name
3	Average leak rate	L/day	Infiltration rate	mm/day
4	Number of such leaks in asset	count	Number of such assets (e.g. gardens of same area)	count
5	Pipe diameter	metres	Area of open space	metres^2
6	Depth of pipe	metres	Depth of root zone	metres
7	Effluent temperature	degrees C	Soil water temperature	degrees C
8	Ground water pH	pH unit	Ground water pH	pH unit
9	Chemical oxygen demand COD	mg/L	Chemical oxygen demand COD	mg/L
10	Biological oxygen demand BOD	mg/L	Biological oxygen demand BOD	mg/L
11	Dissolved organic carbon DOC	mg/L	Dissolved organic carbon DOC	mg/L
12	TN	mg/L	TN	mg/L
13	Cl^-	mg/L	Cl^-	mg/L
14	Na^+	mg/L	Na^+	mg/L
15	K^+	mg/L	K^+	mg/L
16	NH_4^+	mg N/L	NH_4^+	mg/L
17	NO_3^-	mg N/L	NO_3^-	mg/L
18	BO_3^-	mg B/L	BO_3^-	mg/L
19	PO_4^{3-}	mg P/L	P	mg/L
20	SO_4^{2-}	mg S/L	SO_4^{2-}	mg/L
21	Polio virus	pfu/ml	Polio virus	cfu/ml
22	E. coli	cfu/ml	E. coli	cfu/ml
23	Giardia	cfu/ml	Giardia	cfu/ml

2.3.6 Implementation

The software is available in two forms. Batch versions of POSI and SLeakI are used in conjunction with the other programs in the AISUWRS suite of programs. In addition, interactive programs are available for both POSI and SLeakI. The interactive version can be used when data for the other components are not available. It also enables sensitivity analyses to be undertaken - for instance to estimate the effect repairing specific leaks may have on the contamination of the aquifer.

The outputs from both POSI and SLeakI provide estimates of the quantity of water leached (litres) and contaminant load (mg or numbers of infective units). This form of output is additive over various SLeakI and POSI outputs, enabling integration of the various sources of contaminants in a diverse urban environment.

2.3.7 Discussion

The theory used in the development of POSI and SLeakI required many simplifying assumptions. Failure of these assumptions could have a large effect on some variables, but less on others. For example, preferential flow (a deviation from the assumption that the soil is uniform) could increase the rate of flow and thus decrease the residence time. This would have little effect on the conservative inorganic contaminants but would greatly influence pathogen reduction efficacy. There could also be an effect of preferential flow on other contaminants such as phosphate.

Neither model deals with the effects of uptake of water and nutrients by vegetation or of evaporation from the soil surface. This is not considered a limitation of SLeakI, because sewers are generally deep and beyond the root zone of vegetation, except where tree roots exploit the sewer leak. However, for public open space irrigation these effects will be very important for water and nutrients (especially nitrogen, phosphorus and potassium). The results output from POSI should therefore be regarded as upper bounds for recharge and solute fluxes to the water table.

As in all models, the accuracy of the result is limited by the data quality. The inputs require rates of exfiltration from leakage in pipes and rates of infiltration into public open spaces. Furthermore, data are required on the concentration of the various contaminants in the sewage.

There are few data available on the effectiveness of the colmation layer in pathogen removal, both concerning its effectiveness when it is intact and in the proportion of time that it is effective. Further studies are required to validate the two-log reduction assumption used in SLeakI as it could be very important in any risk assessment.

Overall, it was concluded that contaminants leaking from sewers are conservative. There are exceptions – phosphate is likely to be reduced due to sorption, and the form of nitrogen is likely to change to nitrate.

For pathogens, the sensitivity analyses undertaken indicate that soil type and the thickness of soil layers between the leak and the water table are most important, followed by leak rate. This has practical implications in asset maintenance as it indicates that a useful strategy would be to concentrate maintenance (perhaps by asset replacement) on the most sensitive neighbourhoods (those with high soil conductance and short travel distance to a shallow water table) rather than identifying and repairing individual leaks.

The results discussed above are based on assumed models of water movement that neglect preferential flow beneath sewer leaks and other contaminant sources. Actual field data would be valuable in confirming whether the results obtained from the models described in this report are representative of field conditions.

2.3.7.1.1 Appendix 1: Notation

α, β, n = soil properties of various soil types as listed in van Genuchten (1980)
COD = Chemical oxygen demand
DOC = Dissolved organic carbon
g = gravitational acceleration
h = head of water
k = permeability [m^2/Pa]
K = $k\rho g$
λ = constant used in estimation of soil attenuation of contaminants
p = pore pressure [Pa]. $p = 0$ corresponds to atmospheric pressure, $p < 0$ implies suction and $p > 0$ implies positive pore pressure
q_0 = volumetric flux at surface
ρ = density of water [k/L]
t = residence time of a contaminant in the soil profile
θ = volumetric water content [m^3/m^3]
θu = volumetric flux (m/s)
u = average velocity [m/s]
x, y, and z are horizontal and vertical coordinates.

2.4 SIMPLE APPROACH FOR BALANCING TRANSIENT UNSATURATED SOIL PROCESSES IN URBAN AREAS BY THE ANALYTICAL MODEL UL_FLOW

2.4.1 Introduction

An integrated urban water balance is based on the fluxes within water supply and waste water discharge systems as well as on surface and subsurface water fluxes in urban areas. The consideration of the subsurface water fluxes in urban areas, i.e. groundwater flow and seepage in the unsaturated zone, is becoming more and more important in the context of water resources management due to the increasing scarcity of fresh water of potable quality. One major part of an integrated urban water balance is the water balance in unsaturated soils, as it forms the quantitative linkage between urban water systems, surface water and urban groundwater. Groundwater recharge is generated by water that infiltrates into soils from different kinds of sources. The related solute loads of those water fluxes affect groundwater quality.

In urban areas several kinds of sources infiltrating water to the soils are important. The major natural source is the infiltration from the ground surface of the balance of precipitation remaining after evapotranspiration and surface run-off processes. As part of the management of storm water discharge from large impervious surface areas in cities, more and more rain water infiltration ponds are being built. Such ponds act as strong local infiltration sources, whose importance is determined both by the hydrologic conditions and the constructed collecting system that feeds it. Another, very different kind of infiltration source comes from the pipe infrastructure, as leaks in water supply and sewer pipes. These act as point infiltrations and are mainly determined by the usage and quality of those urban water service facilities. In some more dispersed urban settlements, septic tanks are installed locally to dispose of waste water, and these also discharge via local drainage arrangements into the subsurface and eventually to groundwater.

Capillary and gravity forces are mainly driving the water flow and storage processes in unsaturated soils and are quite difficult to quantify due to the strong non-linearity and heterogeneity of the hydraulic soil properties, which are even important at the local scale. Particularly, the near subsurface in urban areas can be found radically modified by human activities stretching back over centuries. In general, different geologic units are present in urban areas with additional substantial variability of their internal structure. This generates large heterogeneities of the soil properties determining flow and transport processes in horizontal as well as vertical directions on the urban scale. Furthermore, the infiltration sources show a large variability both in their location in relation to the soil heterogeneity and temporally in the variations of the infiltration rates and associated contaminant mass loads specific for each single source location. This increases the complexity of the whole system that determines flow and mass transport in the unsaturated zone from the infiltration sources to the saturated aquifer.

Modelling unsaturated flow and groundwater recharge on an urban scale becomes a major undertaking because of the high variability in soil structure and boundary conditions as well as the non-linearity of the unsaturated water flow process. These aspects make it impossible to set up a large scale three-dimensional model for the unsaturated zone of the entire urban area. Instead there is a strong requirement to simplify the problem in an appropriate way, so that important principal components and their dependencies can still be meaningfully represented. In order to be predictive the developed methodology had to be based on physical approaches.

The assessment of the impact of urban areas on urban groundwater has been investigated in recent years by simplified large scale approaches dealing with effects on groundwater levels (Lerner et al., 1990; Foster et al., 1993; Eiswirth, 2001) or groundwater quality (Barrett et al., 1999). Another approach was to estimate exfiltrations from sewer networks by simple water balance methods (Härig & Mull, 1992) or detailed laboratory and field investigations (Dohmann et al., 1999; Vollertsen and Hvitved-Jacobson, 2003). Less effort has been spent on detailed studies on unsaturated processes in this context. Within the AISUWRS concept a feature of the multiple infiltration sources is the dimensionality and complexity of the flow patterns, i.e. regional areal source, local areal sources and local point sources. For each individual source an independent balance model was defined taking into account different unsaturated zone profiles, here referred to as soil profiles, different distances between source and water table and time-dependent infiltration rates (Mohrlok & Bücker-Gittel, 2005). For areal sources a one-dimensional flow pattern was assumed. Under steady state conditions analytical solutions for layered soils were derived and used for the estimation of solute residence times. These solutions together with a simplified volume balance for transient conditions were implemented in the analytical model UL_FLOW (Mohrlok, 2005).

2.4.2 Balancing unsaturated zone processes

The calculation of groundwater recharge is based on the soil water balance in the unsaturated zone between the infiltration source and the water table. Considering a transient infiltration rate this balance is determined by flow and storage processes. Mainly, the inflow into a certain soil volume is increasing the volumetric water content θ of that soil and the outflow is determined by capillary and gravity forces, which are depending on the water content θ itself by non-linear relationships. Water flow in the unsaturated zone is generally balanced by Richards' equation (Richards, 1931):

$$\frac{\partial \theta}{\partial t} = \vec{\nabla}\left[K(\theta)\left(-\vec{\nabla}\psi(\theta)+\vec{e}_z\right)\right] \quad (1)$$

where the hydraulic conductivity $K(\theta)$ and the water tension $\psi(\theta) = -p/\gamma \geq 0$ (given in positive values) due to capillary forces are depending on the water content θ. The suction head p/γ is defined negative with respect to the reference atmospheric pressure $p = 0$. γ is the specific weight of water. Several authors derived empirical formulae for those relationships (Gardner, 1958; Brooks & Corey, 1964; Mualem, 1976; van Genuchten, 1980), that describe the hydraulic properties of the specific soils. For simplicity in mathematical expressions the effective saturation

$$S_e = \frac{\theta - \theta_r}{\theta_s - \theta_r} \quad (2)$$

has been defined. Most common are the relationships from van Genuchten (1980) and Mualem (1976):

$$\psi_{vG}(\theta) = \frac{1}{\alpha}\left(S_e^{-\frac{1}{m}} - 1\right)^{\frac{1}{n}} \quad (3)$$

$$K_{vG}(\theta) = K_s \left[1 - \left(S_e^{-\frac{1}{m}} - 1\right)^m S_e\right]^2 S_e^{\frac{1}{2}}, \quad (4)$$

where θ_r, θ_s are the residual and saturated water content, respectively, K_s is the saturated hydraulic conductivity. The soil-specific van Genuchten parameters α and n are usually used to characterize the soil hydraulic properties and have been summarised for several soil types by Carsel & Parish (1988). The parameter $m = 1 - 1/n$ is usually determined by the parameter n. Here, the relationships from Brooks & Corey (1964) are used for their mathematical simplicity:

$$\psi_{BC}(\theta) = \psi_0 \, S_e^{-\frac{1}{\lambda}} \quad (5)$$

$$K_{BC}(\theta) = K_s \, S_e^{\frac{\eta}{\lambda}}. \quad (6)$$

The parameter $\psi_0 = 1/\alpha$ is related to the van Genuchten parameter α and defines the air entry pressure. The parameters $\lambda = 1 - n$ and $\eta = 3n - 1$ are related to the van Genuchten parameter n. The differences in retention relationships for corresponding parameter values are obvious for low water tension (Figure 2.4.1). The van Genuchten relationship (eq. 3) is continuous down to $\psi = 0$, whereas the Brooks & Corey relationship (eq. 5) reaches the air entry value $\psi = \psi_0$ at saturation $\theta = \theta_s$ and no values $\psi < \psi_0$ exist. However, the hydraulic conductivity $K_{BC}(\psi) = K_s$ is defined for $\psi < \psi_0$.

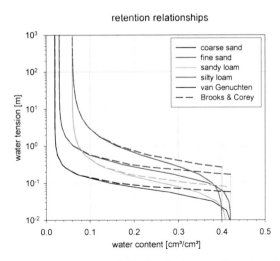

Figure 2.4.1. Retention relationships for different soils (see Table 2.4.1) after van Genuchten (eq. 3) and Brooks & Corey (eq. 5).

Analytical solutions of the Richards Equation (eqn. 1) can be obtained only for very simple cases. The simplest case is the vertical one-dimensional infiltration from areal sources under steady state conditions. For a homogeneous soil layer the remaining non-linear ordinary differential equation:

$$q_{inf} = K(\psi)\left(-\frac{d\psi}{dz}+1\right) \quad (7)$$

has to be solved. The upper boundary condition is given by the infiltration flow rate per unit area q_{inf}, which is assumed to be the excess of the soil water balance in the upper most soil layer (0.5 to 1m) due to climatic conditions. The lower boundary condition is naturally the water table, $\psi = p = 0$.

Assuming constant water content θ over the depth, and therefore constant water tension ψ, existing far above the water table or any other prescribed water tension condition, only gravity is the driving force. This situation is denoted as unit gradient condition and equation (7) simplifies to:

$$q_{inf} = K(\psi) = K(\theta). \quad (8)$$

Hence, the infiltration flow rate per unit area q_{inf} defines the associated water content θ. Using Brooks & Corey relationships (eq. 5 and 6) the water tension and water content are given respectively:

$$\psi = \psi_0 \left(\frac{q_{inf}}{K_s}\right)^{-\frac{1}{\eta}}$$

$$\theta = \theta_r + (\theta_s - \theta_r)\left(\frac{q_{inf}}{K_s}\right)^{\frac{\lambda}{\eta}}. \quad (9)$$

In the case of a water tension condition existing at the lower boundary at z_0 equation (7) has to be integrated to obtain a solution. Using Brooks & Corey relationship (eq. 5) the integration could be obtained by approximation of the denominator as power series expansion for $q_{inf} \ll K(\psi)$, which is valid far from unit gradient conditions close to the lower boundary:

$$\int_{z_0}^{z} dz' = \int_{\psi_0}^{\psi} \frac{d\psi'}{1 - \frac{q_{inf}}{K(\psi)}} \cong \int_{\psi_0}^{\psi}\left[1 + \frac{q_{inf}}{K_s}\left(\frac{\psi}{\psi_0}\right)^{\eta}\right]d\psi' \quad (10)$$

The solution can be given in implicit form:

$$z = z_0 - \psi_0\left[1 - \frac{\psi}{\psi_0} + \frac{q_{inf}}{K_s}\frac{1}{\eta+1}\left(1 - \left(\frac{\psi}{\psi_0}\right)^{\eta+1}\right)\right] = z_0 - \psi_0\left[1 - \left(\frac{\theta-\theta_r}{\theta_s-\theta_r}\right)^{-\frac{1}{\lambda}} + \frac{q_{inf}}{K_s}\frac{1}{\eta+1}\left(1 - \left(\frac{\theta-\theta_r}{\theta_s-\theta_r}\right)^{-\frac{\eta+1}{\lambda}}\right)\right] \quad (11)$$

This solution holds for $z < z_{max}$ defined by:

$$z_{max} = z_0 - \psi_0\left[1 + \frac{q_{inf}}{K_s}\frac{1}{\eta+1} - \left(1 + \frac{1}{\eta+1}\right)\left(\frac{q_{inf}}{K_s}\right)^{-\frac{1}{\eta}}\right] \quad (12)$$

above which the influence of the capillary rise is negligible and equation (8) holds. The full water content distribution over the entire soil layer depth in accordance with a steady state infiltration rate q_{inf} can be described by equation (11) for $z_0 < z < z_{max}$ and by equation (9) for $z > z_{max}$.

This distribution can be used directly to estimate the residence time of a conservative solute within a homogeneous soil layer of depth L by:

$$T = \int_{z_0}^{z_0+L} \frac{\theta(z)}{q_{inf}} dz \qquad (13)$$

These solutions are applicable also to layered soils. For each homogeneous layer i the upper boundary condition is the steady state infiltration rate q_{inf}. The lower boundary condition is defined by the continuity of the water tension ψ at the layer interface z_{i-1} and the water table for the lowest layer respectively. Hence, two cases need to be distinguished with respect to the water tension at the layer interface; $\psi_{i-1}(z_{i-1})$ calculated within the lower layer $i-1$ and the water tension due to unit gradient conditions $\overline{\psi}_i$ (eq. 9) in layer i. In the case $\psi_{i-1}(z_{i-1}) < \overline{\psi}_i$ equation (11) has to be modified and gives for layer i:

$$z = z_{i-1} - \psi_{0,i} \left[\frac{\psi_{i-1}(z_{i-1})}{\psi_{0,i}} - \frac{\psi_i}{\psi_{0,i}} + \frac{q_{inf}}{K_{s,i}} \frac{1}{\eta_i + 1} \left(\left(\frac{\psi_{i-1}(z_{i-1})}{\psi_{0,i}}\right)^{\eta_i+1} - \left(\frac{\psi_i}{\psi_{0,i}}\right)^{\eta_i+1} \right) \right] \qquad (14)$$

The case $\psi_{i-1}(z_{i-1}) > \overline{\psi}_i$ leads to $q_{inf}/K(\psi) \gg 1$ and the approximation in equation (10) has to be adapted. The obtained solution for layer i is:

$$z = z_{i-1} - \frac{K_{s,i}}{q_{inf}} \frac{\psi_{0,i}}{1-\eta_i} \left[\left(\frac{\psi_i}{\psi_{0,i}}\right)^{1-\eta_i} - \left(\frac{\psi_{i-1}(z_{i-1})}{\psi_{0,i}}\right)^{1-\eta_i} \right] - \frac{K_{s,i}^2}{q_{inf}^2} \frac{\psi_{0,i}}{1-2\eta_i} \left[\left(\frac{\psi_i}{\psi_{0,i}}\right)^{1-2\eta_i} - \left(\frac{\psi_{i-1}(z_{i-1})}{\psi_{0,i}}\right)^{1-2\eta_i} \right] \qquad (15)$$

If the upper layer interface $z_i > z_{max,i}$ unit gradient condition occurs within layer i above $z_{max,i}$, the solution for the complete layered soil profile has to be calculated layer by layer, starting at layer 1 at the water table. By usage of Brooks & Corey relationship (eq. 5) the respective calculated water tension values can be transformed into water content values. The residence time within each layer can be estimated from these solutions by applying equation (13).

These analytical solutions were applied to a two-layer soil profile for residence time estimations. One layer was selected as fine- to medium-grained sand and the other was coarse-grained sand. The layer sequence, the layer thicknesses and the infiltration rates were varied. In each case the discontinuity of the water content at the layer interface is obvious (Figure 2.4.2). The influence of the water table on the water content distributions vanished at lower levels for high infiltration rates due to the higher water content defined by unit gradient flow. Residence times became larger for low infiltration rates and high water contents. Even in the case of a thick coarse-grained soil layer and high infiltration the residence time was at least 1 day.

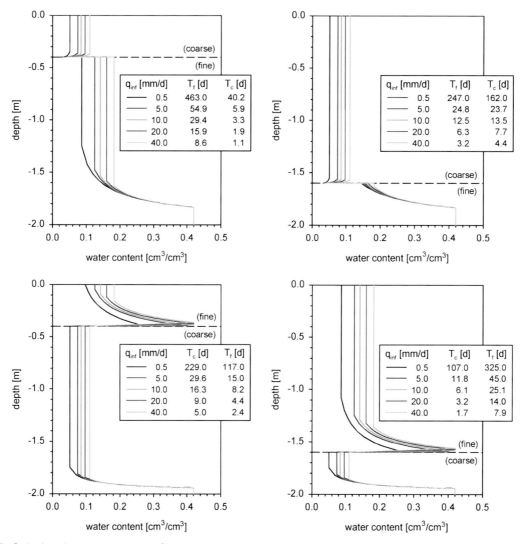

Figure 2.4.2. Calculated water content profiles and residence times in soil profiles with a fine and coarse grained layer of different thicknesses and sequence and at different infiltration rates q_{inf}.

2.4.3 Modelling approach UL_FLOW

The large temporal variability in infiltration rates from the individual sources required the development of a simplified approach to balance the transient water storage processes and transport behaviour of conservative solutes in the unsaturated zone in the heterogeneous urban subsurface. The above-described steady state solutions were combined with a simple water volume balance on a time step basis. Assuming equilibrium conditions defined by those steady state solutions, for a certain infiltration rate q_0 in the previous time step a change in water content $\Delta\theta$ occurs due to changing infiltration rate $\Delta q_{inf} = q_{inf} - q_0$. Furthermore it is assumed that the actual infiltration rate q_{inf} results in the respective equilibrium conditions within a depth L' at the top of the soil profile. Hence, a simple volume balance can be defined for the time step Δt:

$$\Delta q_{inf} \, \Delta t = L^* \, \Delta\theta \, . \tag{16}$$

In order to apply the developed steady state solutions, this volume balance is replaced by using an effective infiltration rate q'_{inf} and an effective change in water content $\Delta\theta'$ considering the whole profile depth L (see Figure 2.4.3):

$$\Delta q'_{inf} \, \Delta t = L \, \Delta\theta' \, . \tag{17}$$

As a consequence, from this consideration the effective infiltration rate q'_{inf} accounts for an intermediate storage in the unsaturated zone and can be calculated by:

$$q'_{inf} = K_s \left[\frac{q_{inf} - q_0}{\theta_s - \theta_r} \frac{\Delta t}{L} + \left(\frac{q_0}{K_s} \right)^{\frac{\lambda}{\eta}} \right]^{\frac{\eta}{\lambda}}. \tag{18}$$

Calculating this effective infiltration rate q'_{inf} for each time step enables the usage of the steady state solutions (eq. 9, 14, 15) for transient calculations of the water flow and the estimation of residence times. Also, multi-layered soils can be considered by this simple balance approach if it is assumed that the infiltration affects only the uppermost layer within the considered time steps (Figure 2.4.3).

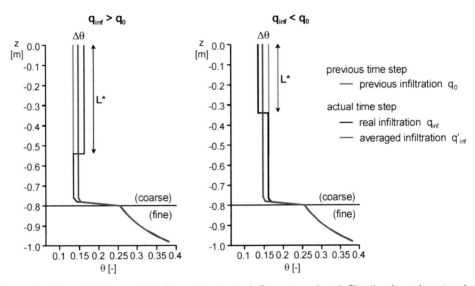

Figure 2.4.3. Concept of the effective vertical flow rate q'_{inf} to define a transient infiltration by using steady state solutions on time step base.

Based on that approach the steady state solutions can be used for computation of the residence time for each time step, the single time step residence time, within each layer by using equation (13). Adding the value of each layer simply gives the single time step residence time of the entire soil profile. Hence, this single time step residence time of each time step is directly inverse correlated with the effective infiltration rate q'_{inf} under the steady state conditions assumed for the respective time step (Figure 2.4.4). Tracer residence times are not given directly by this quantity. However, these single time step residence times define the progress of the migration of a conservative tracer solute within the respective time step. The residence times for the advective transport of such a tracer through the considered soil profile can be computed by adding all migration steps until the bottom of the profile is reached. These computed tracer residence times are related to the time step when the migration has been started and are an integrated quantity over the whole migration period (Figure 2.4.4). Values are only computed when the bottom of the entire profile is reached within the considered simulation time period, which is usually not given for tracer migration started at late time steps.

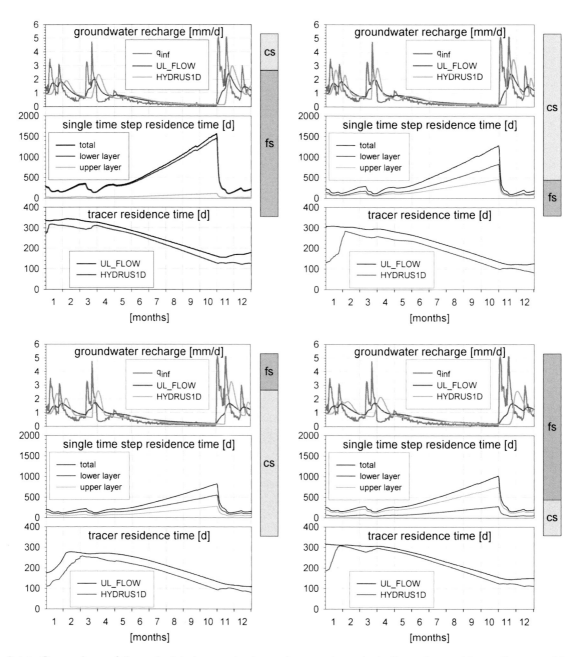

Figure 2.4.4. Comparison of the calculated groundwater recharge rates, single time step residence times and tracer residence times by UL_FLOW (Mohrlok, 2005) and HYDRUS1D (Simunek et al., 2005) for different two layered soil configurations (fine- and coarse-grained sand).

In order to validate this approach implemented in the analytical model UL_FLOW (Mohrlok, 2005) numerical simulations of one-dimensional vertical infiltration have been calculated for a certain set of soil profiles using the numerical model HYDRUS1D (Simunek et al., 2005) using a two years time series of infiltration rates defining the upper boundary condition. The water table, $\psi = 0$, was assumed as the lower boundary condition. Additionally, to validate the estimation of the tracer residence time, tracer migration was simulated by injecting tracer at each time step and analysing the peak arrival of the tracer breakthrough curve at the lower boundary condition of the model domain.

The groundwater recharge rates calculated by the simple UL_FLOW approach and by HYDRUS1D were very similar. A certain time shift occurred for the time series calculated by HYDUS1D since the infiltration moved as a front through the profile whereas the calculation of effective infiltration rates q'_{inf} in UL_FLOW routed the infiltration partly during several time steps directly to the groundwater (Figure 2.4.4). This routing also underestimated the peak recharge rates in the cases of the upper layer consisting of fine-grained sand and the cases of thick coarse-grained sand layers. The single time step residence times are a virtual quantity introduced by the UL_FLOW approach and therefore not computed by HYDRUS1D. However, the tracer residence times calculated by the two different models could be compared and were found rather similar. Due to the already described infiltration routing in the UL_FLOW approach the computed tracer residence times by that approach were slightly higher than those computed by HYDRUS1D. The integrative nature of the tracer residence time together with this result and the delay of the outflow in the HYDRUS1D simulations resulted in the earlier increase observed for the residence times computed by UL_FLOW. In the presented example tracer residence times were computed for all presented time steps since the maximum values were less than one year and only results of the first year of the two year time period are plotted (Figure 2.4.4).

2.4.4 UL_FLOW Application to case study city

The analytical model UL_FLOW (Mohrlok, 2005) has been applied to the data sets from the case study cities of the AISUWRS project. For the major soil types reported from these case study cities the soil hydraulic parameters due to van Genuchten (1980) were defined by different methods (Table 2.4.1). As far as given by Carsel & Parish (1988) the parameters were taken from their publication. If not directly given there, for some loamy soils the parameters were estimated by related soil types from Carsel & Parish (1988). Some parameters of the very fine grained soils have been changed slightly due to recent investigations with these soil classes. For till, the soil parameters were estimated on the basis of a sandy clay taking into account large gravel-sized particles. The parameters of sandy and gravel soil types were derived from drainage experiment analyses of soil samples from Rastatt.

Table 2.4.1. Hydraulic parameters due to van Genuchten (1980) of the reported soil types of the case study cities; (a) investigations by drainage experiments; (b) taken from Carsel & Parish (1988); (c) silt loam from Carsel & Parish (1988) with lower porosity and hydraulic conductivity; (d) estimations according to sandy clay (partly sandy loam) from Carsel & Parish (1988); (e) clay (combined partly clay loam) from Carsel & Parish (1988).

soil type	α [1/cm]	n [-]	θ_r [-]	θ_s [-]	K_s [m/s]
gravel (a)	0.350	4.30	0.01	0.42	1.0e-3
coarse sand (a)	0.175	2.85	0.02	0.42	2.0e-4
fine sand (a)	0.060	2.40	0.03	0.42	3.0e-5
loamy sand (b)	0.124	2.28	0.06	0.41	4.0e-5
sandy loam (b)	0.038	1.89	0.06	0.40	1.0e-5
silty loam (c)	0.020	1.41	0.09	0.41	5.0e-7
till (d)	0.028	1.35	0.07	0.40	3.0e-6
loamy clay (e)	0.011	1.12	0.07	0.40	5.0e-7

In order to assess the possibility of describing flow and transport within the unsaturated soils in urban areas by steady state conditions a parameter study was conducted. The groundwater recharge rates and the single time step residence times as well as the tracer residence times were computed for a two-year infiltration period by UL_FLOW using daily time steps. The influence of storage effects were investigated by varying the soil depths from the areal infiltration source at the surface to the water table for all soil types (Table 2.4.1). As expected, coarse-grained soils and lower soil depths showed much larger variations in the computed quantities than fine grained soils and large soil depths (Figure 2.4.5).

According to the soil type hydraulic conductivity, the residence times are much larger in soils with low conductivity. For instance, in gravel soils the residence times were several days for small soil depths and up to one year for large depths, whereas in silty loam the residence times were several months even for small soil depths. In general, the variability of the tracer residence time was found to be much lower than the variability of the single time step residence time. In general, time series of tracer residence time break down when the computed residence time exceeded the remaining time within the considered period.

The application of the presented approach to the Rastatt case study city was based on the database for the other models within the AISUWRS project (Klinger et al 2006). Mainly, the discretization units were the neighbourhoods defined for the UVQ model (Mitchell et al., 2004). These neighbourhoods were treated as homogeneous units also with respect to their soil properties and boundary conditions for areal infiltration from precipitation. This means that the unsaturated zone related to each neighbourhood was represented by a representative soil profile derived from the geologic units (Wolf et al 2005b), with a certain ground surface and water table elevation defining the soil depth. No water table fluctuations were considered. The time series of the fluxes released to the subsurface calculated by UVQ for the considered base line period in the case studies (see chapter 4.1) were used as the areal infiltration time series.

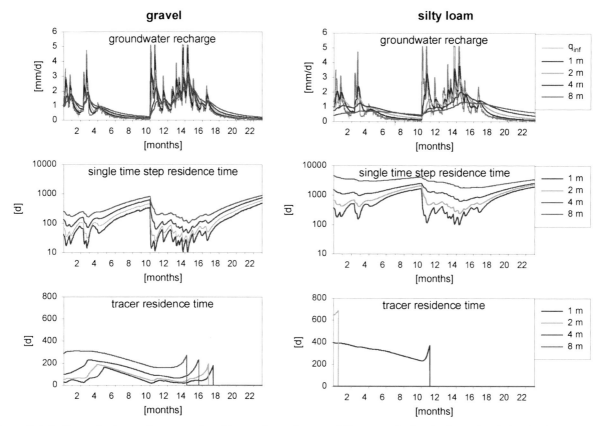

Figure 2.4.5. Groundwater recharge and residence times for two different typical soils from the AISUWRS case study cities calculated using UL_FLOW (Mohrlok, 2005) for a two year infiltration period and different soil depths.

In order to assess the simplified assumption for the AISUWRS model chain, which is to select only a few representative time steps on order to balance unsaturated flow and transport, the influence of time step size on the calculated residence time was investigated. This simplified approach accounts for the transient behaviour of the system much less than the UL_FLOW approach. For comparison with those selected representative time steps, different statistically representative residence times were derived from the computed time series. These statistical values of the residence time were the minimum and the maximum values as well as the 10, 50 and 90 percentiles, threshold values that have not been exceeded in the respective portion of all time steps. This time series analysis was statistically correct only for the presented example with daily time steps. For the analysis of the monthly and yearly time step examples the considered period was too short, and therefore the number of time steps too low, that the analyses can be considered correct in a statistical sense. However, results for all time step lengths are presented to demonstrate the applicability of the methodology.

According to the UVQ results large differences occurred in the infiltration time series of the individual neighbourhoods. For discussion here a neighbourhood with a high infiltration of about 300 mm/a was considered in order to be able to present also the analysis of tracer residence time results (Figure 2.4.6).

In general, the results from using daily or monthly time steps were rather similar. The temporal variations of groundwater recharge rates were similar for the daily and monthly time step results except the time resolution. However, the yearly time steps provide only an average over all that variation. Comparable observations could be made for the residence times. Important differences remain between the computed single time step residence times for daily and monthly time steps. From the monthly time step computation, the residence time values are slightly but significantly higher. This had a strong influence on the derived 50 and 90 percentiles and maximum statistical values but minor influence on the minimum and 10 percentile values. On the other hand no differences were found for the tracer residence times in the time series and the statistical values as well. These time series break down after month 15 when the computed residence time exceeded the remaining time within the considered period. This means that the statistics of daily and monthly time steps might be similar regarding unsaturated soil processes, which can reduce computational effort establishing integrated urban water balances.

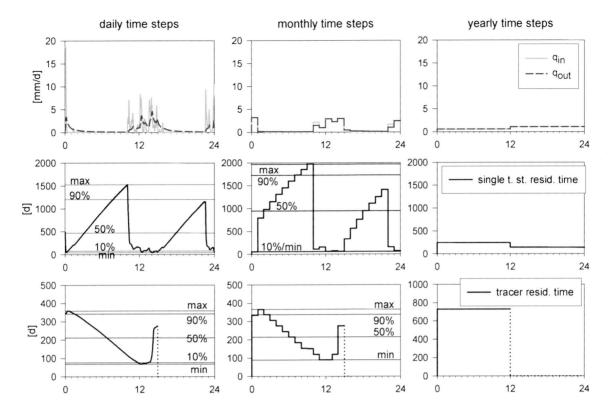

Figure 2.4.6. Comparison of groundwater recharge and residence times time series with their representative statistical values estimated by UL_FLOW (Mohrlok, 2005) for different time step lengths for one of the neighbourhoods in Rastatt case study city, applying the base line infiltration rate calculated with the UVQ model (Mitchell et al., 2004) for Rastatt (Klinger et al, 2006).

In conclusion, the yearly time steps, i.e. annual recharge rates, should not be used if temporal variations are of importance for the assessment of transport behaviour through the unsaturated zone. In particular, the calculated residence times could not be related even to any of the statistical values of the time series with daily and monthly time steps. Furthermore, some of the derived statistical values from the monthly time step time series of the single time step residence times are not representing the short term variations even in a statistical sense as obtained from the daily time step time series.

Considering the whole urban area of Rastatt there was found quite large spatial variability of simple quantities like the mean annual groundwater recharge and the statistical values of the residence times (Figure 2.4.7, Figure 2.4.8). Low recharge rates were correlated with high minimum single time step residence times (Figure 2.4.7) in the respective neighbourhood as expected. Although there is almost no difference for the results of the computations using different time step length the distribution of the minimum single time step residence times show slight differences between daily and monthly time step computation and large differences between daily and yearly time step computations (Figure 2.4.7). In general there was found an increase of residence times with the used time step length.

Comparing the spatial distribution of the several statistical values of the two different residence times shows strong differences (Figure 2.4.8). The minimum single time step residence times were generally lower than the minimum tracer residence times for each neighbourhood. In contrast, the 50 percentile of the single time step residence times was significantly higher than that of the respective tracer residence times. Additionally, the spatial variability of the single time step residence times decreased significantly with increase of the statistical value, whereas only slight changes were observed for the tracer residence times. However, only tracer residence times less than 24 months could have been computed for the considered two-year period. The larger they are the shorter the respective time series has to be (compare Figure 2.4.6). Therefore, for a large number of neighbourhoods no tracer residence times could have been computed (Figure 2.4.8).

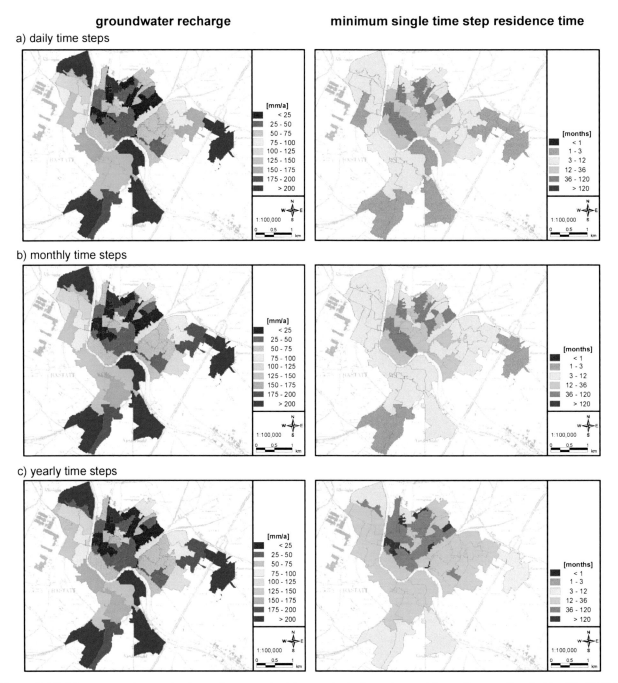

Figure 2.4.7. Mean annual groundwater recharge and minimum single time step residence times estimated by UL_FLOW (Mohrlok, 2005) on different time step length basis for all neighbourhoods in Rastatt case study city applying the base line infiltration rate calculated by Klinger et al (2006) using UVQ (Mitchell et al., 2004) and representative statistical values of the residence times time series.

The application of the UL_FLOW approach demonstrates the possibility to assess transport processes in the unsaturated zone of an entire urban area. Particularly, it was possible to avoid complex transport modelling by introducing single time step residence times for the derivation of tracer residence times representing purely advective transport. Considering the large scale of an urban area, GIS methods are required for the data flow to the model and for the graphical presentation of the result. However, using GIS the computed time series has to be represented by representative statistical parameter values. In the presented work only simple time series analysis was used for that purpose, in order to obtain extreme values and specific percentiles.

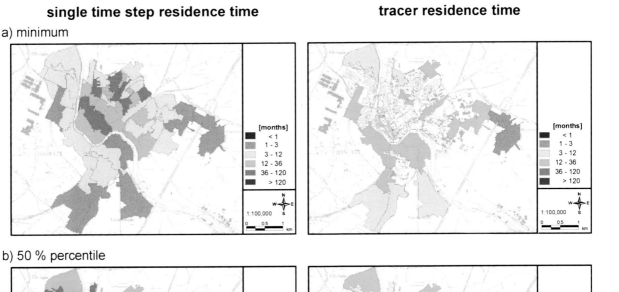

Figure 2.4.8. Comparison of the minimum and 50 percentile single time step residence times and tracer residence times from daily time step computation for all neighbourhoods in Rastatt case study city; no tracer residence times larger than 24 months could be calculated for the applied two year infiltration time period (uncoloured neighbourhoods).

2.4.5 Discussion and Conclusions

The requirement of assessing unsaturated soil processes on large scales with additionally large spatial variability (like an entire urban area) led to the development of the presented approach. This approach uses a one-dimensional steady state solution and a simple volume balance for quantifying transient water flow and conservative tracer residence time in layered unsaturated soil profiles from an areal infiltration source to the water table. Considering steady state in a one-dimensional system defines the volumetric outflow as equal to the inflow due to continuity, since no intermediate storage is available. The derived analytical solution describes the water content distribution within the soil layers (eq. 11, 14, 15). Additionally, it provides the possibility of estimating residence times related to purely advective transport within each layer (eq. 13). This residence time together with the given water content can be used as the basis for estimating transport of reacting solutes, for instance by means of attenuation probability such as that implemented in the AISUWRS models POSI and SLeakI (see chapter 2.3).

The one-dimensional steady state solution has been combined with a simple way to account for the additional balance of storage processes generated by transient inflow conditions. This combined concept has been implemented into the analytical model UL_FLOW (Mohrlok, 2005) by distributing the inflowing volume over the whole depth of the profile within each predefined single time step and deriving an effective steady state infiltration rate q'_{inf} (eq. 18). This procedure provides the possibility to calculate directly the single time step residence time based on that effective steady state infiltration rate q'_{inf} for each time step. Using this single time step residence time it is very simple to route a conservative tracer through the whole profile and compute the cumulative tracer residence time for the entire profile. Both quantities could be computed very fast in comparison with the numerical computation of solute transport in unsaturated soils using HYDRUS1D (Simunek et al., 2005). The groundwater recharge and tracer residence time obtained by UL_FLOW can be regarded as a good estimate as they are similar to the respective results obtained by HYDRUS1D (Figure 2.4.4).

This approach was also used to assess the steady state assumption for the unsaturated soil processes within the AISUWRS modelling concept. The strong daily variations in the infiltration rates can be represented to a certain extent by monthly but not by yearly time step averages. Monthly averages could be used because under certain circumstances the processes in the unsaturated soils are running on that time scale (Figure 2.4.6, Figure 2.4.7). In addition, the statistical time series analyses provide the information about which time steps are appropriate to select to properly represent hydrological conditions and the corresponding unsaturated zone processes.

Within the presented work some important unsaturated zone processes are not yet considered. The most important one is preferential flow leading to fast breakthrough of solutes to the groundwater as a consequence of very short residence time. Even reactive solutes would not be significantly attenuated in such circumstances. However, preferential flow can be considered simply within this presented approach if a further approach will have been developed that describes the amount of the infiltration that is related to preferential flow for the specific soil type. Unfortunately, the developed UL_FLOW approach cannot be applied to another important kind of infiltration sources in urban area, the point source infiltration from pipe leaks. The three-dimensional hydraulics of the unsaturated processes related to point sources are very different from the one-dimensional flow that can be assumed as a reasonable approximation for areal sources. The analytical solution for steady state flow from point sources to the water table through a two layered soil developed by Robinson (2005) and implemented in SLeakI (chapter 2.3) could be used for that purpose. For balancing transient infiltration the simple volume balance as implemented in UL_FLOW might be adapted in the future.

2.5 URBAN GROUNDWATER MODELLING

2.5.1 Introduction

2.5.1.1 Objectives of urban groundwater modelling

Urban groundwater models describe the functioning of the subsurface water system and allow predictions regarding future quantity and quality of this water resource. In most geological settings, numerical groundwater models are applied but some very complicated systems, such as heavily karstified limestones, may be better suited to a risk assessment method than a deterministic model. Urban groundwater models provide information relevant to a range of typical urban issues, such as:

- Predicting the impact of groundwater abstraction resulting from construction-stage dewatering, public or private water supply or geothermal groundwater circulation schemes;
- Delineation of catchment zones;
- Prediction of groundwater quality deterioration as a result of accidental spillages or ongoing urban activities;
- Planning of remediation measures for contaminated sites; or
- Prediction of groundwater quality deterioration from defective sewer systems.

The AISUWRS project applied different model approaches in the case study cities. These reflect both the requirements of the hydrogeological setting and also the pre-existing know how in model operation. Finite difference models based on MODFLOW were applied in Doncaster and Ljubljana, coupled with MT3D for contaminant transport (section 2.5.2). For pre- and postprocessing, Groundwater Vistas (GV©) was used in Doncaster while Visual Modlow© was applied in Ljubljana. A slightly different approach followed in Rastatt where the finite-element code Feflow was applied (section 2.5.3). Feflow includes flow and transport as well as pre- and postprocessing tools in one software package. A saturated zone risk assessment was utilised in Mount Gambier (section 2.5.4).

The aim of this section is not to provide specific details of the groundwater modelling techniques utilised, as extensive literature is already available. Instead, it focuses on the issues that arise when applying such methods and models to urban aquifers and their link with results cascading down from the other AISUWRS models. For further reading on general groundwater modelling, the works of Anderson and Woessner (1992), Rausch et al. (2005), Kovar (2003) or Diersch (2005) are recommended.

2.5.1.2 Key role of the conceptual model

The first step in any modelling exercise is to formulate a conceptual model. Such conceptualisations are needed to simplify the field problem because a complete reconstruction of the field system is not feasible (Anderson and Woessner 1992). However, the closer the conceptual model approximates the field situation, the more faithful the numerical model is likely to be. In groundwater terms a conceptual model describes how water enters or leaves an aquifer system and how it flows through it. Such conceptualisations can be simple sketches that are based on generalised geological cross-sections, with an explanatory legend, or they can be more detailed, so that three-dimensional geological information is represented and notes detailing supporting field information are included. .

As the nature of the conceptual model influences the structure of the numerical model and the features that it has to incorporate, it may also control the selection of the software used to build the numerical model. Conceptual models will evolve as understanding of the natural system develops and as a result the numerical model has to be modified accordingly. This may require alternative software to build the numerical model if the package selected at the start of the project fails to represent new features included in the revised conceptual model.

In general, the solute transport model is based on the same conceptual model used to build the flow model. However, additional conceptualisations may be needed to describe the solute sources and the mechanisms affecting the movement of the contaminant while present in the aquifer, such as dissolution/precipitation or ion exchange.

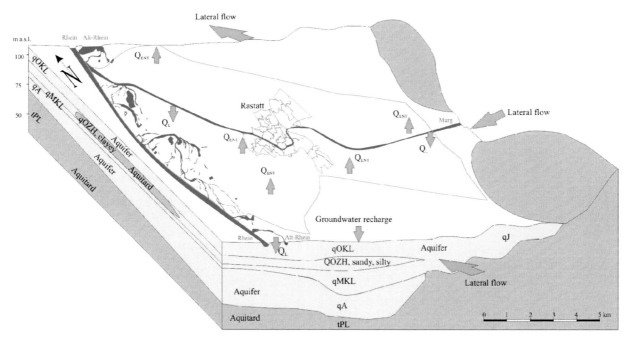

Figure 2.5.1. Conceptual hydrogeological model of Rastatt (modified from Eiswirth 2002). Q_{ww} = Waterworks, Q_L = Leakage from River, qOKL = Upper Gravel Layer, qOZH = Upper Dividing Horizon (silty, sandy), qMKL = Middle Gravel Layer, qJ = not differentiated Younger Quarternary, qA = Older Quarternary, tPL = Pleistocene.

2.5.1.3 Data requirements for groundwater models

The study area is represented in a numerical model by specifying the proper boundary conditions along a geometrical feature. The aquifer characteristics such as the hydraulic conductivity, storage coefficient, aquifer conditions (i.e. confined or unconfined), etc. must be defined. In addition, stresses such as recharge or abstraction boreholes or gains from/losses to rivers or lakes must be included. Anderson and Woessner (1992) list the ideal data requirements for a groundwater flow model, adapted from Moore (1979), these are:

- Geological map and cross sections showing the areal and vertical extent and boundaries of the system.
- Topographic maps showing surface water bodies and divides.
- Contour maps showing the elevation of the base of the aquifers and confining beds.
- Isopach maps showing the thickness of aquifers and confining beds.
- Maps showing the extent and thickness of stream and lake sediments.
- Water table and potentiometric maps for all aquifers.
- Hydrographs of groundwater heads and surface water (river, lake) levels and discharge rates.
- Maps and cross-sections showing the storage properties of the aquifers and confining beds.
- Hydraulic conductivity values and their distribution for stream and lake sediments.
- Spatial and temporal distribution of rates of precipitation, evapotranspiration, groundwater recharge, surface-groundwater interaction, groundwater pumping, and natural groundwater discharge.

When a solute transport model is built the background concentration, chemical reactions, dispersivity, etc. must also be defined. For a solute transport model the additional data requirements are:

- Maps showing the aquifer background concentration, physico-chemical setting and if possible the location of sources of contamination.
- Historical spatial and temporal observations of contaminants.
- Temporal distribution of rates of contaminant application.

For a groundwater flow and solute transport model built to study an urban area's impact on the underlying aquifers, additional information may need to be taken into consideration. For example, geological maps and cross-sections need to be extended upwards beyond the saturated part of the aquifers to the ground surface to include all stratigraphical characteristics of the area. This is because an impermeable layer, present for instance as a thin horizon in superficial deposits or otherwise in the unsaturated zone may occur between the ground surface and the water table and may transfer the water leaking from urban areas laterally to remote locations.

On the other hand, urban geographical characteristics such as the extent of the built area and the characteristics of the different districts within the city may alter the structure of the numerical model. Large variations in characteristics from one neighbourhood to another (residential, commercial or industrial usage, housing density, wastewater disposal practice, etc.) may require use of a fine grid mesh to represent these variations accurately, whereas small variations may permit use of a large cell size with correspondingly simpler computation and data manipulation demands.

Finally, unlike the contaminants generated from a spillage or accident, those originating from everyday urban activities are neither instantaneous in effect nor necessarily severe in their early stages. Often derived from a myriad of small sources rather than a single large incident, such pollution tends to result not in a distinctive concentrated plume but rather in a more diffuse increase (Figure 2.5.2) with locally enhanced 'noise' close to sources (leaks, septic tank drains, etc.). In such areas the contaminant load is approximately proportional to, and the concentration independent of, the urban area size, and develops over time. Contaminant concentration would then increase towards the mean load value unless preventive measures are undertaken (load reduction, recharge process modification, engineering interventions). Historical investigations into the way that older urban areas have evolved are therefore often helpful in identifying the legacy of previous activities and water management practices.

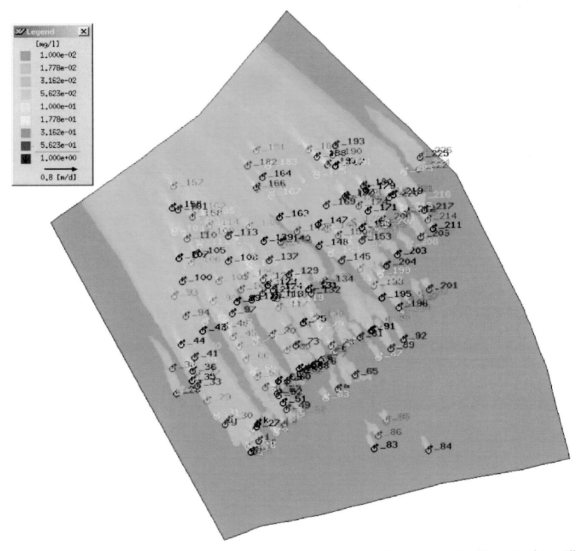

Figure 2.5.2. Leaks from many small sources can create an effect on urban aquifer water quality akin to that from diffuse agricultural pollutants. This theoretical example from the Rastatt case-study shows the very modest effect on downgradient shallow water quality from the injection of 1m³/d of 1mg/l conservative tracer through 226 sewer leaks into an aquifer with a 0.01 mg/l background concentration (Wolf 2004).

The issue of scale is one that affects all urban groundwater modelling. Water resource models tend to be regional in scale, enabling establishment of boundary conditions for the aquifer system that are logical and defensible in geological and hydrogeological terms. These aquifer-scale models are typically used for water resource management purposes, such as location and regulation of major borehole abstractions or planning control measures for resource threats such as saline intrusion, and they are the natural precursor for more detailed studies on those parts of the aquifer that underlie urban areas. However, they can easily encompass a land area of several hundred km² or more and practical constraints on maximum node numbers or run-times tend to lead to mesh spacings of the order of approx. 1 km. Such a coarse scale is typically quite inappropriate to the urban setting because of the complexity of land use across relatively small land areas. When simulating neighbourhoods with different water use characteristics, one is typically dealing with residential, commercial or industrial zones measured in hectares, in which even a grid spacing of 100m could be considered very coarse. Yet it is two orders of magnitude more detailed than a typical regional model would attempt.

The groundwater flow model code that is selected will dictate how this scale issue is resolved in detail, but unless a completely new model is being established for an aquifer under development, the general solution will be to establish one or more sub-regional sectors of the regional models with re-discretisation to reduce the grid size. This permits the continued use of the regional model, although further modification may be necessary if the sub-regional sector is located very close to one of the established boundary conditions, as these can dominate head response.

2.5.1.4 Specific problems in urban areas

While the types of data needed for a groundwater flow model that includes an urban area are broadly the same as those for other settings, there are some special considerations for integrating a groundwater flow and solute transport model array with upstream models in the AISUWRS model chain:

- *Conceptual model:* whether the groundwater flow model is derived from a pre-existing regional resource model or devised especially for the purposes of urban water management, a conceptual model that considers the urban water cycle is an absolute necessity. The urban water balance is a synthesis of several complex processes in which compensating loops can occur that affect heads and contaminant fluxes in sometimes counter-intuitive ways. For instance, summer rainfall on warmed roofs and paved surfaces can be rendered hydrologically ineffective by evaporation until a threshold duration/intensity is reached, when diversion to the subsurface via roof soakaways will be more rapid and effective than infiltration through the adjacent green space. An initial conceptual model, updated as required by the results of field studies or surveillance monitoring, can help to unravel where the model simulation may be most sensitive to uncertainties in the input parameters.
- *Unmodelled processes:* the conceptual model should also identify where parts of the urban water balance are neglected, or only very approximately represented. One example would be the compensating effect of sewer gain. At present, if the modelled scenario results in a rising water table so that increasingly large parts of the pipe network are within the saturated zone, the AISUWRS model chain does not have a procedure for handling sewer gain, as the NEIMO and the groundwater model are not coupled. The result would be unrealistically high groundwater heads, because in reality the losing reaches of sewer and pluvial drain would become gaining reaches, acting as land drains that would mitigate the rate and extent of water table rise.
- *Role of superficial deposit:* for many European cities, Quaternary superficial deposits are a major feature of the geological setting. In some cases the main aquifer itself could be located within a complex array of clays, silts, sands and gravels of glacial, fluvioglacial or alluvial origin, while in others these deposits may mantle the older bedrock aquifer beneath. Rastatt is an example of the former and Doncaster an example of the latter. In either situation the way that the layer at the surface has been modified through urbanisation and is subsequently modelled needs to be considered. For example, a clay stratum a few metres thick in a rural area would be hydraulically significant, impeding rainfall recharge, increasing surface runoff, providing a degree of protection against rapid infiltration of contaminants and possibly even confining heads. Once the land overlying that layer is urbanised, the clay may be punctured by foundations, access chambers or service ducts, excavated for laying sewer, mains or pluvial drainage networks, or physically removed for land contouring around road and rail cuttings. Under such circumstance the modification of the integrity of the clay stratum might mean that its sealing capacity would be sufficiently compromised as to affect how or even whether it is modelled as a separate entity. In older cities the existence of worked ground has to be considered; many European cities trace their origin to medieval or even Roman times and have up to two millennia of constant occupation including a phase of industrialisation. For example, the centres of London and York (Roman Londinium and Eboracum respectively) bear a mantle of anthropogenic detritus several metres thick.
- *Abstraction and recharge:* the fact that all inputs and outputs are cell-related needs to be taken into account. Thus point sources (e.g. a pumping borehole), linear features (e.g. gains/losses from a river/canal/drainage channel) or

area inputs (e.g. the recharge from a given UVQ neighbourhood) are handled alike, as inputs or outputs from an addressed cell. Consequently, the cell dimension defines the resolution of the model, and the temptation to over-interpret results for too small an area in relation to the mesh size is real and needs to be avoided.

- *Recharge assessment:* while the AISUWRS model chain provides insights into urban sources of recharge, the quantification of rainfall infiltration through green space is treated only simplistically. Various methods of assessing rural recharge are adopted in traditional regional water resource models; lysimeter measurements for grassland, soil moisture budgeting for cropped areas (arable and pasture), chloride throughput balances for arid zone catchments, environmental tracers, well hydrograph analysis etc. For various reasons, most of these techniques become inapplicable in the urban environment, so opportunities for corroboration or calibration of the computed recharge from incident rainfall (as opposed to runoff directed to the urban piped drainage network) are limited. Yet, depending on the climatic and geological setting, green space recharge can still be the most important component of the urban water balance.

2.5.2 Using MODFLOW and MT3D for the groundwater flow and contaminant transport modelling tasks

2.5.2.1 Principles

MODFLOW (Macdonald and Harbaugh, 1988) is a three-dimensional groundwater model that simulates the movement of groundwater of constant density through a porous medium. This model uses the finite difference method, wherein the continuous medium is replaced by a finite set of discrete points in space and time and the partial derivatives of the elliptical mathematical equation describing the groundwater flow movement are calculated using the head values at these points. The aquifer is spatially discretised with a mesh of blocks called cells and within each cell there is a "node" at which head is calculated. MODFLOW uses the block-centred formulation to define the configuration of these cells. Cell-to-cell flows are calculated using the head differences between the nodes and the conductance terms. A conductance is defined as the product of the hydraulic conductivity and the cross-sectional area of flow divided by the distance between nodes. Constant head and no flow cells are used to represent conditions along various hydrologic boundaries. Other boundary conditions, such as areas of constant inflow or areas where inflow varies with head, can be simulated through the use of external source terms or through a combination of no-flow cells and external source terms.

MODFLOW uses a modular structure in which similar program functions are grouped together in packages. Several packages have been developed to incorporate the different groundwater flow processes. The packages relevant to the modelling exercise in the AISUWRS project are:

- The River Package, which was developed to simulate the effects of flow between surface-flow features and groundwater systems.
- The Recharge Package, which was designed to simulate areally distributed recharge to the groundwater system.
- The Well Package, which simulates features such as abstraction or injection wells.
- The General-Head Boundary Package, which was designed to represent flow from an external source.

Alongside the MODFLOW groundwater flow model, the MT3D solute transport model is employed to simulate changes in concentration of contaminants in groundwater. Advection, dispersion and some simple chemical reactions with various types of boundary conditions and external sources or sinks can be considered. The MT3D model uses the hydraulic heads produced by MODFLOW and automatically incorporates the sink/source terms and the specified hydrologic boundary conditions defined in MODFLOW. Once the hydraulic gradients between nodes are established, MT3D implements a mixed Eulerian-Lagrangian method for the numerical solution. The Lagrangian part of the method can use a hybrid of the forward tracking characterisation method and the backward tracking modified characterisation method to solve the advection term. The Eulerian part is used in an explicit approach to solve the dispersion and chemical reaction terms. This explicit approach imposes restrictions on the time step size required to solve the numerical equations. The model program uses a modular structure similar to that implemented in MODFLOW. This makes it possible to simulate advection, dispersion, source/sink mixing, or chemical reactions independently hence saving computer memory space. In addition new packages involving other transport processes can be added to the model.

For the Doncaster case study the Groundwater Vistas© (GV©) software was used as a pre-processor to build and run MODFLOW and MT3D and as a post-processor to display the results and to export them as files that are compatible with other post-processing programs or visualisation software. GV© is a model-independent graphical design system that facilitates groundwater modelling in the Microsoft Windows™ environment. Model results are represented using contours, shaded contours, and velocity vectors on both plan and cross-sectional views (see contaminant concentration contour example from Doncaster regional model in Figure 2.5.3). Contaminant concentration time-series data can also be exported to spreadsheet and associated graphical applications, permitting trends to be plotted at key cell locations e.g. at nodes representing observation wells or sensitive public supply boreholes.

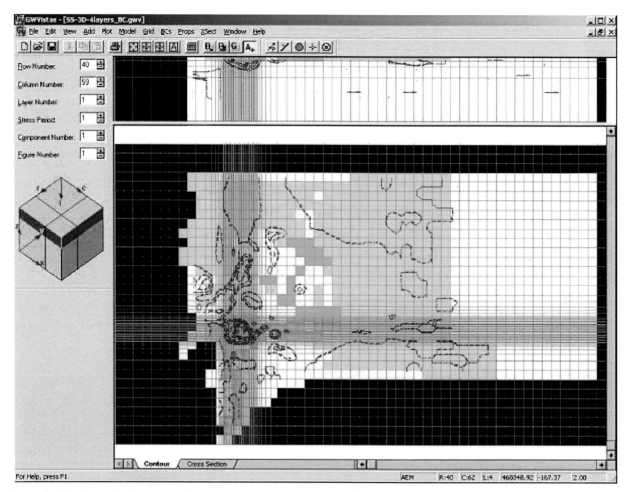

Figure 2.5.3. Groundwater contaminant contouring example from Doncaster case-study using MT3D and GV© post-processor.

2.5.2.2 Model calibration

Model calibration is the most complex part of the groundwater modelling exercise. Target values, to which the numerical model results are compared, must be defined. Target values can be groundwater heads or drawdown measurements, groundwater fluxes, river flows, concentrations, etc. Generally, calibration target values such as drawdown and river flows take a time series form and as such, they are suitable to calibrate time-variant models. In some cases, hydrogeological maps, which provide long term groundwater head contours, are used to calibrate steady-state models.

The quality of the calibration inevitably determines the reliability of any conclusions and recommendations made using the simulation results. There are some generally accepted methods of evaluating and interpreting the model calibration using both qualitative and quantitative measures. Visual MODFLOW provides a comprehensive selection of model calibration analysis tools for evaluating, interpreting and presenting the model calibration, including:
- Calculated versus observed scatter graphs
- Calibration residuals histograms
- Calculated and observed values versus time
- Calibration statistics versus time.

2.5.2.2.1 Doncaster Model calibration

Groundwater flow modelling in the Doncaster case study began with a pre-calibrated regional groundwater flow model prepared by Brown and Rushton (1993), employed by the Environment Agency for England and Wales (EA) for water resource management. The translation process is described in Section 3.2.3.4 and illustrated in Figure 2.5.4.

Original 2-D Doncaster Fortran model (transient)
↓
Translation into 2-D MODFLOW model (transient)
↓
Refined 2-D Modflow model (transient) with some re-discretisation
↓
Refined 2-D Modflow model (steady-state version)
↓
Refined 3-D Modflow model (steady-state four-layer version)

Figure 2.5.4. Different stages in the evolution of the groundwater flow model used for Doncaster case study.

The reasoning behind the decision to use an existing groundwater model was two-fold; firstly to demonstrate that the methodology developed in the AISUWRS project could be applied to pre-existing groundwater models (a likely situation for some future users of the AISUWRS model chain) and secondly to take advantage of a regional model that is regarded as well-calibrated and an adequate representation of the aquifer conditions in the Doncaster area.

As the already-calibrated regional model has been converted into a MODFLOW version, the process was one of checking rather than calibration. The GV© software feature that permits calculation of calibration statistics for head, drawdown, concentration or flux was found to be a useful feature for this purpose. The conversion to steady-state mode adopted 27-year averages of rainfall recharge and urban (non-Bessacarr) recharge, pumping volumes and water levels in the Quaternary cover, together with regional boundary inflows and outflows derived from the original transient mode simulation period of 1970-1997. The groundwater heads produced, the change of storage over time and total leakage between aquifer and overlying stratum (including rivers and drainage channels) were compared with the original model data. Minor differences were expected for several reasons such as the initial conditions, the different grid structures and numerical issues (such as rounding values).

The biggest problem was the representation of pumped withdrawals within the re-discretised area, where cell sizes were progressively reduced by up to two orders of magnitude, from 1 km x 1 km to 100 m x 100 m. In such cells, the wells withdrew the same amount of water as before, but as the cell area was much smaller steeper cones of depression were created. This did affect the Bessacarr-Cantley area because the pumping stations of large public supplies are close by (at Rossington Bridge and Nutwell).

Together these differences prevented a completely faithful representation of the features in the original model, but the groundwater head contours were checked to ensure that the new version represented the same hydraulic gradient as that produced by the original model. An acceptable interpretation was thus obtained so that the steady-state version could be employed for the purposes of scenario comparison. When the solute transport model was built, the indicator solutes were compared to field values at selected monitoring wells. The numerical values were close enough to the observed ones to provide a model acceptable for scenario simulations.

2.5.2.2.2 Ljubljana Model calibration

Model calibration is the process whereby selected model input parameters are adjusted within reasonable limits to produce simulation results that best match the known or measured values. For the Ljubljana case study, calculated versus observed scatter graphs were used to check the model calibration and applied to piezometric data available from 1970 to 2004. These data were used to calculate the hydrological statistics defining the piezometric head fluctuations in groundwater observation wells for the last 35 years. This process provided sufficient data to map the piezometric heads. As the detailed field investigation work took place just over the three years, 2002-2004, it was decided that this period was an acceptable approximation of the hydrological situation being modelled. The groundwater model was built step by step to avoid problems with numerical convergence, a procedure that made the calibration process much easier. An additional aid to the calibration process was the decision to model a significantly wider area than the actual study area. This allowed the model boundaries with fixed conditions to be adequately separated from the scenario simulation area in the city centre, thus avoiding the suppression of the induced changes. Free boundary conditions were represented using the general-head boundary (GHB) package for the entrance and exit zones of the model area. The GHB package was selected instead of the constant head boundary (CHB) again to assist in better management of the model calibration phase. There were two main problems affecting this process of calibration:

- As the model area was several times bigger than the study area, a solution to the representation of pumping drawdown was needed; and
- The piezometric head elevations at the borders near the hilly areas were persistently too low.

To resolve the first problem, the cell sizes were progressively reduced by up to two orders of magnitude, from 100 m x 100 m to 5 m x 5 m in the zones representing the pumping station. In such cells the wells withdrew the same amount of water as before, but as the cell area was much smaller, steeper cones of depression were created. This reduced the gap between calculated and measured piezometric values, although an error of a few metres remained. To solve the second problem, the hydraulic conductivity in this peripheral zone to the main basin was drastically reduced by 1 to 2 orders of magnitude. This decision, although arbitrary at the time, does seem to be validated by new data emerging from recent drilling work and slug tests performed on the margins of these hilly areas. The results of the steady-state groundwater model calibration are shown in Figure 2.5.5, from which it was concluded that the steady-state groundwater model could be employed for the purposes of scenario comparison and also that it represented an adequate foundation for the solute transport modelling requirements of the scenario simulations.

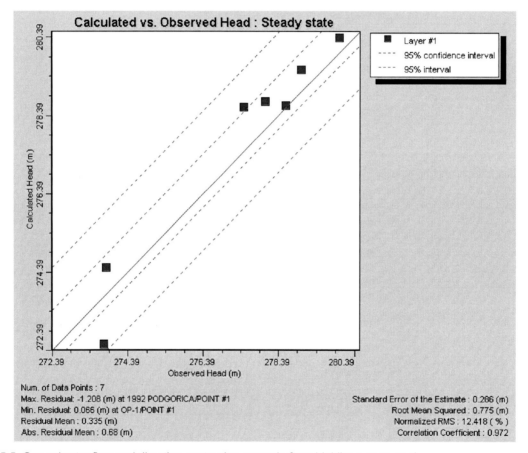

Figure 2.5.5. Groundwater flow and direction contouring example from Ljubljana case-study.

The calibrated groundwater flow model was then used as the basis for transport modelling with MT3D model. Two additional scenarios were added at this stage for comparison with the base case scenario. The first was to assume that all the study area was an industrial neighbourhood and the second assumed it was entirely residential. The calibration of these models indicated that some loads coming from POSI and SleakI were exaggerated. A closer calibration with observed groundwater concentrations would require lower contaminant inputs, but there were insufficient detail in load estimates for Ljubljana to take this further.

2.5.2.3 Model assessment

This section does not provide an exhaustive assessment or a general overview of the quality of MODFLOW or MT3D; instead it discusses the applicability of these software packages within the project context and provides some notes and comments on the ease of their use to achieve the project goals. As the Doncaster case study also used Groundwater Vistas (GV©) as pre- and post-processor, some generic comments on this software are also included:

- The groundwater flow model MODFLOW was capable of representing all the flow processes that were thought important to produce an acceptable simulation of groundwater flows in the study area. Rivers, abstraction wells, general head boundary, injection wells, regional variations in aquifer storage and hydraulic characteristics, layering, etc. are some of MODFLOW features that were used in this study.
- GV© provides the user with the ability to insert rows and columns between the rows and columns of the regular mesh, i.e. the mesh with uniform row and column widths. This is one way of grid refinement and was an important feature used for the Doncaster case study to improve the representation of the case study urban area, but it gave rise to some inconveniences at the re-discretisation stage. For instance, if more than two urban areas need grid refinement, the resultant grid may become similar to a regular grid with fine cell size. The number of created nodes may then impose a numerical burden and the simulation time may increase significantly. The grid refinement technique provided with MODFLOW exacerbates this problem because nodes are inserted in areas where no refinement is needed. This increases the simulation run time and may affect the efficiency of the numerical model.
- Another generic MODFLOW problem arises when inserting a new row or column to split cells into two new ones. That means if a new row splits a river node, two new river nodes are created.
- Other codes have been developed that can more satisfactorily undertake retrospective local refinement of the grid. For example, the code ZOOMQ3D has been successfully applied to this problem of refining the grid in the area of interest only (Jackson, 2004), and the number of nodes required in the Doncaster case study would be greatly reduced if this approach were applied. This is because ZOOMQ3D has been expressly developed using an object-oriented approach, i.e. each feature is represented with a corresponding object. When refinement is introduced, a river object for instance interacts with the new refined grid object and the river routine is updated in the model accordingly, without duplications. ZOOMQ3D would have been superior to MODFLOW in this respect.
- MT3D is a powerful tool that simulates contaminant movement in a porous medium but only some of its features were used in this study, mainly because only conservative solute indicators have been employed for demonstration purposes in this study. These are affected by the advective and diffusion mechanisms only, so subroutines enabling simulation of chemical evolution processes such as precipitation or ion exchange reactions (which could apply to other constituents) were not explored in this study.
- Although parts of the governing equations are solved explicitly, MT3D simulations were numerically stable and acceptable results were produced even under extreme conditions, i.e. when a large contrast between contaminant concentrations was considered at two adjacent nodes.
- GV© successfully fulfilled the objectives it was built for and the purposes for which it was selected. These are: providing a groundwater modelling environment and comprehensive graphical analysis tools to facilitate use of the MODFLOW and MT3D codes. GV© assisted the establishment of the groundwater model in MODFLOW because it has routines that facilitate setting up of the different processes; these features were very useful at the start of the project.
- As in any model interface tool, care must be taken regarding setting up the locations of the different files related to the project. Setting up the working directory of the project to a location, for example, does not automatically direct all GV© features to refer to and read from that location, especially when more than one model is involved. In addition, the edit boxes, where file locations are entered, are not grouped in one edit menu of the interface. This is understandable since these boxes must be located within the menus of the features they are related to, but a period of some familiarisation ought accordingly to be built into the modelling phase.
- The ability of GV© to interact with ArcView shape files and project them over the model grid was also a helpful feature that facilitated the inclusion of urban areas characteristics into the model. Neighbourhood shape files, for example, were used to group grid nodes into zones that corresponded to the neighbourhoods. In addition, the possibility of exporting data from GV©, manipulating them and then importing them back into the GV© environment was a helpful feature that minimised the efforts required to incorporate the characteristics of these zones.
- In some cases, it was easier to prepare files outside the GV© environment and then import them to the project within GV©. This should not be taken as implying that GV© is less helpful in preparing data than other pre-processors; rather that for some applications it is easier to improvise a solution from other software with which many users are more familiar e.g. data manipulation using Excel© or a similar spreadsheet program. For instance, while GV© was useful to set up the basic MT3D model, incorporating additional information about contaminant concentrations into MT3D-compatible files was more easily done outside the GV© environment. The user should therefore become familiar with the structure of MT3D files, especially the feature that GV© produces files with numbers that are next to each other, not separated with spaces, and adapt accordingly so that MT3D routines can read these files.

- There are a variety of analysis technique in GV© for viewing the results. These include plan and cross-sectional plots of contour lines, velocity vector maps, plots of head, drawdown, concentration, and others along a cross-section, time series plots and other options. It was found in practice that the GV© display methods are good for on-screen checks and interpretations but exporting data and results into more dedicated interpretation and display software (such as Surfer© for contouring) was necessary for reporting purposes. For example, the built-in GV© contouring facility had a poor resolution; with upper resolution limits being reached before adequate report-quality contour plots were obtainable.

The next two sections, which identify the more important case study-specific model assessment issues in Doncaster and Ljubljana also contain more general lessons for future users.

2.5.2.3.1 Case study-specific MODFLOW model assessment issues: Doncaster

During the project, it became apparent that using the existing model from the Environmental Agency as a basis had some disadvantages:

- The main one issue was the contrast in scale between models. The source model, being regional, covers several hundred square kilometres, while the AISUWRS project area is comparatively smaller at only a few square kilometres. It became apparent, that both the UVQ and the NEIMO models were intended to simulate small urban areas, at the level if individual households and streets. The chosen suburb Bessacarr-Cantley occupies a very small area within the regional model and lies close to its southern boundary. The result was a mismatch of scales: detailed site-specific models (UVQ and NEIMO) feeding into a comparatively coarse-gridded regional model. This was overcome by rediscretisation but the chosen mesh size (100m grid) remained a scaling compromise between the desire to have nodes of similar size to the neighbourhoods and the practical need to keep the flow model relatively simple with computation run-times of a practical length (≤ 1 hour).
- During translation into the MODFLOW code, it became apparent that the original model included many boundary conditions. Besides specifying the areal boundaries of the model and the recharge, the abstraction and the rivers, the model also needed to emulate the regional model from which it was derived by using boundary conditions for all areas covered by superficial deposits and by the Mercia Mudstone Group confining layer. These together comprised about ¾ of the model domain. Both superficial deposits and the Mercia Mudstone were represented by a leakage term, with the head in those formations being fixed. Bessacarr itself is however not covered by drift, i.e. no general head boundary conditions were specified for that area. It is probable that the good fit of the both original and translated models owes much to these boundary constraints.
- Very little information was available on the way recharge was calculated in the original model. This was important for the climate change scenarios because amendments to the climate files were the cause of the changes in recharge (see section 3.3.8 on scenario modelling).
- The conversion of the model from 2-D to 3-D by inserting layers improved the representation of the effect that deep abstraction from the public supply boreholes would have on the vertical movement within the urban 'composite plume'. This is because abstraction could be specified at nodes in the lower layers of the model. Together with the applied recharge and the effects of water leaking from superficial deposits, abstraction from deep wells established the vertical hydraulic gradients and downward leakage could be simulated as well as lateral movement. This improved the representation of contaminant movement within the aquifer.
- Some care is, however, needed in interpretation of the results when layering is introduced. It was found for instance when checking the contaminant concentration results at the different grids of the multi-layered model that one could easily gain the impression of vertical movement of the contaminant being comparatively more significant than the horizontal advance, because the contaminant appeared at the nodes of the second layer before any horizontal movement to adjacent cells was observed. This was an artefact of the cell geometry, being a result of the relative magnitude of the horizontal and vertical distances separating the grid nodes. In the focus study area, the minimum nodal spaces were 100 m while the vertical distance between nodes in the first and second layers was only 20 m. Mass balance and velocity cross-checking calculations were helpful in elucidating relative contaminant movement rates; downgradient transects at each quadrant of the study area for instance quickly revealed that lateral movement along the upper layer to a monitoring well in an adjacent cell could be measured in terms of a few months whereas vertical movement from the first to the third layer was measured in tens of years. This is another facet of the scale difference problem that users need to be aware of when deciding how to discretise the urban area.

2.5.2.3.2 Case study specific MODFLOW model assessment issues: Ljubljana

- For groundwater flow and pollution transport the decision was made to model a wider area than just the immediate environs of the study area. The main reason was to keep the model boundaries with fixed conditions adequately separated from the scenario simulation area in the city centre is that it avoids these fixed head or fixed flow boundaries suppressing the induced changes from the application of different scenario conditions. This decision then required mesh refining in the study area itself. This helped achieve results for head calibration that fell in the 95% confidence interval.
- The decision to build the groundwater model step by step provided much more control on the calibration because the effects of individual model features could be identified and adjusted step by step. This approach for instance helped resolve the problem of low piezometric head elevations on the basin margins near the hilly areas. This led to the realisation that the study area was also a part of the model border, a situation rather similar to the Doncaster study area. The transition from a one layer to a two-layer model was also easier with the step-by-step approach. At the beginning the same average hydraulic conductivity value from pumping tests was selected for the entire model, but during the calibration phase these were modified in accordance with the ratio of calculated versus measured heads.
- Careful cell mesh design was needed to avoid it influencing the results of MT3D and producing a misrepresentation of the pollution spreading. A good mesh design was also needed for better numerical stability of ModFlow and MT3D. In general, changes in mesh size were smooth in all directions, adopting the rule that transition from one mesh spacing to another should only increase or decrease by a factor of about two per unit of area.
- To calculate the influence of extra recharge (infiltration scenario) and the 10% reduced recharge scenario, two transient groundwater models were built. Although the changes were very small, this does not mean that they are not an influence, but rather that the study area is too small to have a discernible effect on the piezometric levels of the entire model.

The data exchange between the AISUWRS DSS /or model chain has not been specified for the finite difference modules.

2.5.3 Using Feflow for the groundwater flow and contaminant transport modelling tasks

This section deals with a number of difficulties and problems encountered during the saturated flow and transport modelling with Feflow® in the case study city of Rastatt. It is intended to provide additional advice on possible pitfalls for users planning to employ Feflow® in a similar modelling task. For more detailed information on the technical setup of the saturated flow and transport models in Rastatt, see chapter 3.1.

2.5.3.1 *Principles*

Feflow® is a finite element groundwater simulation code by Wasy GmbH Institute for Water Resources Planning and Systems Research, Berlin. As a subsurface flow and transport simulation system, Feflow® is capable of modelling both groundwater dynamics and the associated mass or heat transport processes. The governing equations are based on Darcy's law for fluid flow in porous media. Flow and transport simulations can be run either steady state or transient mode in 2D (vertical or horizontal plane) or 3D. The study domain is discretised into a finite element mesh by alternative triangulation or quadrangulation algorithms. The resulting mesh is built up from discrete elements which are separated by mesh nodes. For each time step, the underlying governing equations are solved for all nodes, delivering individual values for hydraulic head, temperature or mass concentration, depending on the problem class. Feflow® offers multiple options for analysing the simulation results. It features an integrated budget analyser which quantifies water and transport balance terms for all model boundaries and the groundwater recharge. Control parameters like hydraulic head, mass concentration or flow velocity can be visualized as isolines, as colourful fringes both in 3D as well as in 2D plan or as cross-sectional plots. It is possible to produce difference maps from the comparison of computed values with a given reference distribution or to record parameters of interest at any number of specified observation points and wells.

2.5.3.2 *Data requirements*

A three-dimensional Feflow® model consists of a number of layers. These can represent geological units or simply be inserted in order to ensure sufficient discretisation. Each volume element of a layer holds information on material parameters like hydraulic conductivity or porosity, but also groundwater recharge is spatially assigned on an element basis. In the very common case that data need to be interpolated between a limited number of field measurements, several regionalisation options are available in Feflow® (kriging, Akima interpolation, inverse distance weighting).

Communication to GIS data is allowed for by, for example, assigning values from a .dbf file of an ESRI shape file to each model node, or by interpolating between a given number of points in a .dbf file. Feflow® layers are separated by a top and a bottom slice where each node of a slice can hold information on boundary conditions, which opens the model for groundwater flow or mass transport at such a node. Nodes without a specification of boundary conditions are automatically considered as no flow boundaries. Boundary conditions can be constrained by certain maximum and/or minimum values for hydraulic head or mass concentration. This is especially useful for 1st kind mass transport boundaries at nodes where inflowing water is supposed to have a certain constant mass concentration, whereas contaminant plumes with increased concentrations or contaminated water at a pumping well should be able to leave the model freely. Such a situation requires a minimum constraint of zero for the mass boundary. Whenever the respective threshold value for the constraint is exceeded, the boundary condition is set inactive. Consequently, the model is not open at this node any more, but hydraulic head or mass concentration can reach their values freely without imposed levels. Constant flux boundaries can be chosen to be either integral over an entire cross-sectional area (not changing with water level) or to change depending on the water level.

2.5.3.3 Model calibration

Model calibration in Feflow® can be done either manually or by employing the PEST module. PEST is an automatic parameter estimation tool which finds the numerically best solution for the distribution of a specific material parameter in order to approach a set of target values of hydraulic head at a given number of observation points.

In the Rastatt case, the flow model was calibrated by changing the hydraulic conductivity. The PEST module requires a polygon (.ply) file of the areas which are going to be assigned a uniform hydraulic conductivity. Such a polygon file can be created by generating and exporting a supermesh. The file ending needs to be changed into .ply before loading it into the PEST module. The target values need to be assigned in the format of a .trp-file. Again, this can be obtained by changing the ending of, for example, a .txt-file with x and y co-ordinates as well as the corresponding hydraulic heads. In the Rastatt case, it was planned to force the vertical hydraulic conductivity to be 15 % of the horizontal conductivity. The procedure was to include three calibration parameters. Kxx was an independent parameter with a lower boundary of .00001 m/s and an upper boundary of .008 m/s. The PEST module offers the possibility to tie one calibration parameter to another one. Kyy was therefore globally tied to Kxx which means that the same changes are performed on both parameters. Kzz however was assigned an initial value, upper and lower boundary of 15 % of the Kxx values. For each polygon, Kzz needed to be locally tied to the Kxx of the corresponding polygon. Certain areas can be set to be inactive during the PEST optimisation, as was done for example in the Rastatt case for an area of known hydraulic conductivity in the form of a less conductive interlayer. The success of the PEST optimisation can be visualised by generating a reference distribution as an interpolation of the water levels at the observation points and by subtracting the hydraulic head distribution from a simulation run (Figure 2.5.6).

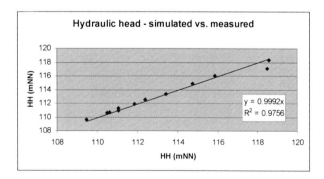

Observation point	Simulated head (m)	Measured head (m)
1	110.95	111.07
2	110.69	110.62
3	111.25	111.07
4	110.55	110.49
5	117.06	118.5
6	116.02	115.84
7	113.36	113.41
8	112.49	112.38
9	114.84	114.73
10	118.34	118.58
11	109.62	109.46
12	111.87	111.84

Figure 2.5.6. Difference of the hydraulic head at twelve reference points in Rastatt.

In order to prevent the dewatering of elements and its associated problems at boundary nodes of a model with a free and movable surface on the first slice (which means that the mesh moves with the water table), it might be necessary to choose the option 'unconstrained' in the Menu Options – Specific options settings in order to avoid possible numerical instability.

Some transport problems in Feflow® have the tendency to produce oscillating mass distributions, or even negative mass concentrations. An effective procedure to overcome this problem is to choose one of Feflow's upwinding options, streamline upwinding or full upwinding. These operations prevent oscillations by deliberately introducing a 'numerical error' into the solution process (also known as 'numerical dispersion').

2.5.3.4 Integration with the AISUWRS model chain

Through the connection of the upper level models in the AISUWRS chain and the saturated flow and transport model in Feflow, the impact of different urban water management strategies – represented through characteristic urban water scenarios – on the net water and contaminant balance as well as on water levels and contaminant concentrations at specific observation points can be assessed. In a first step, a transport model is set up with an initial groundwater recharge and a uniform 'no impact' background concentration of the contaminant in question. Afterwards, the DSS calculates groundwater recharge and contaminant input on a neighbourhood basis for all desired scenarios. The interface between the DSS and Feflow is basically an ESRI neighbourhood shape file. This shape file contains the spatial information on neighbourhood geometry as well as values for recharge and contaminant loads and can be joined to the Feflow simulation. The joining action can be performed in such a way that the original data are changed only in the area of interest. After the simulation run, the results for several scenarios can be compared.

One of the more challenging tasks during this procedure occurred during the setup of the transport model. This was the creation of a uniform background concentration. Several approaches were tested:

(1) The assignment of a constant mass boundary with the desired background concentration for all nodes on the top slice. This option implies also that all nodes on the top slice are open for mass transport. It can occur that mass leaves the model at places where there are no wells or rivers or any of the other features which, in reality, allow the removal of contaminants from the system. One way of preventing this undesired loss of mass is to assign an additional boundary constraint to all nodes, which allows the nodes to reach a higher concentration but actually prevents mass from exiting the model. However, for a large model, the assigning of a boundary constraint on all top slice nodes significantly increases computation time.

(2) The assignment of a constant mass flux, which is determined in accordance with the groundwater recharge and the desired mass concentration for all nodes of the top slice. The problem with this approach is that groundwater recharge is an elemental property, which is accurately stored only at the centre point of an element. The nodal quantities however are always subject to interpolation. This fact leads to the peculiarity that exporting groundwater recharge values (In-/Outflow on top) as nodal quantities and then importing them again leads to a distribution deviating from the original one. Subsequently, using the same nodal quantities for the calculation of the constant mass flux leads to a biased result. On the other hand, the correct groundwater recharge values can be exported for all element centre points, but the Feflow® transport boundary menu (for constant mass flux) is only able to assign nodal quantities. These circumstances complicate the correct determination of constant mass fluxes. In addition, the assignment of transport boundaries can indeed be easily done along line segments, which could for example be imported from ESRI shapefiles. However, the spatial input of boundary values on a polygon basis is not supported in the boundary menu. When working with constant mass fluxes, it is therefore difficult to change mass fluxes from the DSS output on a neighbourhood or polygon basis later on.

(3) The use of the transport material parameter source/sink. The unit of this parameter is $g/d/m^3$. During the simulation, the input values are multiplied by the saturated volume of each element, resulting in a contaminant flux in g/d. However, the saturated volume of an element directly under the free and movable surface at each time step can only be determined by using the IFM programming platform. One solution approach is to insert a layer of known thickness (e.g. 1 m) at a depth which stays fully saturated throughout the entire simulation and to assign the source/sink term to the top slice of this layer. The source/sink values can then be determined by the simple operation: Source/sink $[g/d/m^3]$ = In/outflow on top $[.0004\ m/d] \cdot .0004 \cdot$ desired background concentration $[g/m^3]$ / thickness of the respective layer $[1m]$.

The parameter In-/Outflow on top contains the groundwater recharge values. They are material parameters, which are stored on an element basis and should be exported at the centre points in order to be accurate. Before multiplying the groundwater recharge by the background concentration, all negative recharge values (for example at lenticular water bodies) should be replaced by zero in order to prevent a negative mass flux. Concerning the thickness of the layers, all layers need to be specified in a way that ensures that the thickness of the layer underneath the slice with the mass flux input remains the same throughout the solution process. One solution is a layer configuration. In the case of a five layer model for instance, slice 1 is equal to the groundwater level, the mass source is assigned to slice 2 which could lie 20 cm under slice 1 and layer 2 between slices 2 and 3 is 1 m thick: It must be assured that slices 4, 5 and 6 keep a sufficient distance from slice 3 in order to avoid unintentionally 'pushing' slice 3 upwards. This can be achieved in the menu 'Modify slice distance', bearing in mind the 50 % rule (see Feflow help for this menu).

The last step to produce a uniform background concentration is to assign constant mass boundaries at all nodes that are open for convection (e.g. 1st 2nd and 3rd kind flow boundaries, wells). Whenever mass is supposed to be able to leave the model freely, like at a pumping well inside the urban focus area, a minimum constraint of zero needs to be applied to this boundary (see WASY 2002).

For the scenario testing, an ESRI shapefile is required which contains:

(1) Groundwater recharge values. They are delivered by the DSS and need to be converted to Feflow units. They are applied to the first slice as the flow parameter In-/Outflow on top.
(2) Contaminant flux. The DSS delivers contaminant loads in g/y/Neighbourhood. The following data preparation is necessary:
Feflow units [$g/m^3/d$] = DSS output [g/y/Nbhd] / Area Nbhd [m^2] /365 / thickness of the layer [1 m]. The resulting values can be applied to the slice which is underlain by a layer of known thickness which stays fully saturated.

2.5.3.5 Model assessment

Within the AISUWRS modelling tasks Feflow has proven its general applicability to complex mass transport problems. The possibility to refine the mesh at points with large gradients offers an effective countermeasure to numerical dispersion. Furthermore, the model can be put to other uses in a very straightforward way: once set up for a large scale, it can also be applied to local problems with contaminated sites just by locally increasing the mesh density. In this manner it is possible for the municipality or environmental agency to set up a model for planning at a large regional scale but promote improved environmental risk assessments with numerical models by local consultants at the same time.

However it must be pointed out that the attachment of mass fluxes to spatially distributed groundwater recharge patterns was difficult at the time of writing. The visual display of water and contaminant balance results in Feflow® is clear and easy to retrieve. However, in order to compare the exact balance term for individual compartments of the water and contaminant budget, it is necessary to manually write down the numbers and to reproduce them in a specifically created scenario assessment file. Even more cumbersome is the comparison of hydraulic heads or contaminant concentrations in observation wells as they are not directly available from the Feflow® user interface. They need to be exported and analysed externally.

Different numerical solutions like "full upwinding" or "streamline upwinding" have a major influence on the result. For every case it needs to be decided which option is most suited for the problem and consequently it is recommended to perform both solutions for comparison. Additionally, the implementation of correct mass transport boundaries of rivers introduces uncertainty in the model; this occurred in the study area with the Murg river.

Despite the pitfalls encountered, the generated numerical groundwater flow and transport model remains the most precise tool for the estimation of impacts on quantity and quality currently available. Feflow encompasses all necessary pre- and postprocessing functionalities and its user friendliness has been upgraded continuously during the last decade with the feedback of the user community. Nevertheless it is recommended for first time users to secure access to the Feflow support team if complex mass transport problems need to be solved.

2.5.4 Use of a Risk Assessment method for water quality management in karstified aquifers

2.5.4.1 Principles

A risk-based approach to management and protection of drinking water quality, based on the Hazard Analysis and Critical Control Points (HACCP) framework previously applied internationally in the food industry (CAC, 2003), has recently been adopted in the Australian drinking water guidelines (NHMRC and NRMMC, 2004) (Figure 2.5.7), and also in the draft guidelines for water recycling (NRMMC and EPHC, 2005).

One of the elements contributing to the HACCP framework is hazard identification and risk assessment. Thus, assessing the impact of urban water systems on a receiving aquifer or water body can be approached in a manner consistent with the risk assessment component of the HACCP framework (Figure 2.5.8). Furthermore this method can incorporate potable water quality guidelines (NHMRC and NRMMC, 2004) and aquatic ecosystem water quality criteria (ANZECC and ARMCANZ 2000, SA EPA 2003). The measurement of risk in HACCP can range from a simple qualitative assessment through to a fully quantitative assessment (for example, using the @Risk® software). Furthermore, the risk assessment can be iterative, adding complexity as the supporting input data develops.

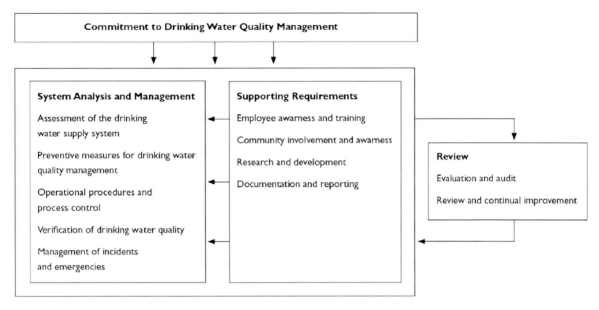

Figure 2.5.7. Framework for management of drinking water quality (NHMRC and NRRMC, 2004).

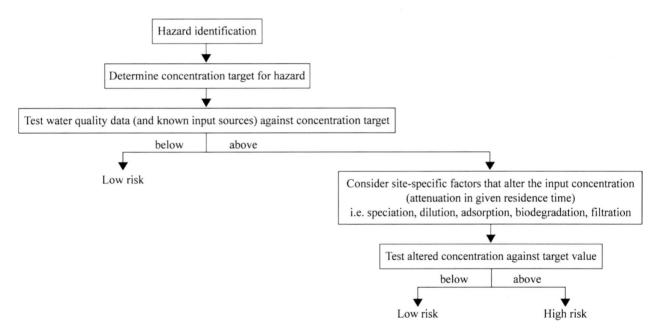

Figure 2.5.8. Framework to assess the impact of urban water systems on water quality. Target values can be based on potable and environmental water quality guidelines.

2.5.4.2 Data requirements

The risk assessment can accommodate varying detail in input data (Table 2.5.1). The initial risk assessment assumes a very conservative approach in response to data limitations. However for a quantitative assessment, the degree of confidence increases as the uncertainty associated with input parameters decreases. Knowledge of attenuation processes appropriate for the saturated zone is essential to this approach. It can be applied to contaminant attenuation both in groundwater and in receiving surface water bodies.

Table 2.5.1. Description of the required risk assessment input variables based on a qualitative or quantitative application.

Qualitative	Quantitative
Identify hazards related to urban water systems and contaminants associated with these hazards. Some contaminants may be predetermined and carried through the model chain. Alternative information sources include water quality data and land use information.	Concentrations of potential contaminants. .
Target values of surrogate parameters (guideline or historical values)	Applicable drinking water and/or environmental guidelines.
Identified attenuation processes (where applicable)	Solubility; Dilution; Background concentrations; Distribution coefficient for partitioning between aqueous and solid phase; Henry's constant for partitioning between gas and water phase (where groundwater discharges into an open water body); Degradation or pathogen inactivation half-life; and Aquifer residence time.

2.5.4.3 Attenuation processes

2.5.4.3.1 Solubility

The concentration of a contaminant may be limited by its solubility, under the conditions encountered in the groundwater or the receiving water body. Solubility limits can be applied to examine the worst case scenario, such as shock loadings of organic chemicals.

2.5.4.3.2 Dilution factor

Dilution factors alter the concentration from the urban water system inputs, by allowing for mixing with the groundwater.

2.5.4.3.3 Degradation half-life

For species subject to degradation, the final concentration is given by C_t (mol/L):

$$C_t = C_o e^{-kt} \tag{1}$$

where C_o is the initial concentration (mol/L), t is the minimum residence time (day) and k is the decay rate constant (day^{-1}). Half-lives from the literature (where $C_t/C_o = 0.5$) or one $\log_{10}$ removal times (where $C_t/C_o = 0.1$) are used to calculate decay rate constants. Half-lives also incorporate the half-life of degradation byproducts where possible. For calculating degradation, the minimum aquifer residence time is used in Eq.1. This is consistent with a conservative risk assessment and represents the highest risk event.

2.5.4.3.4 Distribution coefficient

When applying attenuation parameters such as distribution coefficients, it is assumed that these processes are independent of residence time and have an unlimited capacity. For species retarded by sorption, the partitioning between the solid and aqueous phases for linear isotherms (and often approximated for non-linear isotherms) is given by the distribution coefficient, K_d:

$$K_d = \frac{C_s}{C_{aq}} \tag{2}$$

where K_d is the distribution coefficient for a linear isotherm (L/kg), C_s and C_{aq} are the concentrations adsorbed to the solid phase (mol/kg) and in the aqueous phase (mol/L) respectively. K_d can be estimated from the adsorption coefficient related to the organic carbon content, K_{OC} (m^3/kg OC) and the fraction of organic carbon present in the porous media, f_{OC} (%). K_d and K_{OC} can be determined by laboratory studies, or calculated from the octanol-water partition coefficient (K_{OW}) (Oliver et al. 1996a;b).

For species subject to volatilisation, the partitioning between the gas and aqueous phases is given by the Henry's law constant, H (dimensionless):

$$H = \frac{C_{aq}}{C_g} \quad (3)$$

where C_{aq} and C_g are the concentrations in the aqueous and gas phases (mol/L) respectively. The Henry's constant can also be expressed as K_H (M/atm):

$$K_H = \frac{H}{RT} \quad (4)$$

where R is the gas constant (0.082057 dm³.atm/K.mol) and T is the temperature (°K).

For volatile species (H<10), the final concentration in a receiving body following time dependent volatilisation is given by C_t (mol/L):

$$C_t = C_o e^{-(vA/V)t} \quad (5)$$

where C_o is the initial concentration (mol/L), t is the residence time (day), v is the mass transfer velocity (m/day), A is the surface area of the water body (m²) and V is the volume of the water body (m³). This assumes the water body is well mixed and transfer can occur through the entire depth. Note that half-lives given for surface water bodies may incorporate both volatilisation and degradation.

2.5.4.3.5 Aquifer residence time

Minimum aquifer residence time (t_{min} in days) can be determined from groundwater modelling or be determined empirically using tracer tests. It may also be calculated, for example, using the approach used by Miller et al., (2002) for determining the minimum travel time between two wells:

$$t_{min} = \frac{\pi D n_e L^2}{3Q} \quad (6)$$

where D is the aquifer thickness (m), n_e is the effective porosity, L is the distance between the point of injection and recovery (m) and Q is the rate of injection and recovery (m³/d). A negligible ambient hydraulic gradient is assumed.

The risk assessment for the organic species is based on the minimum time required for degradation to reduce the contaminant concentration to the target value.

2.5.4.4 Integration within Decision Support System

Currently the risk assessment operates independently from the Decision Support System (DSS). Contaminant concentrations for stormwater were based on measured data. Outputs from the DSS via the Urban Volumes and Quality (UVQ) model were consistent with the medians of the measured data. However, a wider contaminant suite was tested than was available through the DSS. Groundwater contaminant concentrations for the sewer leaks were provided from the DSS via the SLeakI and POSI model components. These concentrations were incorporated into the range of groundwater concentrations used in the risk assessment.

2.5.4.5 Model assessment

The choice of a risk-based assessment is consistent with the current Australian framework for managing drinking water systems (NHMRC and NRMMC, 2004). This generic framework can be used in a qualitative sense, where quantitative input data are poor, but can also accommodate greater complexity when input data allow the use of a quantitative risk assessment. HACCP provides a framework that can incorporate this type of risk assessment, along with several other elements, for management of drinking water. For fully quantitative risk assessments, the use of sensitivity analysis would indicate which areas of uncertainty are the greatest. This allows the outputs from the risk assessment to be configured to focus on the parameter of highest priority. For example, if aquifer residence time is deemed as the highest priority area of uncertainty, the risk assessment output can deliver critical residence times for incorporation into the HACCP framework.

2.5.5 Summary

Two different numerical groundwater model codes (one finite difference and the other finite element) and a risk assessment procedure have been successfully applied as the end-members in the AISUWRS model chain. Each has been able to utilise the data provided by the upstream models in the array to demonstrate the impact on four quite diverse aquifer settings of urban recharge and its associated contaminant load. Their operation has had to be deployed in different ways, but together they show that the AISUWRS concept, of using linked models to understand, simulate and then predict the effects of complex urban hydrological processes, is a valid and practicable option for urban water resource planners.

Given the conceptually different approach of MODFLOW and Feflow®, the operational details of their implementation will differ, and some of the key features have been described in the preceding sections. However, at the heart of their application is the groundwater modelling truism; successful models start with a considered conceptual model so that the numerical expression of the model can be set up with identifiable and defensible hydrological boundaries. In Rastatt, Ljubljana and Doncaster alike this led early on to the decision to model much larger areas than the districts where detailed studies were concentrated, permitting the construction of defensible hydrologic boundaries at a sufficiently distance to avoid suppression of head effects induced by scenario changes in the study area. Where the geometry of the aquifer did not entirely permit this (as in the case of Doncaster where the aquifer edge was close by), the scenario simulations suffered.

Another general feature to consider in applying a groundwater model is the decision to progress from the steady state to the transient mode. This has to be made from a study-specific assessment of the data available to support the more demanding requirements of transient mode simulations. These demands relate as much to the data inputs to the upstream models as they do to the time-series water level, hydrologic and climatic data requirements of the groundwater models themselves. In Doncaster, Rastatt and Ljubljana, this led to the decision to undertake the main scenario comparisons with steady state flow fields but transient mass transport. One set of model runs with transient flow field between 1960-1990 was performed in Rastatt but it proved the applicability of the steady state assumption.. The option always remains with such simulations to progress to transient mode once the underlying data permit.

A third general feature that has emerged from the case-study applications is the need for sensitive selection of layering options, crucial to a credible representation of downstream water quality changes in the transport model. This may lead to the definition of layers that do not necessarily coincide with major changes in the geological strata, but may instead in some circumstance be necessary just to divide the continuum of a thick aquifer, to better represent the downgradient movement of contaminants in a vertical as well as horizontal direction.

Finally, it has to be accepted that many cities are unavoidably sited on aquifers that are inherently difficult to simulate with a numerical model, and those located on karstic limestones form a major class. For such cities, risk assessment offers a practical alternative that is consistent with management practices. The Mount Gambier study provides a demonstration of this procedure in operation.

2.6 INTEGRATING IT ALL: THE DECISION SUPPORT SYSTEM

2.6.1 Introduction and Overview

The Decision Support System (DSS) developed within AISUWRS is the integrating software that supports the selection and comparison of predefined urban water scenarios. This is achieved by linking the individual models together to allow them to be run as one entity, rather than individual stand-alone models. Figure 2.6.1 depicts a conceptual flow diagram of the DSS components and their linking in the AISUWRS model chain. The DSS provides the ability to track water and associated contaminants emanating from urban areas from the source until they reach the aquifer, enabling prediction of the impact of different scenarios on groundwater contamination and assessment of its implications for urban development. The most important function of the DSS is to provide a common platform with a graphical user interface that controls a number of complex models and allows for the passing of information between models. The interface enables the user to set key parameters for each of the models, to run the model chain and then to view a summary of the output. It was envisaged at the beginning of the AISUWRS project that it might be possible to develop a tool that identified preferred strategies for the sustainable management of urban groundwater resources. However, the research undertaken has shown that the complexity of an urban water system interacting with a heterogeneous natural system requires a very detailed site-specific analysis and understanding of the involved processes, and consequently this proposed optimisation component of the DSS was not developed.

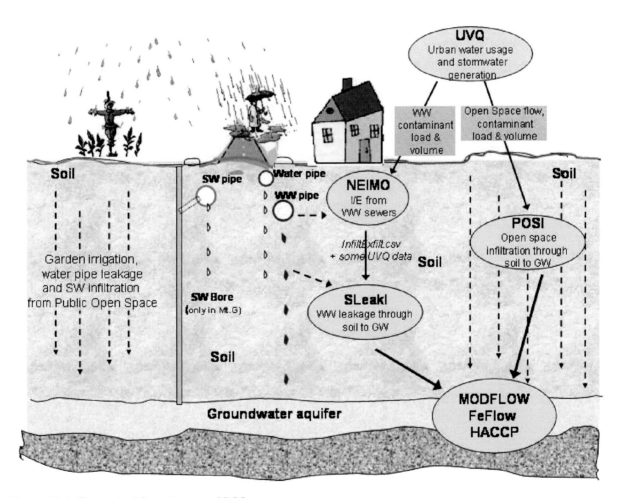

Figure 2.6.1. Conceptual flow diagram of DSS.

The DSS provides users with a tool to track water volumes and the associated potential groundwater contaminants from a source, such as a leaking sewer, and allows the assessment of the movement of water volumes and contaminants and their attenuation through the unsaturated zone until they reach the aquifer. The groundwater models deal with significant complexity in the subsurface (such as layering, 3-dimensional flow, transient flow) which makes it difficult to run them remotely, so the DSS was not designed to run these models and instead was developed to just provide the input files to them. The major benefit from the application of the DSS was to link the urban water scheduling capabilities of UVQ with the process models that estimate pipe leakage and unsaturated zone flow and transport, providing a holistic urban water system tool that enables water services to be represented in a flexible manner. This affords the opportunity to represent and investigate the implications of a wide range of conventional and emergent techniques for providing water supply, stormwater and wastewater services. Users are also able to explore the likely implications of critical uncertainties such as alterations to water consumer behaviour or climate change. The technical details and requirements of the DSS are described in more detail in the following sections.

2.6.2 Scenario Generation

A critical requirement of the DSS is to control the models and to allow the impact of individual urban scenarios on groundwater contamination to be assessed. The future is inherently uncertain, yet there is a need to take decisions and develop strategies that will impact on this uncertain future. The running of scenarios provides water managers with a powerful tool for dealing with future uncertainties and developing strategies that are robust to a range of potential future scenarios. A scenario is not a prediction of how the future will turn out, but rather a proposition about how it may unfold and by permitting "what if?" analysis, their use in the AISUWRS DSS allows water managers to explore the likely nature and impact of potential changes to the urban water system in terms of changes both in groundwater level and quality.

In order to assess the applicability of the AISUWRS approach, the DSS has been applied to four case study cities with a range of supply and disposal scenarios that have been developed for each city. The scenarios themselves were devised to represent typical problem situations that might evolve over the next few decades. Typical problem scenarios analysed by the project teams included:

- Baseline: the current situation- used as a starting point to compare other scenarios against and to explore the likely impacts of current management practices.
- Sewer replacement: Simulates the replacement or rehabilitation of sewers (for example, those that are more than 40 years old)
- Climate change: Explores likely implications of global climate change on groundwater resources using local climatic change predictions, such as decrease in rainfall
- Decentralised water systems: Simulates the implementation of decentralised water supply and wastewater treatment/disposal technologies
- Enhanced infiltration of urban surface runoff
- Demand change: Explores the impacts of increased water consumption driven by rapid population growth or changes in usage patterns
- Groundwater contamination: Simulates groundwater quality parameters exceeding World Health Organisation guidelines (either through an accident or systematic pollution)

The precise form of the scenarios differed, depending on the individual characteristics of each case study city, with the final selection developed in consultation with key stakeholders from each city. In each case, the range of scenarios was defined in consultation with the Technical Reference Group established for the city in question. In order to keep the process illustrative, this typically meant that simplified assumptions were applied. For instance, pipe rehabilitation was assumed to be completed instantaneously. Two types of scenarios were used. Firstly, 'action' scenarios focussed on potential strategies or changes that could be implemented by a water authority, such as rehabilitation of sewers of more than a specified age. The second type of scenario focused on drivers of change that were outside the control of the stakeholders, such as climate change or population increase. Water managers cannot control such exogenous scenarios, but they could nonetheless significantly impact on the urban water system. Not all scenarios were appropriate for each of the case study cities; for example Mount Gambier did not consider a sewer rehabilitation scenario, as their pipe infrastructure is less than forty years old.

The stages in setting up and running scenarios in the DSS are:

(1) Definition of the parameters of the urban water system in the Urban Volume and Quality Model (UVQ). This includes climate inputs, land use, contaminant profiles of different flows, water usage patterns and population distribution
(2) Selection of appropriate UVQ project from within the DSS
(3) Selection of the model time period
(4) Running of the DSS and viewing of the outputs

The scenario running enables the potential impact of a single factor on groundwater to be identified, then the factors can be combined to include a range of responses including water re-use strategies, proactive pipeline rehabilitation, relocation of current potable water extraction systems and upgrading current water restriction policies and treatment systems. Different factors can either compensate for or be additive to the effects of others and it is an advantage when devising a water management policy to be able to disentangle and quantify the relative effects of each component on water levels or recipient aquifer water quality.

As the DSS supports the selection and comparison of predefined problem scenarios, it allows the end-user to choose between decisions in order to find a best-practice response e.g. should operational strategy be to accept some contamination and reinforce treatment for sensitive water uses or should system improvements that prevent contamination be the optimum investment decision. The DSS generates a set of data that can be used to provide sustainability indicators that use internal and external environmental, social and economical drivers. However, the detailed analysis of the socio-economic context is highly specific for each case study and therefore is performed outside the automated modelling chain. For instance the comparison between German and Australian case study cities has revealed similar attitudes towards water-related problems, but significant differences in the perception of the implications of groundwater contamination arise because of the Blue Lake's key role in supporting Mt Gambier's tourism industry. The scenario outputs have been used in a separate socio-economic model to assess the socio-economic implications of different strategies (see Chapter 5).

2.6.3 Running the DSS

The DSS provides a Graphical User Interface (GUI) in which the user can set up key parameters and run the AISUWRS model chain (Figure 2.6.2). In order to assist model calibration and testing, the user can select partial running of the model chain, providing flexibility to facilitate scenario assessment. The following sections describe some of the functionality contained within the DSS.

The UVQ screen in the GUI enables users to locate the UVQ executable file and also select the time period for which UVQ, and hence the DSS, will run. The scenarios are selected by choosing a UVQ project file to run. As discussed in Chapter 2.1, UVQ provides the ability to represent and investigate the implications of a wide range of conventional and emergent techniques for providing water supply, stormwater and wastewater services.

The control of the Network Exfiltration and Infiltration Model (NEIMO) by the DSS was more difficult because in many cities, particularly older ones, sewer pipes carry both wastewater and stormwater. However, increasingly cities are using separate systems to remove stormwater runoff and wastewater in order to prevent the overflow of combined systems during high rainfall events. Both types of systems need to be represented. The user can select either the wastewater or stormwater network, so the impact of leakage on groundwater quality both from stormwater and wastewater pipes can be considered where there are separate systems.

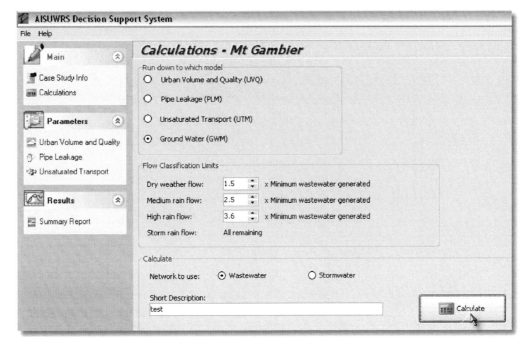

Figure 2.6.2. Front page of AISUWRS DSS.

Nevertheless, for each scenario, analysis requires the groundwater contamination potential be simulated over a period of years to decades and the computational resources required by NEIMO for a typical sewer network (total length greater than 100 km) for this time frame are very high. It was not practical to perform the calculations for these periods at a daily time step and an alternative solution, which has been adopted is to design the DSS to initially analyse the UVQ output data and identify four days that represent typical weather events. The weather event days are classified into the following groups:

- Dry weather flow
- Medium rain flow
- High rain flow
- Storm rain flow.

The event day selection is based on flow levels in wastewater pipes, with data taken from the first neighbourhood in the UVQ model and then applied to subsequent neighbourhoods with the simplifying assumption being that the impact of a rain event is consistent across all neighbourhoods. The minimum wastewater flow is identified for the year and then each day of the year is assigned a weather event flow, for example dry weather flow conditions could be assigned to days where wastewater flow is ≤150% of the minimum flow. The DSS then performs the NEIMO calculations on medians of these representative 4 days, allocated on a yearly basis over the time frame required. It then passes the output to the Sewer Leakage Index model (SLeakI) to assess the attenuation of contamination emanating from pipeline leaks. The user can manually select the wastewater flow classification limits that are used in determining the weather event days. The DSS also gives the user the option of selecting specific dates to represent weather event flows.

The NEIMO screen (referred to by its former name Pipe Leakage Model in the DSS) allows the user to navigate to the correct input files for both wastewater and stormwater networks (see Figure 2.6.3). There are also a number of key parameters that can be set within the DSS for the NEIMO model. This includes specifying a value for Manning's roughness coefficient, which relates to the roughness of the pipe's internal surface and is used in Manning's equation to determine the water level in the pipe. The user can also specify the hydraulic conductivity of different soil types and the delta L, which relates to the thickness of the colmation layer. The presence and thickness of the colmation layer are important factors in slowing the leakage rate and also in the attenuation of potential contaminants. The DSS also provides a user interface for locating and modifying text input files for NEIMO. This functionality is important to ensure users were easily able to locate input files and also to prevent errors by making it clear which input files are being used for each DSS run.

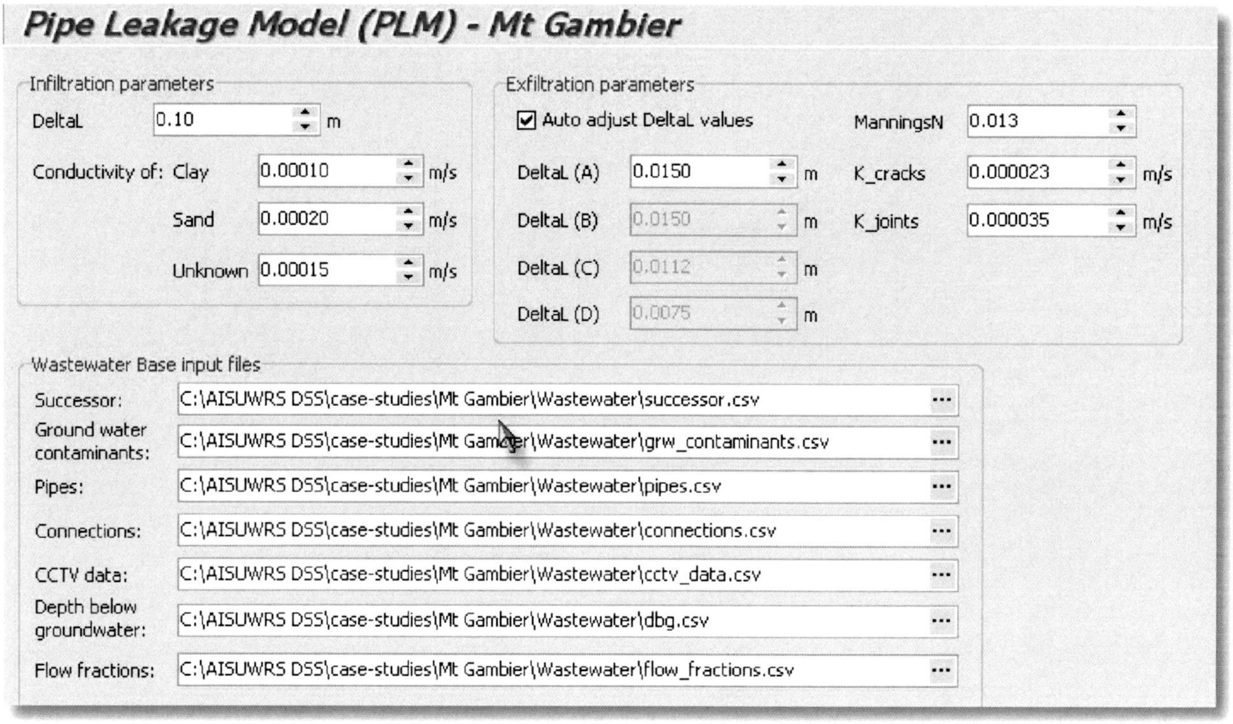

Figure 2.6.3. NEIMO Input Screen.

The impact of pipe leakage can be simulated using either best case or a worst-case situation. The volume of leakage for each asset is passed to SLeakI from NEIMO, with the best and worst-case scenario referring to the number of leaking defects for each asset. The best-case scenario is considered to be when the leakage volume is exfiltrating from multiple defects along the asset, while worst case is taken as one leak at the end of the asset. The number of leaks that the exfiltration volume is spread across implies a defect size, so it influences the travel time through the colmation layer and the unsaturated soil zone and consequently also the attenuation and decay of contaminants. The colmation layer clogs up the soil on the outside of the defect and has a limiting influence on the exfiltration rate. Research has shown that leaks predominately occur at asset joints. Therefore the asset length was divided by the standard length for the material used in order to calculate the number of leaks for the best-case situation. For example, the average length of a PVC pipe is 6 metres, so in the best case scenario an asset of 20 metres will have three leak points.

The best and worst case scenarios only differ significantly in impact for those groundwater contaminant loads that attenuate or decay (non-conservative contaminants). For example in the Mount Gambier case study there was a significant difference between the scenarios in the phosphorus load predicted to reach the groundwater. The phosphorus load on average for the best-case scenario was less than 1% of the worst-case situation.

In addition to the parameters for the SLeakI model, key parameters for the Public Open Space Index model (POSI) are also set from within the DSS. Such parameters as the soil characteristics for instance are critical for determining the residence time and attenuation of contaminants as they move through the unsaturated zone. The values needed for SLeakI are soil type, thickness, temperature and pH, while POSI requires values for depth of the root zone, groundwater temperature and pH.

Once the DSS has run the user can view a summary of the results, which can also be exported in a variety of formats (.pdf, .html, or .rtf) or printed out. The summary report presents the key outputs from each of the models for the individual scenarios accessed, with further detailed information for each run available from the corresponding MS Access® database.

2.6.4 Data Management Pathways

The primary role of the DSS was to facilitate the passing of data between the models in the chain. Figure 2.6.4 provides a summary of the primary data pathways.

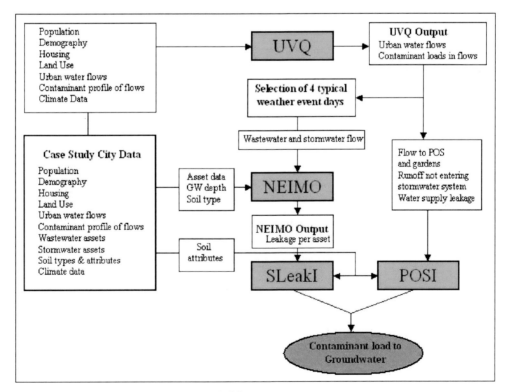

Figure 2.6.4. Primary data management pathways in DSS.

The individual models use different units to quantify water flows and contaminant loads. Therefore, a critical task in the DSS was to ensure the transfer of data in a consistent form between the different models. Appendix 1 lists the key input and output files in the DSS and the unit conversions between the models.

2.6.5 Database Interactions

The AISUWRS DSS is linked to a Microsoft Access® database that stores the output of each model in the DSS, and also stores input files and parameters used in each run. The direct linking of the database content to geographical information systems is possible but is not included in the standard version.

Figure 2.6.5 is a conceptual overview of the interactions between the DSS and the database. The DSS is the central component, as it passes information between the models and also writes output files to the database. A key feature of the database is that for each run of the DSS the input files and parameter settings for each model are saved along with the outputs. This feature is useful when evaluating different scenarios, as it provides a clear linkage between specific model outputs and the input parameters and files used. This functionality is useful for undertaking a sensitivity analysis of a particular parameter.

The DSS combines the outputs from the unsaturated zone models (SLeakI and POSI) to produce the groundwater output file that is then written to the database. This file records the contaminant load and water quantities reaching the aquifer for each day in the DSS run. The database stores this information in a format that can then be used as an input for a number of commercially available groundwater transport models, such as Feflow and MT3D.

The Access® database has been designed with a 'Form' view (default display when opening the database manually). These forms allow the user to easily step through the data via the Project Number that represents each DSS run.

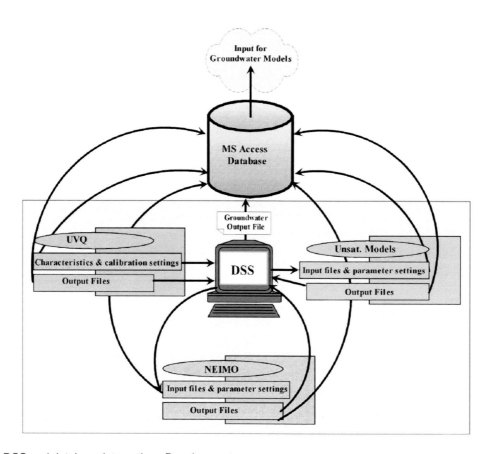

Figure 2.6.5. DSS and database interactions Development.

Borland Delphi 2005® (service pack 3) was used as the primary development environment for the AISUWRS DSS. To provide an attractive user interface, a third-party component collection (ExpressQuantumGrid Suite®) from Developer Express Inc. (http://www.devexpress.com/) was used.

Summary report functionality was also developed within the DSS. This provides a digest of the key outputs for each of the models. The reporting functionality was developed using report package Rave Reports® v6 from Nevrona Designs Inc. (http://www.nevrona.com/). The package allows users to view the reports immediately, print out or save to various formats (.html, .rtf, .pdf).

To help in the debugging and remote user error reporting, the software package madExcept® (http://www.madshi.net/) was implemented within the DSS. Whenever there is a crash/malfunction in the DSS, madExcept® will automatically catch it, analyse it, and give the end user the possibility to send an email to the developer containing a full bug report along with a message explaining what they were doing when the error occurred. This greatly speeds up the debugging process by making it clear at which point the DSS crashed.

2.6.6 System Requirements

The DSS is installed by running setup.exe and by default is installed into c:\AISUWRS DSS, although the user can change this if required. The user also needs to manually install Python® v2.3 (http://www.python.org/) and UVQ for the program to function. It is mandatory to install both Python® and the DSS in the same directory. SLeakI and POSI are automatically installed.

The installed DSS requires 55Mb of hard disk space (includes base data for the 4 case study cities), but once the DSS is running, the database for a city will grow at about 10Mb per run. This growth is very dependent on the size of the network being modelled. A Pentium 4® PC with at least 512Mb of memory and at least 500Mb of free space on the hard disk is strongly recommended. The DSS has been developed and run on Windows XP® and should also run on the Windows 2000® operating system.

Run time for the DSS is dependent upon the size of the pipe network (number of assets) and also the number of neighbourhoods defined. In the AISUWRS case study cities the run times varied significantly, which was related to both the characteristics of the study area and also the computing set-up. Running the DSS over a two-year period using a computer with a Pentium 4® 3Ghz processor for most case studies would take less than 1 hour with some as quick as 10 minutes.

2.6.7 Conclusions

A Decision Support System has been described that controls a number of complex models to predict the transport of water and its associated contaminants from source to ground water. The tool integrates individual models of urban water balance, pipeline leakage and unsaturated zone transport. It supports the selection and comparison of problem scenarios, which allows the implications of the effect of these scenarios on groundwater contamination to be assessed. When combined with appropriate groundwater flow and transport models and a socio-economic methodology, the Decision Support System provides a comprehensive tool that allows the implications of different urban water practices on ground water sustainability to be assessed.

2.6.8 Appendix 1: Transfer of data between models

UVQ Outputs

PlmUVQWWinput.txt: Wastewater flows and contaminant loads passed to NEIMO as uvq_nbrhoods.csv with no change in units

UFMGardenToGW.txt: Infiltration volume and contaminant load to groundwater from gardens, includes rainfall, irrigation and stormwater overflow. Passed to POSI.

UFMPOSToGW.txt: Same as above, but groundwater infiltration from public open space

UFMSWInfiltrationBasinToGW.txt: Pervious stormwater storage infiltration to groundwater. Passed to POSI.

UFMTapToGW.txt: Infiltration of water supply pipe leakage. Passed to POSI.

NEIMO input (uvq_nbhoods.csv)

Volumes: Wastewater generated (m^3/connection/day/neighbourhood)
Area: Neighbourhood area (m^2/neighbourhood)

NEIMO output (InfiltExfilt.csv)

Volumes m^3/day/asset: exfiltration, infiltration
Contaminants: grams: boron, potassium, sodium, chlorine, nitrogen, phosphorus, sulphate, suspended solids, total organic carbon, endocrine disruptors
    pfu/100ml: virus
    cfu/100ml: E.coli, faecal streptococci

SLeakI input (leakdata.txt)

Average leak rate per site/asset (litres/day) taken from NEIMO exfiltration and multiplied by 1000 to convert to litres.
Contaminants: mg/litre: boron, potassium, sodium, chlorine, nitrogen, phosphorus, sulphate, suspended solids, total organic carbon, endocrine Disruptors (UVQ output multiplied by 1000)
    pfu/100ml: polio virus (no change from NEIMO)
    cfu/100ml: E coli, faecal streptococci (no change from NEIMO)

POSI input (POSIInfiltrationData.txt)

Average infiltration rate per site (mm/day) calculated from UVQ output files using the following calculation:
    (Flow_kL / (Area m^2 No. of Areas)) 1000
Contaminants: mg/litre: boron, potassium, sodium, chlorine, nitrogen, phosphorus, sulphate, suspended solids, total organic carbon, endocrine disruptors (UVQ output multiplied by 1000)
    pfu/100ml: polio virus: (no change from UVQ)
    cfu/100ml: E coli, faecal streptococci (no change from UVQ)

SLeakI output (load.txt)

Average leak rate per site/asset (litres/day)
Contaminants: mg/litre: boron, potassium, sodium, chlorine, nitrogen, phosphorus, sulphate, suspended solids, total organic carbon, endocrine disruptors
    pfu/100ml: polio virus
    cfu/100ml: E coli, faecal streptococci

POSI output (load.txt)

Average infiltration rate per site (mm/day)
Water output (litres/day)
Contaminants: mg/litre: boron, potassium, sodium, chlorine, nitrogen, phosphorus, sulphate, suspended solids, total organic carbon, endocrine disruptors
    pfu/100ml: polio virus
    cfu/100ml: E coli, faecal streptococci

GWM output (GWoutputYYYY.csv)

Produced by combining SLeakI output + POSI output
> Best: GroundWaterRecharge per NBHD (L/day) = (SLeakI best water output + POSI water output)
> Worst: GroundWaterRecharge per NBHD (L/day) = (SLeakI worst water output + POSI water output)

Contaminants (Best Case & Worst Case): mg/litre: boron, potassium, sodium, chloride, nitrogen, phosphorus, sulphate, suspended solids, total organic carbon, endocrine disruptors
> pfu/100ml: polio virus
> cfu/100ml: E coli, faecal streptococci

Total load: units/day

3
Application to real world problems

3.1 Jochen Klinger[1], Leif Wolf[1], Christina Schrage[1], Martin Schäfer[1], Heinz Hötzl[1]

3.2 Brian Morris[2], Joerg Rueedi[3], Majdi Mansour[2], Aiden Cronin[3]

3.3 Petra Souvent[4], Goran Vizintin[4] and Barbara Cencur Curk[4]

3.4 Stephen Cook[5], Joanne Vanderzalm[5], Stewart Burn[5], Peter Dillon[5] and Declan Page[5]

[1] Department of Applied Geology (AGK), University of Karlsruhe, Germany
[2] British Geological Survey (BGS), Wallingford, UK
[3] Robens Centre for Public and Environmental Health, EIHMS, University of Surrey, Guildford, UK
[4] Institute for Mining, Geotechnology and Environment (IRGO), Ljubljana, Slovenia
[5] CSIRO, Melbourne, Australia

3.1 A POROUS AQUIFER: RASTATT, GERMANY
3.1.1 Background
3.1.1.1 Physical and geological setting

The study area of Rastatt has approximately 50,000 inhabitants and is located 30 km south of Karlsruhe, close to the eastern border of the Upper Rhine Valley, SW-Germany. The city extends from north to south for 6.7 km, and from east to west for 7.9 km. The city area is characterised by a rather flat topography with an elevation of 115 m a.s.l. in the centre, a minimum of 110.5 m a.s.l. at the location "Große Brufert" and a maximum of 130 m a.s.l. at the castle "Favorite".

The geology of the Upper Rhine Graben in the area of Rastatt can be divided into four main tectonic units: a graben block, a down fault block, a marginal block (fore hills), and an outlier zone (basement) (Figure 3.1.1). The graben is bordered by the Black Forest and the Vosges. During the Holocene epoch, the River Rhine excavated the Lower Terrace Formations within the present Rhine depression and filled it up with recent material of varying thickness. With a generally steady slope, the Rhine depression borders the valley terrace that in this region corresponds to the uniform Lower Terrace floodplain. Along the mountain border, the Rivers Kinzig, Murg, and others coming down form the mountains have deposited a Holocene channel. As shown in Figure 3.1.1 in the region of Rastatt the Pleistocene sediments are some 60 m thick with thickness increasing continuously towards the north.

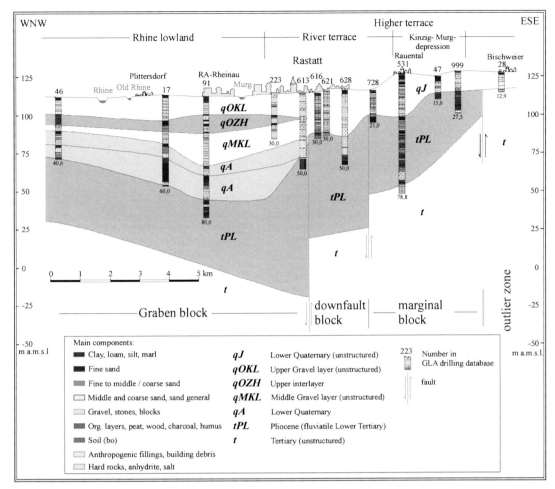

Figure 3.1.1. Geological cross section (NW-SE) through the Upper Rhine Valley in the area of Rastatt. Modified after (Eiswirth et al. 2003).

3.1.2 Hydrogeological setting

As shown schematically in Figure 3.1.1, up to four aquifers can be distinguished in the Rastatt area: the Upper Gravel Layer (OKL), the Middle Gravel Layer (MKL), the Older Quaternary (qA) and the Pliocene aquifer (tPL). The aquifer sequence is overlain by Holocene cover sediments of varying grain sizes and thicknesses. The Upper Gravel Layer and the Middle Gravel Layer are separated by the Upper Interlayer, which consists of more fine-grained sediments. The Upper Interlayer does not occur across the whole study area and so constitutes only a partial hydraulic barrier. Therefore the Upper Gravel Layer and the Middle Gravel Layer are frequently grouped together as the Younger Quaternary aquifer. The names Upper and Middle Gravel Layer suggest that there would also be a Lower Gravel Layer and a Middle Interlayer present but this system stems from other parts of the Upper Rhine Graben, where more frequent fine-grained horizons mean these subdivisions can be recognised. In the Rastatt area, the Middle Gravel Layer is underlain by the Older Quaternary aquifer, which is distinguished by a finer-grained lithology. The Older Quaternary is itself underlain by the Pliocene aquifer.

3.1.3 Urban development setting

Rastatt was established on the lowlands of the River Rhine and was historically mentioned the first time in the year 1084 under the name "Rastetten". Due to politically unstable periods between 1100 to the end of the 19th century (Palatine war of succession, occupation, revolution in Baden, etc.) the town was burned down several times, right down to the foundations of the walls (Ebeling 1991).

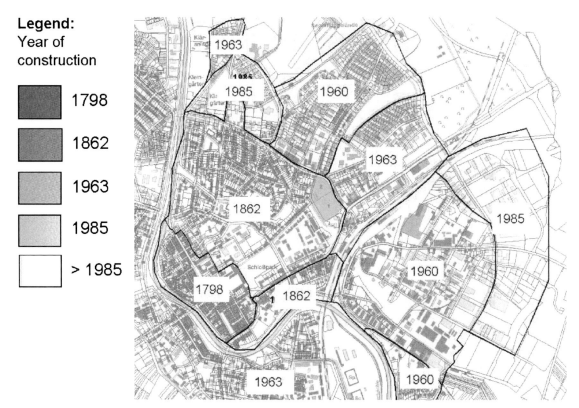

Figure 3.1.2. Settlement structure of Rastatt and the main construction periods of the different boroughs (Schweinfurth et al 2002, modified).

The most significant changes were noticed in the late 19th and the beginning of the 20th century. From 1890-onwards new industrial settlements came to Rastatt and brought the city prosperity. Also the new building of the synagogue of Rastatt falls into this time. The First World War and the world economic crisis suddenly stopped this economic upswing. Unemployment, housing shortage and poverty dominated the life of the border town in the coming decades. From the mid-1920s increased economic growth was accompanied by population growth and in 1930 Rastatt had ca. 15,000 inhabitants and 38 factories with more than 20 employees. At the end of the Second World War (WW II) Rastatt was hit by air attacks and partly destroyed. During the reconstruction of the city in the decade following WW II, Rastatt started to take on its present-day appearance. From 1950 an economic upswing started. Together with industry the population grew continuously. Completely new road alignments and boroughs were established adjacent to the historical centre of Rastatt. The population of 12,000 in 1945 increased to more than 23,000 by 1958. From 1971 – 1975, several neighbouring communities (Ottersdorf, Wintersdorf, Plittersdorf, Rauental and Niederbühl) were suburbanized due to a territorial reform of the federal state of Baden-Württemberg and the population increased to 39,000. In the mid-1970s, town redevelopment started with the classification of a traffic-calming area and the replacement of older one-storied houses through new architecture, with the aim to increase the quality of life in the city. In Figure 3.1.2 the age of construction of the different boroughs is illustrated.

The industrial area of Rastatt developed in a region that is nowadays considered to be one of the strongest in Europe. Enterprises of almost all sizes found good conditions and established factories in the urban area of Rastatt. The economy is characterized by long-established metal processing enterprises like Stierlen-Marquet and Diana and through automobile component suppliers like Berga. In recent years Daimler Chrysler expanded its gearing factory and started the production of the small car series "A-Klasse". Additionally, BWR Bizerba, an automobile assembly facility started with a factory in Rastatt.

Besides the large-scale enterprises there are additionally numerous middle and small-size enterprises as well as craft workshops, all of which contribute to good economic development. In the promotion of economic development Rastatt also offers active support with the "office for middle-size development Rastatt", intended to attract small and middle-scale enterprises (Ebeling 1991).

3.1.4 Urban water infrastructure

Water leaking from drinking water mains is known to be a major source of groundwater recharge in urban areas. The public water supply in Rastatt is provided by the star.Energiewerke GmbH. The company emerged from the local public sector in January 2003 and supplies the city not only with water but also with gas, electricity, heat and telecommunications. The public drinking water supply network consists of 188.8 km of pipes and serves 7,851 household connections. The water mains are mainly made of ductile cast iron with diameters up to 600 mm. To protect them from corrosion, the inside of the pipes is lined with cement mortar. The connecting pipes consist of polyethylene. Around 1,650 hydrants ensure water supply for fire fighting. The evolution of drinking water supply volumes is listed in Table 3.1.1.

Table 3.1.1. Drinking water supply characteristics in Rastatt as specified in the socio-economic questionnaires (see work package 12 of AISUWRS).

		1985	1990	1995	2000	2004
Drinking water production	m³/y	2,636,000	2,857,900	2,745,174	2,730,623	2,530,165
Drinking water imports	m³/y	-	19,820	20,830	19,190	18,570
Production + imports	m³/y	2,636,000	2,877,720	2,766,004	2,749,813	2,548,735
Total water consumption	m³/y	2,402,000	2,402,000	2,550,836	2,475,933	2,456,000

It is evident that there have been only minor changes from the 2.7 million m³ supplied on average, despite the growth in inhabitants from c. 39,700 persons in 1985 to c. 46,000 persons in 2000. This reduction in water demand per capita has been achieved by more efficient technology in industry and households as well as raised public awareness. In recent years the consumption has decreased even further. In an internal report of the star.Energiewerke GmbH, the leakage rates have been specified. Table 3.1.2 lists the losses and relates them to the supply area. This results in groundwater recharge rates from drinking water system leaks equivalent to between 10.2 mm/y and 17.2 mm/y.

Table 3.1.2. Leakage rates determined by star.Energiewerke and the corresponding groundwater recharge averaged over the estimated supply area.

Year	Water supplied			Water losses	
	m³	mm/y equivalent	%	m³	mm/y equivalent
1996	2,579,000	152.7	11.32	291,943	15.8
1997	2,498,000	147.7	9.27	231,565	12.5
1998	2,434,000	144.5	7.76	188,878	10.2
2000	2,475,933	147.0	12.84	317,910	17.2

While the leakage rate averaged over the supply area is known, the distribution of leaks in the drinking water mains has not yet been determined in detail in Rastatt. Extensive condition monitoring campaigns relying on sector wise pressure measurements are planned. However, as the total losses of the system are considerably below the German average, the water supplier is not under high pressure. It may be assumed that the losses depend on the supplied volumes and the density of pipes in the network. The groundwater recharge from leaky drinking water supply pipes is based on the assumption of a 10 % loss of the supplied water volume. It can be noticed that the densely populated parts in the city centre exhibit leakage volumes equivalent to 55-74 mm/y groundwater recharge. Combined with the high degree of surface sealing, the drinking water losses can become the largest component of the groundwater recharge total beneath these areas.

3.1.5 Data sources

For the set up of the AISUWRS model chain more than 350 input parameters can be implemented. The data serve on the one hand to provide a detailed description of the settings and the boundary conditions. On the other hand they have been acquired for the calibration of each model that has been employed and for the adjustment of the whole model chain to real world measurements, in order to minimise reliance on model default values that may not be appropriate to the urban/aquifer setting. The following tables list the data requirements and their sources. A big effort was made by stakeholders and the local departments of the city of Rastatt. In Table 3.1.3 the data gathered for the set up of the UVQ model for the study area can be seen. In the context of the AISUWRS model chain UVQ is certainly the approach that needs the most input data.

Table 3.1.3. Source of the data used for the set up of the UVQ model.

Approach	Data	Source	Format	Additional information
UVQ	Cadastral	CED Rastatt*	GIS layer	Data have been exported from the Smallworld GIS ™ to the ArcGIS® shape file format
	Surface sealing	CED Rastatt*, Arcadis Trischler & Partner	Paper format	Data have been transferred manually to the digital landparcel layer
	Demographic data (inhabitants per area)	Dept. of town marketing	GIS layer	Data where provided in number of inhabitants per km²
	Sewer network, discharge direction	CED Rastatt*	GIS layer	Essential data for the definition of the catchment areas
	Mains network	star.Energiewerke	GIS layer	
	Mains losses	star.Energiewerke	Dig. Table	No spatial data available, losses determined over the total area supplied due to iits sensitivity
	Water use data per capita, area in which water was used (toilet, kitchen, etc.)	star.Energiewerke	Dig. Table	Latest release of the data 2003
	Loads per capita, waste water contaminants	Literature, field investigations	Dig. Table	Wide range of variations given in literature, adjustment by using own field measurements
	Industrial water use	-	-	Gained through analysis of flow meter measurements at the WWTP and at the outlet of industrial catchments
	Contaminants from impervious surface run off (roof, paved, road)	Literature study	Dig. Table	
	Waste water flows	WWTP Rastatt, own field investigations	Dig. Table	The total discharge is measured at the WWTP, this includes however the waste water of the suburban areas too. Additional calculation have been perfomed to receive an area related discharge. Own measurement have been used for smaller catchments
	Soil properties	LGRB**, field investigations	GIS layer	Master thesis Osswald (2002)
	Climate data	LfU***, regional climate model	Dig. Table	Time series for the description of the climate are gathered on the one hand from automatised weather station and additionally from the regional climate model HIRHAM (DMI)

CED Rastatt*: Civil Engineer Department of Rastatt, LGRB**: Regional Authorities for Geology, Resources and Mining, LfU***: Regional Authorities for Environmental Protection of Baden-Württemberg.

In Table 3.1.4 the data sources to meet the requirements of the pipeline exfiltration model NEIMO, for the unsaturated zone modelling with SleakI and POSI as well as the data needed for the modelling of the saturated zone flow and the transport processes with the commercial code Feflow® are listed.

3.1.6 Field investigation programme

3.1.6.1 Objectives

The field investigations in Rastatt have been performed to assess the impact of urbanisation and anthropogenic activities on the urban water cycle. By the detection of typical sewage constituents in the groundwater the influence of exfiltrating sewage through sewer defects was estimated. Additional experiments have been conducted to determine leakage rates of a single sewer defect. The experiments led to a better understanding of the interaction between sewerage and groundwater and delivered basic data sets for the calibration processes of the AISUWRS model chain. A more complete documentation of the field investigation programme is provided in Wolf et al. (2005b).

3.1.6.2 Measured impacts on groundwater quality

Sampling strategy

The Environmental Agency in Rastatt had already collected available information on contaminated sites and created a database of incidents and measurements, and relevant results like the distribution of chlorinated hydrocarbons are documented in Arcadis Trischler & Partner (1999) and Eiswirth (2002). Therefore the focus of the field investigation in the AISUWRS-project was the quantification of the impact of leaky sewer systems. Several different sampling strategies were employed over the course of time on an extensive network of wells (Figure 3.1.3).

Table 3.1.4. Data requirements and source of the NEIMO, SLeakI, POSI and groundwater flow and transport model.

Approach	Data	Source	Format	Additional information
NEIMO	Sewer network (see above)	CED Rastatt*	GIS layer	Data export from the smallworld GIS
	Detailed sewer network properties (material, grade, diameter, etc)	CED Rastatt*	Dig. Table	Data export from the smallworld GIS
	Elevation of the sewerage, position relative to water table	CED Rastatt*, Drillling data (LGRB**)	GIS layer	
	Tree structure of the sewer network	CED Rastatt*		Gained by a back tracking query
	Exfiltration quantities for calibration of NEIMO	Field investigations, Monte Carlo simulation		Test site Kehler Strasse provided detailed information to adjust the NEIMO approach
	Bedding material	CED Rastatt*		It was commonly assumed to be sand and was stated by the CED* of Rastatt
SLeakI POSI	Soil profile Soil properties	LGRB**		
	Water table	LfU***		Water table was interpolated between the gw level of the observation wells
Groundwater flow and contaminant transport model (FEFLOW®)	Geological setting of aquifer	LGRB**	Dig. Table	Drilling profiles
	Properties of the aquifer	LGRB**, literature study		
	Boundary conditions	Literature study, calibration		Lateral inflow from the Black Forest
	Observation wells	LfU***	Dig. Table	
	Production wells	star.Energiewerke	Dig. Table	
	Water levels of surface water bodies		Dig. Table	Time series monitored at river-, lake gauges
	Groundwater recharge	Literature, calibration	GIS layer	See existing ground water flow model of Kuehlers (2000) and regional flow model Upper Rhine Valley (LFU 1996).

CED Rastatt*: Civil Engineer Department of Rastatt, LGRB**: Regional Authorities for Geology, Resources and Mining, LfU***: Regional Authorities for Environmental Protection of Baden-Württemberg.

As the sampling campaigns during the years 2001 and 2002, on a network of existing wells, had already indicated that no widespread influence of leaky sewers could be detected, it was decided to build new observation wells close to known sewer leaks. For this, the sewer defect database was searched for large-diameter sewers with severe defects. In a second step the pre-selected sites were evaluated in terms of their accessibility and potential conflicts with other urban infrastructure (gas or water pipelines, telecommunications and electricity).

Finally, eight new groundwater monitoring wells were constructed. The wells were drilled at a diameter of 225 mm and equipped with 50 mm PVC pipes and screens. The annular space was filled with gravel and sealed with bentonite against contamination from the surface. Three further sewer focus wells have been constructed in summer 2004 at the sewer test site Danziger Strasse.

Citywide sampling campaigns
In order to compare the quality of groundwater beneath the urban area of Rastatt with the surrounding rural area, the observation wells are grouped according to their spatial position (Figure 3.1.3). Well group 0 comprises wells that are situated outside the city area ('Out of town'), either in forests or on agricultural land. Well group 1 is assigned to wells that are situated on the outskirts of the city ('City limits'). Well group 2 comprises wells that are located directly beneath the settlements and is dubbed as urban background. Well group 3 consists of wells that have been specifically drilled to monitor the effects of wastewater exfiltration ('sewer focus').

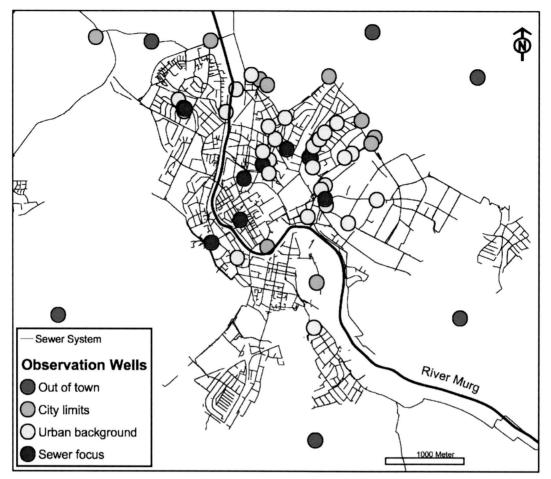

Figure 3.1.3. Classifying observation wells based on their exposure to urban pollution sources. Groundwater observation wells used for sampling during the years 2001-2005 (Wolf et al. 2005b).

The comparison between the different well groups is listed in Table 3.1.5 for selected parameters of the sampling programme. It can be seen that the focus observation wells (Well Group 3) exhibit significantly elevated values compared to the average of the other three well groups for the parameters electrical conductivity, temperature, sodium, chloride, potassium, phosphate, ammonium and boron. The highest contrast can be observed for ammonium (+ 446.5 % in the arithmetic mean) potassium (+ 116 % in the arithmetic mean), sodium (+ 80.5 % in the arithmetic mean), chloride (+ 77.3 % in the arithmetic mean) and boron (+ 65.2 % in the arithmetic mean). An inverse relationship is found for dissolved oxygen and nitrate.

The data set can be further analysed with respect to the straightness of the increase from rural area to sewer focus observation wells. A straight increase is given if the concentrations at the sewer focus wells (group 3) exceed the concentrations of the city background wells (group 2) which exceed the concentrations of the outskirts wells (group 1) which themselves are above the concentrations of the rural background wells (group 0). A straight increase in both arithmetic mean and median values has only been observed for potassium, boron and temperature.

The marked contrast between the sewer focus group and the urban background group shows that the quality deterioration is connected to large defects rather than to the whole network. However, the elevated concentrations of the urban background group compared to the rural group demonstrates the diffuse pollution originating from the superposition of a large number of leaks.

Monthly samplings at focus wells

In order to analyse the seasonal variations of the hydrochemical parameters in more detail, monthly samplings were performed at 10 (later 13) selected wells throughout Rastatt in addition to the annual large sampling campaigns. The water samples were screened for major anions and cations as well as boron and ammonium.

Only eight different stations are displayed in Figure 3.1.4 with the temporal variations of ammonium concentrations as the other stations remained below the detection limit of 0.02 mg/l. The highest concentrations of ammonium were measured at the focus observation well Ottersdorfer Strasse which is located close to a defect branch connection. The well at the Ottersdorfer Strasse also exhibits very high concentrations of the iodated contrast medium iomeprol (max. 1655 ng/l) as well as very high boron concentrations (max. 0.2 mg/l). The increasing ammonium concentrations indicate that the environment around the source has not yet reached stable conditions or that the influence of the defects is increasing.

Table 3.1.5. Selected hydrochemical parameters related to well groups (0 = out of town, 1 = city limits, 2 = urban background, 3 = sewer focus) and identified trends (++ = straight increase from group 0 toward group 3, + = increase from group 0 towards group 3 with one exception, o = ambiguous pattern, - = inverse correlation). Data from own samplings 2001 – 2005 (Wolf et al. 2005b).

Well Group	Category	Distance Nearest Sewer	Distance Upstream Sewer	Cond [µS/cm]	Temp [°C]	Diss. O_2 [mg/l]	Na [mg/l]	K [mg/l]	Cl [mg/l]	NO_3 [mg/l]	PO_4 [mg/l]	SO_4 [mg/l]	NH_4 [mg/l]	B [mg/l]
0	Arith. Mean	841,7	1116,7	484,9	11,1	3,1	10,5	2,1	17,8	27,2	0,8	31,1	0,06	0,032
1		44,0	257,3	556,8	12,5	1,0	14,5	3,3	19,8	5,1	0,7	34,5	0,16	0,056
2		14,8	47,8	489,2	13,1	1,9	13,3	4,7	14,1	10,0	0,8	30,4	0,14	0,061
3		2,6	10,2	706,9	14,0	1,2	23,1	7,3	30,5	12,0	1,1	38,1	0,66	0,082
0	Median	875,0	1000,0	424,0	11,2	2,3	9,1	1,2	15,7	18,6	0,7	31,2	0,03	0,024
1		24,0	270,0	585,0	12,4	0,6	11,3	2,9	17,2	4,8	0,7	32,1	0,08	0,037
2		9,0	37,0	499,6	13,2	2,0	13,0	3,9	14,8	7,9	0,7	33,5	0,04	0,060
3		2,0	3,0	622,5	14,1	0,8	20,7	6,1	26,2	8,6	1,0	37,5	0,38	0,093
0	Min	410,0	1000,0	232,6	9,9	0,1	5,9	1,1	8,3	1,6	0,7	11,5	0,03	0,015
1		7,0	13,0	216,5	9,8	0,1	7,4	1,5	6,9	0,2	0,7	11,6	0,02	0,018
2		1,5	1,5	193,1	8,9	0,4	7,8	1,6	8,1	1,3	0,5	13,0	0,02	0,036
3		1,0	1,8	462,2	12,6	0,4	11,3	3,1	11,7	1,4	0,7	16,1	0,01	0,036
0	Max	1150,0	1700,0	723,0	12,1	7,4	22,0	5,6	28,4	70,1	1,3	55,3	0,13	0,076
1		90,0	620,0	876,0	19,1	3,4	33,2	10,3	53,9	8,9	1,0	73,0	0,78	0,231
2		50,0	190,0	649,0	14,6	3,6	21,3	13,6	21,4	26,9	1,6	43,3	1,14	0,109
3		8,3	80,0	1052,7	15,1	6,2	38,5	16,3	53,6	54,1	2,3	73,1	2,71	0,143
0	Measurements considered	6	6	31	34	30	30	30	20	18	6	19	23	27
1		22	22	35	35	30	35	35	35	31	12	35	15	27
2		23	23	88	87	74	82	82	72	63	28	72	58	67
3		14	14	137	138	121	121	121	79	74	37	78	102	111
0	Increase towards focus wells	Mean	++	++	-	+	++	+	o	o	o	+	++	
1		Median	+	++	o	++	++	+	o	o	++	+	++	
2		Min	+	+	o	++	++	o	o	o	++	-	++	
3		Max	+	o	o	o	++	o	o	+	o	++	+	
0	Focus above average [%]	Mean	38,5	14,4	-38,9	80,5	116,0	77,3	-14,6	42,6	19,0	446,5	65,2	
1		Median	23,8	15,1	-53,3	86,4	128,7	65,2	-17,7	46,0	16,4	616,8	131,7	
2		Min	115,9	32,2	103,0	59,6	122,7	51,4	35,1	10,0	34,2	-52,6	54,2	
3		Max	40,5	-1,1	28,3	51,1	66,0	55,1	53,3	82,0	27,8	296,6	3,3	

Summarizing the monthly sampling campaigns it can be stated that seasonal patterns are not pronounced in the urban groundwater wells. A short-term increase of sodium and chloride concentrations due to road salting could not be confirmed. Strongly sewage-influenced wells showed elevated concentrations of both ammonium and boron.

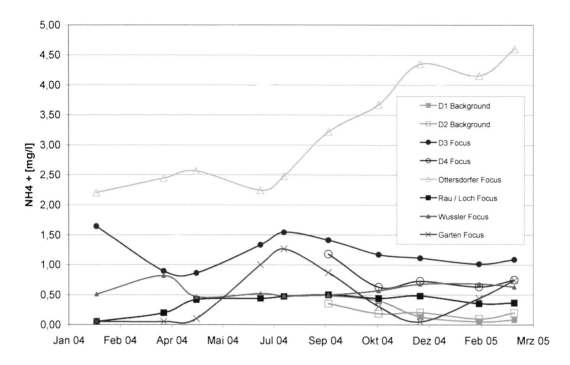

Figure 3.1.4. Temporal evolution of ammonium concentrations in focus observation wells in Rastatt. Detection limit is at 0.02 mg/l (Wolf et al. 2005b).

Test site Danziger Strasse

A very dense groundwater monitoring scheme has been implemented at the test site Rastatt-Danziger Strasse, where four observation wells were constructed along the groundwater flow line. Two defective public sewers are present as shown in Figure 3.1.5.

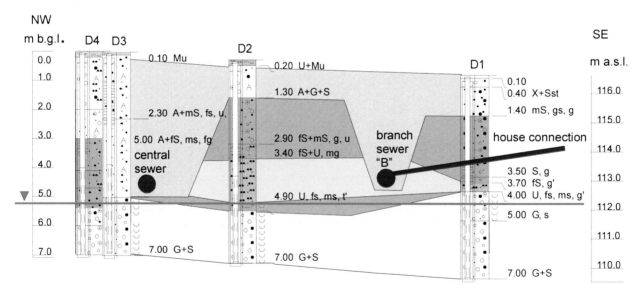

Figure 3.1.5. Geological cross section at the test site Danziger Strasse. Groundwater flow direction is from SE to NW (Wolf et al. 2005b).

Figure 3.1.6 shows a period of three weeks in February 2005 in more detail and also in relation to the water levels measured inside the sewer. Following heavy rainfall on 12th February, the electrical conductivities in the observation wells D3 and D4 declined by 350 µS/cm. During the next two weeks electrical conductivities rose again as the wastewater composition returns from 100 µS/cm during rain events to electrical conductivities around 1500-2000 µS/cm. The next decline on the 28th of February is more difficult to explain as no major changes in sewer fill level were noticed in the sewer. In this case, changed wastewater composition or altered microbiological conditions in the clogging layer could be responsible for this signature. One could also think that the infiltration at the surrounding green spaces during the rain event on the 12th of February produced a wetting front that reached the water table at the 28th of February and leads to a dilution of the groundwater. However, this effect should be noticeable also in the remaining three wells (D1, D2, ElfP1).

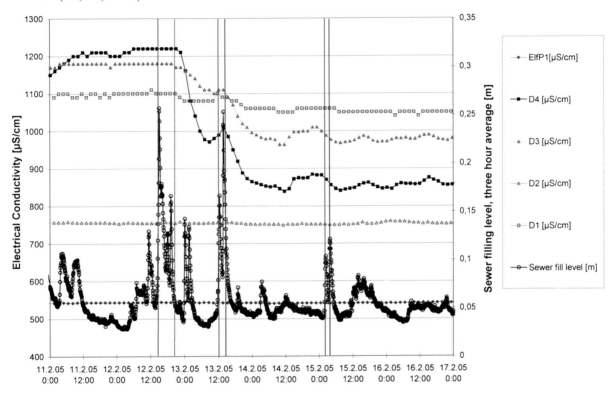

Figure 3.1.6. Detailed observations for three weeks in February 2005 and influence of water levels in the sewer on groundwater quality. A sharp decrease in electrical conductivity of groundwater can be observed in wells D3 and D4 following heavy rainfall on 12.2.2005. The reference wells D2 and Elf P1 remain unaffected (Wolf et al. 2005b).

From the online measurements at the sewer test site Danziger Strasse it can be concluded that:
- Sewers are in close contact with the aquifer.
- Significant water and solute volumes reach the water table.
- Travel times through the unsaturated zone are short (< 1 day).
- Upstream and background observation wells are not affected by the fluctuations.
- Large volumes of wastewater exfiltrate during storm events but result in a drop in electrical conductivity due to the diluting effect of the low dissolved-solids rain water on the wastewater chemistry.

Screening for pharmaceutical residues
The screening for pharmaceutical residues in groundwater and sewage samples revealed the existences of beta blockers (metoprolol, sotalol) in the urban groundwater (see Table 3.1.6). Compared to the state-wide survey undertaken by the LfU (LfU 2002) pharmaceutics have been found with elevated frequency in the Rastatt groundwater samples. These results correspond to other parameters indicating sewer leakage at the test site Kehler Strasse. It can be seen that only limited degradation or attenuation of most pharmaceutics occurs on the 50 cm seepage route (the sampling is performed after the sewage has passed a 50 cm deep, sand-filled collector box beneath a sewer leak). All substances identified in the sewage have also been recorded in the seepage water in a similar concentration range. Occasionally, the concentrations in the seepage water were above the concentrations in the sewage. In order to establish a mass balance in this strongly transient system, a denser monitoring scheme is obviously necessary.

Table 3.1.6. Pharmaceutical residues detected in groundwater, soil water, and wastewater in Rastatt (11.6.2005); Positive findings displayed bold. Values in brackets indicate a clear detection of the substance which falls slightly below the official detection limit (Wolf et al. 2005b). The detection limits vary between 10 and 100 ng/l depending on the substance class and the background matrix of the sampled water or wastewater.

Substance	Groundwater							Seepage water	Wastewater	
	D 1	D 2	D 3	D 4	Garten-Str.	Zay-Str.	Ottersdorfer-Str.	Test site Kehler Strasse	Kehler Strasse	WWTP
	ng/L	ng/L	ng/L	ng/L	ng/L	ng/L	ng/L	ng/L	ng/L	ng/L
Analgesics:										
Phenazon	< 10	< 10	< 10	< 10	< 10	< 10	< 10	< 20	< 50	< 100
Dimethylaminophenazon	< 10	< 10	< 10	< 10	< 10	< 10	< 10	**150**	< 50	< 100
Propyphenazon	< 10	< 10	< 10	< 10	< 10	< 10	< 10	< 20	< 50	< 100
Betablocker:										
Atenolol	< 10	< 10	< 10	< 10	< 10	< 10	< 10	**450**	**620**	**480**
Betaxolol	< 10	< 10	< 10	< 10	< 10	< 10	< 10	< 20	< 50	< 100
Bisoprolol	< 10	< 10	< 10	< 10	< 10	< 10	< 10	< 20	< 50	< 100
Metoprolol	< 10	< 10	< 10	< 10	**30**	< 10	< 10	**1100**	**540**	**500**
Pindolol	< 10	< 10	< 10	< 10	< 10	< 10	< 10	< 20	< 50	< 100
Propranolol	< 10	< 10	< 10	< 10	< 10	< 10	< 10	< 20	< 50	< 100
Sotalol	< 10	< 10	< 10	< 10	< 10	< 10	**920**	**1200**	**1200**	**620**
Broncholytics, Secretolytics:										
Clenbuterol	< 10	< 10	< 10	< 10	< 10	< 10	< 10	< 20	< 50	< 100
Salbutamol	< 10	< 10	< 10	< 10	< 10	< 10	< 10	< 20	< 50	< 100
Terbutalin	< 10	< 10	< 10	< 10	< 10	< 10	< 10	< 20	< 50	< 100
Zytostatics:										
Ifosfamid	< 10	< 10	< 10	< 10	< 10	< 10	< 10	< 20	< 50	< 100
Cyclophosphamid	< 10	< 10	< 10	< 10	< 10	< 10	< 10	< 20	< 50	< 100
Antibiotics:										
Azithromycin			< 10		< 10	< 10	< 10	< 20	< 50	(54)
Dehydrato-Erythromycin A			< 10		(5)	< 10	< 10	**98**	(39)	**240**
Clarithromycin			< 10		< 10	< 10	< 10	**45**	< 50	**150**
Roxithromycin			< 10		< 10	< 10	< 10	**31**	< 50	< 100
Clindamycin			< 10		(6)	< 10	(8)	**230**	< 50	(65)
Ronidazol			< 10		< 10	< 10	< 10	< 20	< 50	< 100
Metronidazol			< 10		< 10	< 10	< 10	< 20	< 50	< 100
Sulfadiazin			< 10		< 10	< 10	< 10	**21**	< 50	< 100
Sulfamerazin			< 10		< 10	< 10	< 10	< 20	< 50	< 100
Furazolidon			< 10		< 10	< 10	< 10	< 20	< 50	< 100
Sulfadimidin			< 10		< 10	< 10	< 10	< 20	< 50	< 100
Sulfamethoxazol			< 10		< 10	< 10	< 10	**50**	**170**	**440**
Dapson			< 10		< 10	< 10	< 10	< 20	< 50	< 100
Trimethoprim			< 10		< 10	< 10	< 10	**41**	**120**	**190**
Amoxicillin			< 10		< 10	< 10	< 10	< 20	< 50	< 100
Penicillin G			< 10		< 10	< 10	< 10	< 20	< 50	< 100
Penicillin V			< 10		< 10	< 10	< 10	< 20	< 50	< 100
Cloxacillin			< 10		< 10	< 10	< 10	< 20	< 50	< 100
Oxacillin			< 10		< 10	< 10	< 10	< 20	< 50	< 100
Nafcillin			< 10		< 10	< 10	< 10	< 20	< 50	< 100
Dicloxacillin			< 10		< 10	< 10	< 10	< 20	< 50	< 100
Oleandomycin			< 10		< 10	< 10	< 10	< 20	< 50	< 100
Chloramphenicol			< 10		< 10	< 10	< 10	< 20	< 50	< 100
Spiramycin			< 10		< 10	< 10	< 10	< 20	< 50	< 100
Tylosin			< 10		< 10	< 10	< 10	< 20	< 50	< 100

The pharmaceutical group of iodated x-ray contrast media received special attention in the sampling programme. 114 samples from 46 wells were analysed and resulted in a total of 51 positive detects of iodated contrast media. The frequency of the positive detects as well as the maximum concentrations exceed the values from the state-wide sampling campaign in Baden-Württemberg (LfU 2002). No x-ray contrast media were found in wells more than 60 m from the nearest upstream sewer. The occurrence of the iodated x-ray contrast media clearly proves the existence of major amounts of sewage in the shallow urban groundwater. Due to their chemical robustness, iodated x-ray contrast media are well suited as a marker species. However they are not distributed evenly over the city area (see Figure 3.1.7).

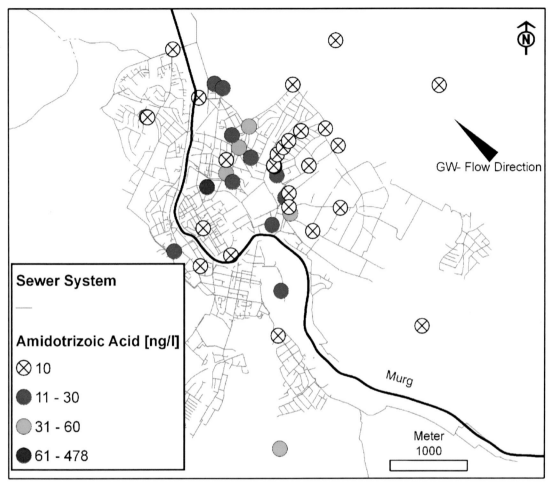

Figure 3.1.7. Amidotrizoic acid (in ng/l) in groundwater of the Upper Gravel Layer in Rastatt. Mean concentrations over samples from 2002 to 2004.

Microbiological parameters

The microbiological sampling programme comprised *Escherichia coli*, total coliforms, enterococci, faecal streptococci, sulphate reducing clostridia, *Clostridia perfringens*, coliphage and *Pseudomonas aeroginosa*. Indicators of faecal contamination have been found in several groundwater samples. The groundwater from 6 out of 12 wells is not suitable for human consumption according to German and international drinking water guidelines. However, no correlation between leak geometry, distance to the leak and microbiological indicators was found. In addition, the correlation between other wastewater indicators (e.g. ammonium, boron, pharmaceutics) and microbiological indicators was very weak. The findings of *Escherichia coli* in well D1 support the idea of a defect house branch connection as an additional sewage source. In conclusion it must be stated that wastewater in Rastatt contains high numbers of all the organisms screened for but an effective removal is taking place within the sediments surrounding the sewer (see Wolf, 2006).

Long-term evolution of groundwater levels

The long-term evolution of groundwater levels was assessed, based on existing LfU-measurements of groundwater levels starting as early as 1913. Only minor changes have been observed in the area of Rastatt during the last 90 years and the anthropogenic impact on groundwater levels in this area has been low. This is due to the very productive younger Quaternary aquifer, the high recharge rates and the exchange with the surface waters of the Rhine and Murg systems all of which serve to damp out other components of the water cycle such as pumped withdrawals or new urban sources of recharge.

3.1.6.3 Sewer test site Kehler Strasse

The new sewer test site Rastatt Kehler Strasse, constructed in July 2004, was set up in order to monitor exfiltration processes in undisturbed environments under real operating conditions. The water level in the sewer, the flow velocity, the composition of the sewage and the exfiltration rate from a *ca.* 120 cm² leak were monitored with a high time resolution over a period of more than 12 months. Figure 3.1.8 displays schematically the set up of the test site.

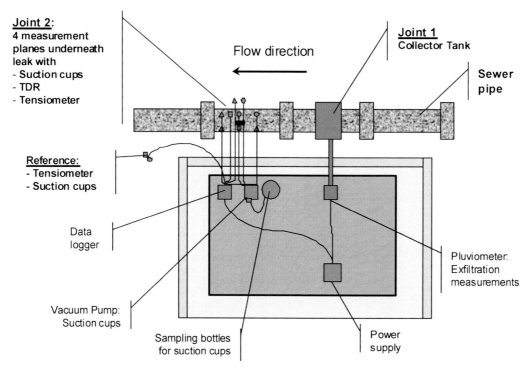

Figure 3.1.8. Schematic view of the test site Kehler strasse.

The experiment can be divided into two phases and is displayed in Figure 3.1.9
- Phase I: During phase I, a cavity which remained from the construction process, was gradually filled with sewer sediments and bio film material. Exfiltration rates up to 230 l/day were measured after storm events. This situation may be regarded as representative for sewers which are subject to alternating infiltration/exfiltration situations dominated by a seasonally changing water table. During the infiltration period, surrounding sediments are washed into the sewer, leaving behind a void around the leak. If groundwater levels are dropping again, the void is slowly refilled with sewage material. At this stage, very high exfiltration rates, as measured during phase I of the Kehler Strasse experiment, are likely to occur.
- Phase II: After six months of continuous operation, the exfiltration rates approached a pseudo-steady state. The exfiltration during dry weather flow mounted to an average of 1.3 l/day and 5 l/day connected to storm events. About 42 % of the total exfiltration volume during the period from January to June 2005 can be attributed to rain events which are affecting the exfiltration rates during 26 % of the balancing period time.

Attempts to derive hydraulic conductivities based on sewer water level, leak size and exfiltration volumes were inconclusive as the correlation between sewer water level and exfiltration rate is quite poor. It is planned to overcome the problem in future with more precise information on the travel time. A recent investigation using the advanced optical inspection technique PANORAMO has demonstrated that the leak geometry is more complex than was anticipated during the construction. A function that links the leak area with the fill levels will have to be introduced. Furthermore, the uneven distribution of the colmation layer along the leak will have to be taken into account. Summarising, it can be stated that the data monitored at the test site Kehler Strasse demonstrate the high complexity and the many factors influencing the colmation process.

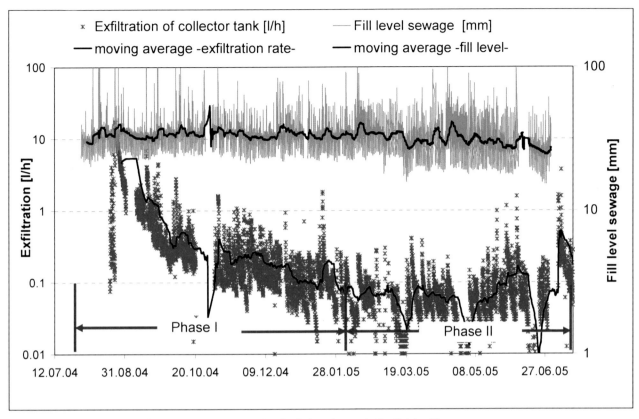

Figure 3.1.9. Long-term exfiltration rate measured at leak 1 at the test site Kehler Strasse.

3.1.7 AISUWRS model setup for Rastatt

3.1.7.1 UVQ (Urban Volume and Quality)

Data requirements – data processing

The model approach in UVQ accounts for all stages in the passage of water through the urban system. However, to meet the model requirements an enormous amount of input data needs to be acquired and processed. The study area is described by the physical characteristics (distribution of house, garden, paved area of a residential property, open space, roads, etc). Additionally demographic data (inhabitants per house) and information about water consumption behaviour of the population is required.

To cover the contaminant modelling, input concentrations from rainfall, surface runoff and indoor water use are needed. In total almost 300 fields cover all input possibilities. The data used to set up the UVQ model for the Rastatt study area were gathered through literature studies, data acquisition for the city itself or direct field measurements. In Table 3.1.7 data used for the set up of the model and their formats are listed.

Table 3.1.7. Data sets used for the set up of the UVQ model.

Spatial Scale	Parameter	Unit	Datasource / Data processing
Land Block Scale	Number of Land Blocks	nb	Digital Landparcel Map, GIS Analysis
	av. Block Area	m^2	Digital Landparcel Map, GIS Analysis
	Average Occupancy	nb	Digital Landparcel Map, GIS Analysis
	av. Garden Area	m^2	Digital Landparcel Map, GIS Analysis
	av. Roof Area	m^3	GIS Analysis of the Building layer
	av. Paved Area	m^4	Digital Landparcel map, GIS analysis
	Percentage of Garden irrigated	%	Classification according to the landuse, Satellite image
	Roof Runoff to Spoondrain	ratio	Digital land parcel map, GIS analysis
Neighbourhood Scale	Total Area	ha	Digital Landparcel Map, GIS Analysis
	Road Area	ha	Digital Landparcel Map, GIS Analysis
	Open Space Area	ha	Digital Landparcel Map, GIS Analysis
	Percentage of Open Space irrigated	%	Digital Landparcel Map, GIS Analysis
	Imported Supply Leakage	%	Water Supply Company Rastatt *
	Wastewater as exfiltration	ratio	Pipeline Leakage Model, Monte Carlo assessement
Waste Water Outputs	Wastewater from Neighbourhood goes to ID	nb	Digital Sewer Map, GIS Analysis
	Storm water from Neighbourhood goes to ID	nb	Digital Sewer Map, GIS Analysis
Indoor Water Usage & Contaminants	Kitchen	L/c/d	Literature with local data
	Bathroom	L/c/d	Literature
	Toilet	L/c/d	Literature
	Laundry	L/c/d	Literature

Spatial scales - Physical characteristics
UVQ operates on three different spatial scales to represent a study area i.e. land block, neighbourhood and study area. On each scale the input parameters have to be determined before the model can be run (i.e. area distribution, water use, population density). The Civil Engineering Department of Rastatt provided a many digital data sets including a layer of the land parcels and the buildings, a thematic layer of the sewer network and a satellite image to distinguish between the different land use types like single residential, commercial or industrial areas. The digital data sets were processed with a GIS system (ArcGIS 9.0) that allowed a fast definition and editing of the data sets.

A land block is the smallest spatial unit in UVQ and represents a single house property. The total land block area is distributed to roof (house or garage), paved and garden area. The digital layers of the land parcels and the buildings were used to define the land blocks in the study area. Through a query prompted in the GIS, the total number of land blocks could be identified automatically. Land parcels containing no buildings were defined as open space areas.

A neighbourhood is commonly represented by a set of land blocks with a similar land use, e.g. residential properties or predominantly industrial and commercial areas can be defined as homogeneous areas. Due to the completely different water consumption behaviour in public utilities, large schools or hospitals can be defined as a separate neighbourhood.

Further criteria for the definition of the neighbourhoods are the sewer system (combined or separate sewer system) and the delineation of the different catchment areas. The use of the GIS allowed the superimposing of the different digital layers (land parcel and sewer network layer) and therefore the determination of the homogeneous areas. Besides the land blocks, roads and shared open space like parks or graveyards occur and have to be quantified. Very large features (e.g. the main graveyard) were assigned to special neighbourhoods while smaller features were integrated into the average percentages of open space.

The study area is the compilation of different neighbourhoods, which may vary with regard to their land use, size and water management setting. They are linked by the sewer system. Waste- and storm water generated in an upstream area is conveyed to the downstream area. However, in UVQ the flow direction from one to another neighbourhood follows ascending numbers. In Figure 3.1.10 the aggregation of the neighbourhoods defined to represent the urban area of Rastatt is shown. In total 74 neighbourhoods have been defined.

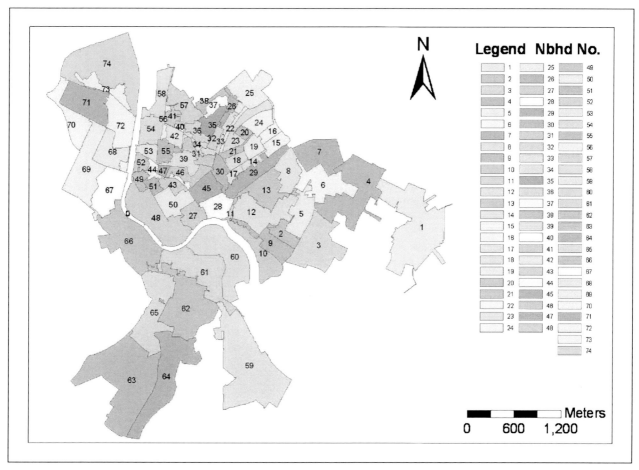

Figure 3.1.10. Neighbourhoods representing the study area of Rastatt for the UVQ modelling.

Climate conditions

To account for the climatic conditions UVQ implements potential evaporation and the precipitation either as rain or as snow. The time series for the modelling of the relatively short time 2002 to 2005 are derived from an automatic weather station which is located in the south eastern part of Rastatt, and were provided from the Institute for Environmental Protection Baden-Württemberg (LfU Baden-Württemberg). The monitoring programme includes measurements of precipitation, temperature, dew point, saturation and vapour pressure that allow the calculation of the potential evaporation used in the UVQ climate file.

For the modelling of a climate change scenario climate data of the regional climate model HIRHAM of the Danish Meteorological Institute (DMI) with a spatial resolution of 12x12 km was used. Fortunately Rastatt, the area of interest lies in the centre of one cell for which the data could be acquired. This simulation delivered data in daily steps as needed by the UVQ model and is an up to date regional climate model with the finest resolution available. Two timelines were made available by the DMI: 1960-1990 as the control period and 2070-2100 as the prediction period (predictions for an earlier time period are accounted as not very reliable (Schaefer 2006).

Water supply data

The drinking water in Rastatt derives from the groundwater of the underlying Quaternary deposits. The three waterworks in the vicinity of Rastatt tap the Upper Gravel Layer that serves as the main aquifer. In 1998 the water works Ottersdorf southwest, Niederbühl south and Rauental east of Rastatt have pumped c. 2.53 Mm^3 groundwater to supply the inhabitants of Rastatt and the industrial enterprises. The water is conveyed over a supply network of 189 km and app. 8000 house connections to the consumers (Eiswirth 2002).

The water supply company in Rastatt is continuously (every year) informing the inhabitants about the water quality and the water use per capita. The amount is split according to the water use in the house. In 2001 the average use per capita was at c. 110 l/d (star.Energiewerke 2002). The water quantity and the proportion used in each usage category is shown in Table 3.1.8.

Table 3.1.8. Water usage per capita statistically determined for the inhabitants of Rastatt (star.Energiewerke 2002).

Kitchen		Toilet		Bathroom		Laundry	Outdoor water use	
Drinking water, cooking	2.20	Toilet flushing	35.20	Shower, bath	33.0		Car wash Other use	11.00
Dish washing	4.40	Cleaning	-	Personal hygiene	6.6	13.20	Garden irrigation	4.40
	6.60		35.20		39.60	13.20		15.40
Total [l/c/d]				110				

In a separate study (Arcadis Trischler & Partner 1999, Eiswirth 2001) the water consumption behaviour in a typical household was investigated. The amount which was determined differs only slightly from the assignments done by the water supply company, distributed across a mean usage of 128 l/c/d.

Industrial water use

A separate data set for the industrial use in Rastatt could not be gathered. Two options for the estimation of the industrial water use do exist:

1. The analysis of flow measurements at the outlet of an industrial catchment. This approach is rather uncertain as it is an integration over the catchment where even infiltration processes can be included.

2. A second method accounts for the proportion of the total amount of water supplied in the urban area of Rastatt by difference. The water supply company records the total volumes supplied per year. These quantities include the water use of the industrial factories. Knowing the number of inhabitants as well as their water consumption the industrial water use can be calculated by difference. Even though these amounts include the irrigation of public open space areas and the water used for extinguishing of fires a quantity which is in a reliable range could be reproduced. Table 3.1.9 shows an extract of the recorded supply quantities and the calculated quantities for the industrial enterprises.

Table 3.1.9. Water supply data in the urban area of Rastatt.

Year	Total amount of water supplied	Inhabitants supplied	Av. water use per capita per day (total)	Inhabitants modelled in UVQ	Total water use for UVQ inhabitants	Av. water use per capita (residetial)	Water use of UVQ inhabitants (residential)	Industrial water use (calculated)
	[Mm³/a]	[nb]	[l/c/d]	[nb]	[Mm³/a]	[l/c/d]	[Mm³/a]	[Mm³/a]
1996	2.579	46264	153	36152	2.015	110	1.452	0.564
1997	2.498	46340	148	36152	1.949	110	1.452	0.497
1998	2.434	46133	145	36152	1.907	110	1.452	0.456
2004	2.456	47517	142	36152	1.869	110	1.452	0.417

Contaminant loads

Contaminant data are derived mainly from literature values gathered during the set up of the AISUWRS model chain. In Table 3.1.10 and Table 3.1.11 the loads of the contaminants are listed. Note that the contaminants in the house are entered as daily loads whereas the outdoor water components contain parameters linked to precipitation events. For these substances concentrations have to be defined.

Table 3.1.10. Indoor water use data and contaminant loads used for UVQ modelling (Eiswirth 2002, Gray et al. 1999, Böhm et al 1999, Rueedi 2005, Klinger et al. 2006).

Parameter	Unit	Compartment			
		kitchen	bathroom	toilet	laundry
Water flow	(L/c/day)	17.05	41.25	36.85	14.85
B	(mg/c/d)				200
K	(mg/c/d)	40		2000	10
Cl	(mg/c/d)	10000	6400		10000
N	(mg/c/d)	238	462	13709	328
P	(mg/c/d)	42	22	1568	52
SS	(mg/c/d)	3990	803	36240	4858
BOD	(mg/c/d)			60000	
COD	(mg/c/d)			120000	
feacal streptococci	(cfu/c/day)	1000	6071875	3000000	3.1E+07
Cu	(mg/c/d)	1.39	8.15	1.20	3.97
Zn	(mg/c/d)	19.23	215.00	10.85	75.77
Pb	(mg/c/d)	0.16	9.37	0.02	7.76

Table 3.1.11. Outdoor water use data and contaminant loads used for UVQ modelling (Eiswirth 2002, Gray et al. 1999, Böhm et al. 1999, Rueedi 2005).

Paramater	Unit	Imported (tap)	Rainfall	Pavement Runoff	Roof Runoff	Road Runoff	Roof first flush	Fertiliser to POS	Ground water	Fertiliser to garden
B	(mg/L)	0.02								
K	(mg/L)			2.5	2.5	2.5	5			
Cl	(mg/L)	19								
N	(mg/L)	0.11	0.85	2.1	4.5	0.82	9	81	0.4	80
P	(mg/L)	0.007	0.2	0.928	0.3	1.5	0.6	12		40
SS	(mg/L)	0.26	10	95	60	264	120	34	74	
BOD	(mg/L)									
COD	(mg/L)									
feacal streptococci	(cfu/L)			34294	294	51067	294			
Cu	(mg/L)	0.00	0.05	0.04	0.31	0.07	0.00		0.07	0.00
Zn	(mg/L)	0.01	0.17	0.02	0.60	0.05	0.05		0.01	0.05
Pb	(mg/L)	0.005	0.021	0.104	0.024	0.200	0.003		0.034	0.003

As contaminant loads can vary from region to region a sewage sampling campaign was performed at the outlet of a small catchment area of 55 ha and app. 1,600 inhabitants. Mainly domestic sewage is conveyed through the sewerage and these measurements therefore provided a concentration that could be adjusted to loads produced by the in-house water use. The concentrations measured and modelled are listed in Table 3.1.12.

For the nutrients nitrogen (N) and phosphorus (P) the model reproduces realistic concentrations with only small deviations. This is a positive outcome as the input loads are based only on literature values and were not modified. The results confirm the suitability of the literature values for the input loads used on household scale for this urban setting. As UVQ is not accounting for dynamic processes like the transformation of ammonium to nitrate the total amount for nitrogen and phosphorus is modelled.

As well as for the nutrients, the input loads for the suspended solids, consisting of organic and inorganic particles could be adequately represented. The difference between modelled and measured concentration is lying in a reliable range and it could be stated that the model is able to reproduce the content of suspended solids for this area.

The input concentrations of heavy metals copper (Cu), zinc (Zn) and lead (Pb) are very low and it is difficult to state whether the results are in a reasonable range as the measured concentrations are below the detection limit. The concentration of the heavy metals has to be handled with care, particularly if an influence of surface runoff to the waste water systems is noticed. Precipitation events can lead to a fast increase of the heavy metal concentration due to flushing of sources such as the abrasion products of automobile brakes and vehicle corrosion products.

Table 3.1.12. Contaminant loads measured and modelled for a small catchment area in the southwestern part of Rastatt (Klinger et al. 2005).

Parameter	Unit	Water usage - cont. loads per capita				Cont. Concentration [mg/l]	
		Kitchen	Bathroom	Toilet	Laundry	modelled	measured*
Water usage	[l/c/d]	56.8	21.0	13.2	44.8	-	-
N (total)	[mg/c/d]	228	462	13709	327	73.43	76
P (total)	[mg/c/d]	42	222	1568	52	8.39	13
Potassium	[mg/c/d]						32
Chloride	[mg/c/d]						125
Boron	[mg/c/d]						1.57
Susp. Solids	[mg/c/d]	3990	803	36240	4858	228	194
Cu	[mg/c/d]	1.4	8.2	1.2	3.9	0.073	0.04
Zn	[mg/c/d]	19	215	10	75	1.6	<0.01
Pb	[mg/c/d]	0.2	9.4	0.0	7.8	0.086	<0.1

3.1.7.2 NEIMO (Network Exfiltration and Infiltration Model)

NEIMO permits calculation of exfiltration and infiltration from gravity flow pipes. The model accounts for the physical characteristics of the pipes (material, diameter, grade, etc.), for the defects (generated or real defects), the wastewater flow in the pipe and the location with respect to the water table.

Data requirements

The set up of a model requires in total eight input files containing the information to represent the sewer network, to account for fill level variations, etc. Table 3.1.13 lists the required files, the content, the origin and the institution from which they were gathered. These needed to be edited to provide the input files in the eight required formats.

Table 3.1.13. Data source of the input files used for the set up of the NEIMO model.

File	Content	Datasource
flow_fractions.csv	holds the diurnal filllevel variations in the pipe	provided by NEIMO, is not specific to a network
generic_curves.csv	contains the defect sizes, distribution, opening of all pipes; used if no cctv data are available	generated by NEIMO, is specific to the network
grw_contaminants.csv	data of the groundwater contaminants; for infiltration modelling	UVQ
uvq_nbhoods.csv	water use and contaminant loads per property related to the neighbourhoods	UVQ
pipes.csv	pipe properties, length, grade etc	Civil Engineering Office Rastatt
connections.csv	number of property connection per asset	GIS Operation
successor.csv	contains up- and downstream relationship between pipe assets	Civil Engineering Office Rastatt
cctv_data.csv	crack records of cctv inspections	Civil Engineering Office Rastatt
dbg.csv	depth below groundwater	GIS Operation

Pipes.csv

The pipes files store the properties of each asset in the network. The data are derived from the sewer network data base in Rastatt maintained by the Civil Engineering Office of Rastatt. The original sewer database operates on the SMALLWORLD™ GIS software and an export using ASCII-files was required to transform the information to the ESRI-shape file format used at the Department of Applied Geology (AGK). Except for the construction date, all fields of the pipes.csv file could be filled with already existing data. Table 3.1.14 shows the parameters serving for the property description.

Table 3.1.14. Parameters stored in the pipes.csv.

Field name	Signifation	Data source
ID	Identification number of the pipe	sewer network database
PTYPE	Reticulated network or presure pipe	sewer network database
DN	Diameter	sewer network database
LENGTH	length of the pipe	sewer network database
GRADE	length until an elevation difference of 1m is reached	recalculated using the slope
MAT	Material of the pipe	sewer network database
GROUND	Bedding material of the pipe	generally assigned (sand)
CDATE	Construction data	settlement structure
NBH_ID	spatial reference	UVQ Model

The following parameters require special attention:
- Grade: is the length of a pipe until a height difference of 1 m is reached. Commonly in European countries a slope given in percent is used. The difference between grade and gradient must therefore be considered. As the gradient and the length of the pipe are known the grade was calculated.
- Ground: the geological environment is generally not stored in the sewer network database. Corresponding to the most common construction practice in Rastatt, the value for all the pipes was set to SAND as the bed of the pipe trench is filled with sand during the installation.
- CDATE: is the construction date, the date when the pipe was installed. In Rastatt the date of construction or rehabilitation has been registered for the last 10 years. These assignments cover only a minor percentage of the pipes as the sewerage network of the city is much older. To meet the NEIMO requirements the construction date or pipe age was assigned according to the main development period of the respective town quarter. In case of the historic city centre the introduction of the sewer system was set as the pipe age. This procedure will overestimate the actual pipe age as it neglects the renovation and rehabilitation works during the last decades but no other reliable information source was available.

Connections.csv

The connections file holds the number of customers or rather the properties connected to an asset. The number of house connections was assigned by the set up of a GIS query. The land parcels defined during the set up of the UVQ model were used. The centre point of every land parcel containing a building was determined and the shortest distance to the next sewer pipe calculated. Land parcels containing no houses were declared as open space and so the determination was done completely automatically. It is recommended to use the land parcel rather than the buildings itself as a residential or commercial property contains often more than one building, i.e. garage or a barrack, which leads to an overestimation of connections.

Successor.csv

The file gives the up- and downstream relation between the pipes of the network. The program code requires a straight tree structure of the assets. Each upstream lying pipe has only one downstream lying asset or only one single outlet. However, a downstream lying pipe can be the collector of several upstream pipes.

The file consists of two columns:

- UPSTREAM: Conveys the sewage from the up-gradient lying pipe
- DOWNSTREAM: Receives the sewage from the upstream lying pipes

The Civil Engineering Office of Rastatt performed a backtracking query for certain sewer branches in the study area Rastatt. The query is prompted at the outlet of the city and all the pipes, which are draining to that final branch, are subsequently listed. This query was performed in total for 6 areas to cover the complete area of Rastatt as the structure is rather complex at places where several smaller catchment areas are interconnected.

Problems arise when the urban area is characterized by a very low topographical gradient and a flat sewer system, so the network structure is not always following consistently the required tree structure. In the case of Rastatt large pumping stations are connecting the suburban areas. Furthermore the sewer network is designed to react dynamically on fill level variations. Temporary watersheds exist only during dry weather flow. Raising fill levels leads to completely different flow paths as the retention basins and combined sewer overflows are activated to prevent flooding of the connected house properties. This is even more important in cities where a combined sewer system is dominating like in Rastatt.

As a consequence of the flat topography, several cases occurred where an upstream lying sewer pipe has two outlets, which relieve the sewerage network during precipitation events. Additional manual work was performed to meet the straight tree structure requirement of the program.

Further problems occurred where single pipes are not connected to the sewerage or to a land parcel. To prevent error messages these pipes have been deleted and are not considered in the calculation. Figure 3.1.11 and Figure 3.1.12 are displaying the typical problems that occurred during the data preparation of the input files.

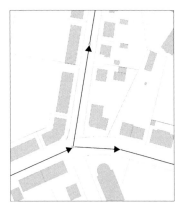

Figure 3.1.11. One upstream pipe and two downstream pipes; the criterion of the tree structure is violated.

Figure 3.1.12. Pipe branches with contradicting flow directions; the result of not properly stored data.

Regarding the urban area of Rastatt 11 CSOs (Combined Sewer Overflow) are installed. Each CSO has its own outlet to the receiving water body, mostly the river Murg traverses Rastatt. Four further outlets drain to natural retention basins. In Figure 3.1.13 an overview of the sewer network and the CSO area is illustrated.

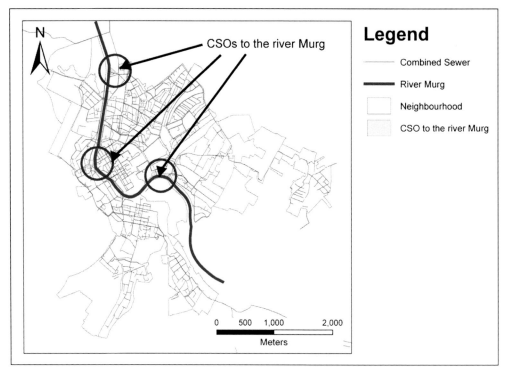

Figure 3.1.13. Sewer network of Rastatt. The red circles are indicating selected CSOs to the river Murg.

As the tree structure requires one single outlet, a virtual pipe was introduced to connect all outlet pipes of Rastatt. This final pipe with a diameter of 2000 mm was defined as a new one (no aging processes), without any defects to prevent extensive exfiltration. Even though the dry weather discharge is not influenced, the effect during rain days is significant. In reality there will be a discharge of stormwater to the river Murg, which is not considered in the UVQ model. Consequently, there will be a large overestimation of the discharge and a smaller overestimation of exfiltration quantities due to the exaggerated hydraulic load of the system.

Dbg.csv (depth below groundwater)

The dbg.csv gives the depth below groundwater and accounts for the position of the pipe relative to the water table. If a pipe is positioned above the water table exfiltration occurs otherwise infiltration prevails.

Note: Even though the NEIMO approach implements the processes of infiltration, variations in groundwater level are not included, i.e. the system of the water table is considered to be static. In Rastatt that can lead to a significant difference as about 35% of the sewerage is lying within the saturated zone during high groundwater level periods whereas only 5% are affected at low groundwater level periods (Wolf et al. 2005b). For the set up of the model the long-term average groundwater level was used and thus only 10% of the considered sewer network is lying below the water table.

The determination of the distance to the groundwater or the water column above an asset was performed by a GIS query. For each manhole the position relative to the groundwater level was determined. If a pipe is below the water table, the values entered have to be positive in the dbg.csv file.

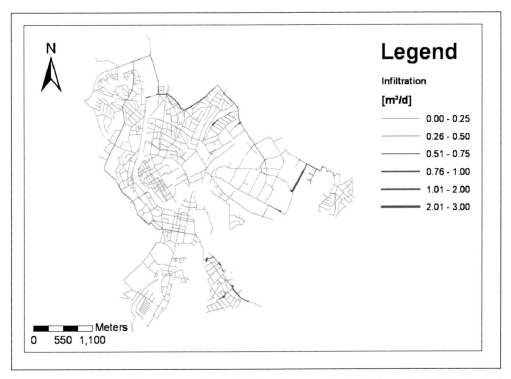

Figure 3.1.14. The main collectors of Rastatt are situated below the groundwater table and subject to infiltration.

CCTV data

NEIMO offers the possibility to implement sewer defects recorded by a CCTV inspection (Closed-Circuit TeleVision). In Baden-Württemberg a regulation orders the inspection of the sewer network at periodic intervals (EKVO 1999). Due to that fact more than 90% of the public sewer network in Rastatt was inspected since 2003. The defects are classified according to a rehabilitation priority scheme (ATV-M149 1999) and are stored in the sewer data base of Rastatt. NEIMO divides the circumference of a pipe into four sectors (see Figure 3.1.15). The defects of the CCTV inspection are also grouped into four sections; only the arrangement differs as they are distributed as quadrants.

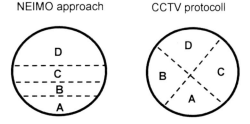

Figure 3.1.15. Division of the pipe circumference in NEIMO and the sewer inspections in Rastatt.

Two kinds of defects are distinguished in the CCTV file: Cracks and joint displacements. Cracks represent real cracks, shards, holes etc. The total defect area of the pipe is given in mm² and is distributed to the sectors (A), (B), (C) and (D). Joint displacements are entered as an opening width in mm and are active over the total profile. Note that the total crack area of an asset includes all defects of a pipe asset (typical length 50 m). On the contrary, the joint displacements specified in the input file are assumed to occur at every joint (the default value for the joint length is set to one meter). Obstructions cannot be represented in the CCTV file. The consideration of defective house connections relies on the availability of a valid estimate for the defect size related to the damage.

Calibration

The international literature encompasses a wide range of possible exfiltration quantities: lower exfiltration rates of 0.2 l/s/km have been determined by Härig (1991) through statistical analysis. Blackwood (2005) and Vollertsen & Hitved-Jacobsen (2000) performed experiments at single leaks and came to exfiltration rates of 0.3 to 2.0 l/s/km. These numbers show that the determination of reliable exfiltration rates is rather difficult. For the validation of the NEIMO approach and to receive reliable assumptions for the physical characteristics, a small-scale stand-alone model was set up to calibrate the relevant parameters (i.e. thickness and hydraulic conductivity). For the adjustment of the measured parameters exfiltration and flow meter measurements of the test site Kehler Strasse (see chapter 3.1.6.3) have been used. The stand-alone modelling is designed by three sequenced pipes (Figure 3.1.16). The two upstream lying pipes receive 75 m³/d from two house connections sequentially. This results in a total sewage flow of 300 m³/d, which corresponds to the measured discharge at the test site. In the downstream pipe a sewer defect of 120 cm² in sector A was introduced and has therefore the same defect area as the artificial defect at the test site.

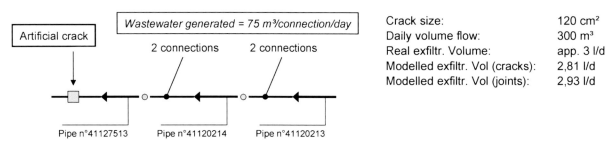

Figure 3.1.16. Scheme of the stand alone model for the calibration of the NEIMO parameters.

In Table 3.1.15 the physical characteristics for the representation of the exfiltration processes determined within the course of a parameter study are listed. The DeltaL (thickness of the colmation layer) in sector (A) was assumed to be as thick as the pipe wall and was constantly kept at 0.02 m. The DeltaL in the upper sectors were difficult to estimate as no real measurements are available. Finally the DeltaL in the sectors (B), (C) and (D) have been adjusted by the comparison of the total water balance of the exfiltrating water. A best guess was given by Wolf & Hötzl (2006) using a Monte Carlo simulation.

Table 3.1.15. Parameters for the exfiltration modelling.

Sector		Default	Calibration Thickness	Parameter		Default	Calibration
DeltaL (A)	[m]	0.02	0.02	ManningsN	[-]	0.13	
DeltaL (B)	[m]	0.02	0.005				
DeltaL (C)	[m]	0.015	0.0025	K_cracks	[m/s]	2.30E-05	1.20E-05
DeltaL (C)	[m]	0.01	0.0001	K_Joints	[m/s]	3.50E-05	6.00E-06

As the location and the symmetry of the artificial leak can be considered either as a normal crack or as a joint displacement two calibration approaches have been set up. The hydraulic conductivities have been determined for the cracks and the joint displacements have been determined separately. The default settings are higher by a factor of two for the k_cracks and a factor of six for the k_joints compared to the calibrated values. Using the default values the exfiltration quantity would be significantly increasing. It has to be stated that the k_values of the colmation layer in the different sectors are always the same, whereas its thickness can vary depending on the frequency of flooding and the supply of particulate matter and nutrients. While the different thicknesses were not measured in the field, the respective parameters were changed in the course of the calibration process to fit the measured exfiltration rates at the test site Kehler Strasse .

Infiltration

The computation of the infiltration is a much simpler approach compared to the exfiltration and it does not consider a colmation layer. It accounts for the groundwater level (hydrostatic pressure), the defect area, the conductivity of the soil and the limiting soil layer. The size of the defect (gathered by CCTV inspections), the groundwater elevation (measured at observation wells and interpolated via GIS analysis) as well as the k-values for the surrounding soil layer are known. The parameter with the highest uncertainty is the limiting soil layer, in which the flow velocity increases. The term limiting soil layer denotes the relevant distance DeltaL along which the hydraulic pressure is reduced to zero. The smaller the limiting soil layer is set, the higher the infiltration quantities become. Table 3.1.16 shows the basic assumptions of NEIMO for the infiltration and the parameters that have been assessed during the set up of the NEIMO model.

Table 3.1.16. Default and calibrated values for the calculation of the infiltration.

		Default	Adjusted
DeltaL	[m]	0.10	2.00
Conductivity of			
Clay	[m/s]	0.0001	0.0001
Sand	[m/s]	0.0002	0.0002
Unknown	[m/s]	0.00015	0.00015

A calibration was only performed in terms of plausibility checking using the results from the entire city sewer network. It remains unknown whether there are errors included in the estimation of the defect sizes or whether DeltaL is not properly adjusted. However, with a better availability of night flow measurements it would be possible to have a more rigorous calibration and validation exercise.

Event days

NEIMO defines different event days with respect to the precipitation and the related varying discharge. This results in dry weather, medium-, high- and storm rain flow. The different event days have to be chosen by the user himself. The reference day is a day of dry weather flow, which will be then multiplied by a factor for each discharge event. It was commonly decided in the AISUWRS project to define all days with less than 1 mm of precipitation as dry weather days. This upper limit was defined to account for the initial loss of the impervious surfaces due to its roughness as it is assumed that surface runoff occurs when more than 1 mm of precipitation has fallen. The analysis of flow patterns in the sewerage showed that the discharge of rain days with ~1mm/d corresponds with the dry weather discharge multiplied by the factor of 1.1.

The "medium rain flow" was determined by the analysis of the climate file for the year 1972 (see Figure 3.1.17). Medium rain flow occurs during rain events of 1 to 4 mm and results in a quantity of 3.8 times the dry weather discharge. High rain flow occurs during precipitation events of 4 to 12 mm. All rainfall events >12 mm have been determined as storm rain flow days.

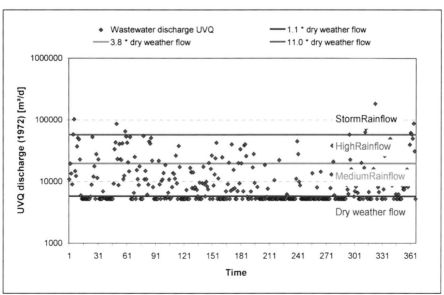

Figure 3.1.17. Classification of the event days for the implementation into the NEIMO model.

In Table 3.1.17 the defined discharge events are listed and the frequency of these events that occurred during the year 1972.

Table 3.1.17. Distribution of the event days for the year 1972.

Eventday	Eventdays	Av. Exfiltration	Sum Exfiltration	Portion of total exfiltration	Portion of time
	[nb]	[m³]	[m³/Y]	[%]	[%]
DryWeatherFlow	195	79	15308	25.35	53.42
MediumRainFlow	125	92	11471	19.00	34.25
HighRainFlow	36	669	24080	39.88	9.86
StormRainFlow	9	1058	9522	15.77	2.47
Total	365		60381	100	100

Even though the flow patterns for the year 1972 are stamped by the dry weather discharge, which is present for more than 50 % of the time, it accounts for only 25% of the total exfiltrating volume leaving the sewers. The exfiltration increases only modestly during medium rain flow. An explanation is given by the sewer network, which is designed with respect to high precipitation events. Medium rain flow leads therefore only to small fill level increase and a low exfiltration rate.

Noticeable conditions are reached during high and storm rain flow. Even though these conditions are present for only 12 % of the time, more than 50 % of the exfiltration occurs. It can be stated that during high and storm weather events a strong impact on the subsoil occurs as large volumes of untreated sewage enter the unsaturated zone. Even though surface runoff is loaded with heavy metals from impervious runoff, the sewage is strongly diluted due to large precipitation quantities. Investigations are still necessary to assess the real risk to the soil and the surface water bodies during heavy discharge events.

3.1.7.3 Unsaturated zone

The distance between sewer bottom and groundwater level also describes the thickness of the unsaturated soil zone, which is responsible for much contaminant attenuation and degradation. Beneath sewers with only a very thin unsaturated zone, the contaminants are directly released into the groundwater. The thickness of the unsaturated zone is therefore a key factor in the risk assessment for leaky sewers. Based on a GIS-analysis using water table and sewer bottom position, a risk map was produced for the Rastatt sewer network (Figure 3.1.18). The map differentiates between four risk classes based on the unsaturated zone characteristics:

- Infiltration: Sewer is always beneath the water table (even during the low water table situation on the 9.11.1991). No exfiltration possible. Low risk
- High: Sewer is subject to intermittent infiltration and exfiltration. Direct contact to the aquifer + possible voids due to sediment transport. Sewer is located below the high water table (9.11.1992) and above the characteristic long term low (11.8.1988). Highest risk class.
- Medium: Constant exfiltration but less than 1 m unsaturated zone. Comparatively medium risk.
- Low: Constant exfiltration and always more than 1 m unsaturated zone. Comparatively low risk.

The travel times through the unsaturated zone were calculated using the UL-Flow model and are described in detail in chapter 2.4 of this book. With regard to the main scenarios, it was assumed that all substances pass the unsaturated zone without being subject to sorption or decay. This is of course a worst-case assumption for most contaminants but it is valid for the considered marker substances of boron and chloride.

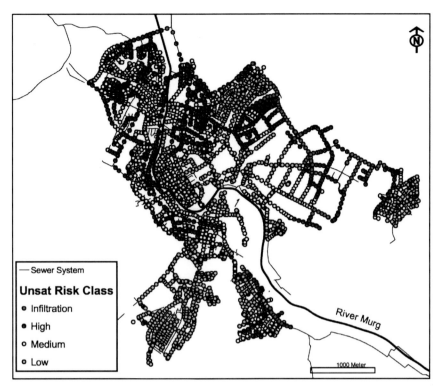

Figure 3.1.18. Indication of the risk class in Rastatt depending of the position of the sewers relative to the groundwater.

3.1.7.4 Groundwater flow model

The following chapters 3.1.3.4 and 3.1.3.5 provide an insight into the saturated zone modelling. They describe model setup, calibration and results specifically for the Rastatt case study. For a more general introduction to the software package Feflow®, which was employed in this exercise, as well as an indication of the strengths and weaknesses of this program within the AISUWRS model chain, see chapter 2.5.

The numerical groundwater model at the end of the model chain in Rastatt employs the commercial finite element package Feflow® by WASY. The location of the model domain, which covers an area of about 88 km², can be seen in Figure 3.1.19. Model set-up substantially draws upon a steady-state two-dimensional model, which has been set up for the same domain by Klinger (2003) and also uses data from a regional well-calibrated transient flow model by Kuehlers (2000). Within the model domain lies the urban focus area, which consists of 74 'neighbourhoods', comprising a total of 11 km².

Figure 3.1.19. Location of the model domain (Klinger 2003) and the focus area.

The 3D-model is discretised by triangulation into a mesh of 155,460 elements (**Error! Reference source not found.**). It consists of five layers representing the Upper Quaternary aquifer (the first three representing the Upper Gravel Layer, the fourth the Upper Interlayer, the fifth the Middle Gravel Layer). The 5 layers are defined by 6 slices. They add up to a thickness between 10 and 40 m and are based on the interpretation of borehole data.

The following boundary conditions are assigned to the model (**Error! Reference source not found.**):

- Dirichlet boundary (Type 1): The northern boundary is occupied by a constant head boundary on the first slice. This boundary represents the River Rhine. Its height ranges from 111 m a.s.l. to 108 m a.s.l. The River Rhine drains about 62,500 m^3/d over this boundary.
- Neumann boundary (Type 2): In the southwest an integral flux is assigned to all slices. This boundary represents the inflow from the Black Forest Mountains. The value was taken from the model by Kuehlers (2000) and adjusted according to the cross-sectional area. It accounts for an inflow of about 14,000 m^3/d.
- Cauchy boundaries (Type 3): Transfer boundaries are assigned to most waterways in the model domain. River water levels and transfer coefficients are adopted from the calibration of the model by Klinger (2003). They lead to the variation of infiltration of about 53,700 m^3/d and an exfiltration of about 57,600 m^3/d.
- Well boundaries (Type 4): A database is assigned to the model containing pumping rates for 33 wells which deliver roughly 19,000 m^3/d. The pumping rates represent the situation of October 1986, the date showing a long-term medium water level. The entire flow model is calibrated to the water levels at this date.

In order to determine the initial amount of groundwater recharge, a number of estimates have been made for the study area. They include an assessment by Pfuetzner (1995) based on time series of precipitation and potential evapotranspiration from 1960-1994. This estimate takes into account the soil type, land use and depth to the water table and was validated with lysimeter data and time series of groundwater levels. It was extended to incorporate different sealing degrees for urban areas by Kuehlers (2000) and finally adapted to the model area by Klinger (2003), also considering the negative recharge of lenticular water bodies. This information is applied to the three-dimensional groundwater model and represents the state of knowledge before the AISUWRS project. With this input, groundwater recharge in the entire study domain amounts to about 77,000 m^3/d, equivalent to an average of around 300 mm/y.

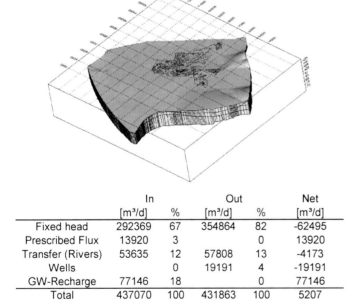
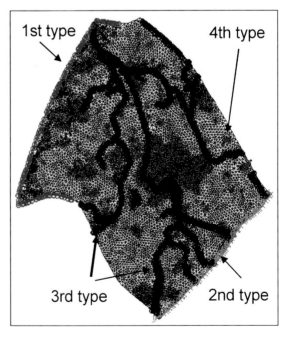

	In [m^3/d]	%	Out [m^3/d]	%	Net [m^3/d]
Fixed head	292369	67	354864	82	-62495
Prescribed Flux	13920	3		0	13920
Transfer (Rivers)	53635	12	57808	13	-4173
Wells		0	19191	4	-19191
GW-Recharge	77146	18		0	77146
Total	437070	100	431863	100	5207

Figure 3.1.20: 3D-view of the model body and overall water balance.

Figure 3.1.21: Top view of the mesh with specification of boundary conditions.

The flow model is calibrated by changing the hydraulic conductivity (expressed as kf-value in the German literature). The initial kf-value distribution was taken from existing calibrated models and interpretations of pumping tests. Since the manual calibration based on this distribution proved to be difficult, it was decided to use the PEST module, which is an automatic Parameter EStimation Tool available in Feflow®. The model area was divided into certain patches where conductivity was going to be uniform. Twelve observation wells throughout the model domain served as target values for the calibration. The area in the vicinity of the river Rhine was not included in the PEST process, because it includes a number of water bodies which were supposed to keep their formerly assigned kf-values. The layer representing the Upper Interlayer was manually assigned a fixed kf-value of lower permeability (Kx and Ky: 4.10^{-6} m/s in the central part, $6 \cdot 10^{-4}$ m/s in the outer part, Kz: $6 \cdot 10^{-7}$ m/s and $9 \cdot 10^{-5}$ m/s) in order to account for the less permeable geological structure, which is known from borehole records. Employing the PEST module brings about the numerically best solution for the model setup at hand. It resulted in a horizontal hydraulic conductivity between $2.4 \; 10^{-5}$ m/s and $8 \; 10^{-3}$ m/s. The vertical hydraulic conductivity was set to be 15 % of the horizontal conductivity. Figure 3.1.22 shows the horizontal conductivity on layer 4. The kf-value distribution described above leads to contour lines in Figure 3.1.23 which show the infiltration from the river Murg in this area and also the bending of isolines in the vicinity of the River Rhine, which mirrors what is observed in reality.

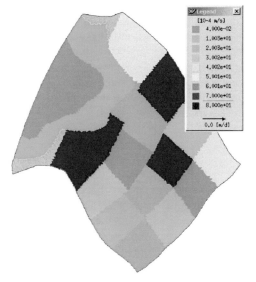

Figure 3.1.22. Distribution of the kf-values in layer 4 as a result of the calibration.

Figure 3.1.23. Display of the calculated groundwater contour lines.

At all reference points, the difference between observed and calibrated water level elevations is less than 50 cm (Figure 3.1.24), and especially in the urban focus area and downgradient of the city, the fit reaches sufficient accuracy.

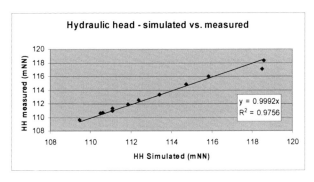

Observation point	Simulated head (m)	Measured head (m)
1	110.95	111.07
2	110.69	110.62
3	111.25	111.07
4	110.55	110.49
5	117.06	118.5
6	116.02	115.84
7	113.36	113.41
8	112.49	112.38
9	114.84	114.73
10	118.34	118.58
11	109.62	109.46
12	111.87	111.84

Figure 3.1.24. Difference of the hydraulic head at twelve reference points.

Regarding water quantity, different water management scenarios, which lead to different amounts of groundwater recharge, can be assessed with the flow model through their impacts on groundwater level and water balance.

For each scenario, the DSS results for groundwater recharge replace the recharge values of the initial model in the urban focus area. To prepare for such a connection, the DSS output files need to be integrated into a database and the values of interest need to be converted to Feflow® units. This database is joined to the attribute table of a GIS shape file, which in turn is linked with the Feflow® project.

3.1.7.5 Groundwater transport model

The calibrated flow model serves as the basis for the transport simulations. The transport model, which is described in this section, was set up for boron as a characteristic conservative urban contaminant. Initial conditions and background concentrations were determined through hydrochemical sampling campaigns (Wolf et al. 2005b). For the case of boron, the background concentration was set to be 0.024 mg/l.

In order to set a uniform background concentration, all water that enters the model – both groundwater recharge and water input at boundary nodes - needs to be assigned the corresponding concentration. For boundary conditions, this can be easily achieved by adding a first kind boundary condition to all nodes representing rivers (constant head at the River Rhine and transfer boundaries along the smaller rivers) and at the southeastern boundary where a constant flux enters the model domain at all slices. However, in order to allow contaminant plumes to freely leave the model at boundary nodes (e.g. at pumping wells or through groundwater exfiltration into rivers), it is necessary to include a so-called constraint of a minimum concentration of zero in the boundary conditions. This prevents the boundary conditions from imposing a low concentration onto areas which otherwise would reach higher concentration levels.

Regarding groundwater recharge, it is necessary to export the groundwater recharge values for each finite element, to multiply them with the target background concentration and to assign the resulting mass flux as a mass source to each element again. During these model setup tasks, it was also tried to assign a different concentration to the river Murg as opposed to the other rivers in the model domain. The river Murg as a Black Forest river has a lower boron concentration than the background in the Rhine Valley and lies at roughly 0.015 mg/l upstream of the wastewater treatment plant in Rastatt and at roughly 0.05 mg/l downstream of there. However, the implementation of these field data in the numerical model led to a dilution of the contaminant plume by the river Murg which was much more prominent than it is observed in reality. For this reason, it was decided to keep the river Murg concentration at the value of the overall background concentration.

With regard to quality issues, a number of predictions are attempted on how different urban water scenarios affect the load and distribution of characteristic urban substances within and beyond the urban aquifer. By applying a neighbourhood-specific urban contaminant input to the transport model for several scenarios and by comparing the extent of contaminant plumes and pollutant concentrations at selected observation points, the contamination risk potential and eventually also the sustainability of different urban water management strategies can be assessed.

After setting a uniform background concentration, the transport model is prepared to test the impacts of different AISUWRS scenarios. In this respect, the DSS was employed to calculate the conservative transport of boron. It must be kept in mind, that in the described case, the unsaturated zone models were not engaged and therefore, the attenuation processes in the unsaturated zone are not accounted for. The DSS output for boron loads replace the values of the initial model in the urban focus area. Five real world observation wells are chosen along a transect parallel to the groundwater flow direction. At these observation points, contaminant concentrations can be recorded.

In order to avoid having to work with UVQ or DSS files of 30 years or more duration, a comparison was done on the differences of the results from using a 'representative year' as compared to the long-term average. A Feflow® model was set up with groundwater recharge values which were determined as the average recharge of the years 1960-1990 in UVQ for each individual neighbourhood. A second model held the UVQ recharge values for the year 1972 with a groundwater recharge of 322 mm/y over the entire urban focus area and therefore comes closest to the 30-year average of 339 mm/y. A comparison of groundwater levels at 50 observation points on three different slices shows an average deviation of 0.1 mm rise in groundwater level in the 'representative year'. Groundwater recharge over the entire model domain increased by 0.76 %. The changes occurring being minor, it was decided to conduct the scenario testing with a 'representative year' instead of long-term time series. For the climate change scenario, the year 2092 with a groundwater recharge of 301 mm/y was determined as the 'representative year'.

3.1.8 Scenario modeling results

3.1.8.1 Baseline scenario

The baseline scenario describes the status quo of the urban water system, relying on data which have either been acquired during the AISUWRS project or which constitute a long term average over the last decades. The parameters describing the infrastructure characteristics and demographic conditions of the study area are representative for the status of the year 2003. The baseline scenario served for the comparison to all other scenarios, which have been set up. In Table 3.1.18 the datasets used in the different model approaches are listed.

Table 3.1.18. Basic input data sets used for the different models.

Model	Basic settings, boundary conditions
UVQ	The physical characteristics represent the status of the year 2003
	Demographic data and water usage patterns are representative of the year 2003
	Climate data deriving from the regional climate model HIRHAM of the DMI. The year 1972 was determined to be the characteristic year within the time line of 1960 – 1990.
NEIMO	Real sewer defects have been used, deriving from CCTV inspection in Rastatt recorded in the period 1995-2000
	Exfiltration and infiltration was calculated for the average meteorological conditions between 1960-1990.
	Event days determined: 195 (dry):125 (medium):36 (high):9 (storm) flow conditions
Unsaturated models	Uniform sandy soil was assumed for SLeakI/POSI whereas the parameterisation of UL_Flow relies on the definition of typical soil profiles as encountered within the drilling programme of the focus observation wells.
Saturated flow and transport modeling (Feflow®)	For the total model area of the groundwater flow model the average groundwater recharge of 1986 was introduced.
	The groundwater recharge inside the Rastatt city domain was calculated by UVQ and NEIMO for the meteorological average between 1960-1990.

Determination of the representative year

In the comparison between the 30 years average and a 'representative' year (groundwater recharge coming closest to the average of a 30 year period) the numerical groundwater model showed only deviations of a tenth of a millimetre in groundwater levels. Therefore, the subsequent scenario tests were conducted only for the year 1972 which exhibits a typical groundwater recharge for the time span 1960-1990.

Calibration of the model chain: Discharge measurements at the WWTP can be used to assess the plausibility of the simulated discharge. In the Rastatt case however, the simulated catchment area is only a part of the entire service area of the WWTP. In order to allow a rough comparison, the discharge volumes at the WWTP (draining c. 46,000 inhabitants) were corrected according to the number of inhabitants in the model area (36,152 inhabitants). However, this correction does not take into account the different percentages of industrial users in the city centre and the surrounding suburbs/villages. Furthermore it is not possible to deduct the significant amount of infiltrating groundwater from the suburb of Rastatt-Plittersdorf close to the River Rhine. It is therefore recommended for future investigations to fit the model area with an area of known discharge characteristics or to install a flow meter directly at the outlet of the model area. In Rastatt, this was done for several sub catchments (Klinger et al. 2005). Nevertheless the comparison in Figure 3.1.25 demonstrates the general plausibility of the model results. Two limitations can be observed:

1. Seasonal effects like varying infiltration rates through groundwater level variations are not present in the model results due to the steady state representation of the groundwater level (see Figure 3.1.25, increased discharge in Jan-Mar 2003 measured at the WWTP).
2. The model array does not account for combined sewer overflows (there are more than 15 in the urban area of Rastatt). This results in extremely high discharge peaks and therefore is an overestimation of the WWTP throughput/discharge during precipitation events.

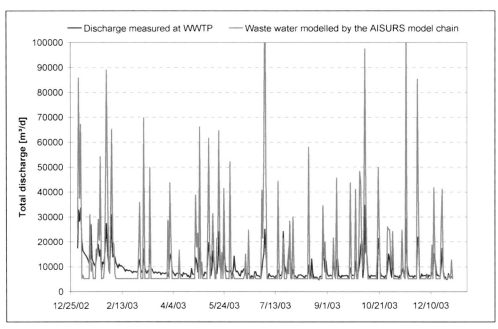

Figure 3.1.25. Discharge quantities determined by the model chain (green) and measured at the WWTP.

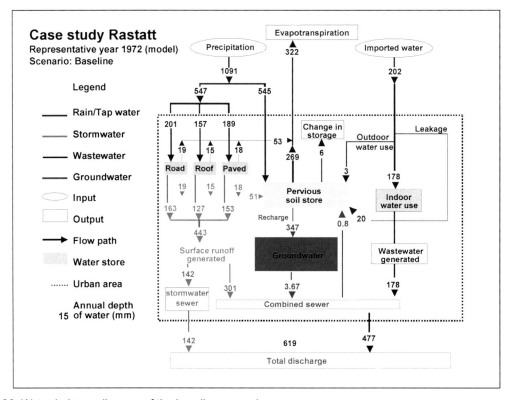

Figure 3.1.26. Water balance diagram of the baseline scenario.

Water balance: As illustrated in Figure 3.1.26 the water balance of the base line scenario shows a rather high groundwater recharge of 347 mm/y. Traditional estimates used in regional groundwater flow models cite the urban recharge to 35.6 mm – 80 mm per year depending on the population density and the sealing degree (Kuehlers 2000, LfU 1996). Nevertheless, the plausibility of Figure 3.1.26 can be checked with a simple example: The degree of land surface sealing of the study area Rastatt reaches app. 50 %. The average natural recharge measured at the lysimeter Rauental is 426 mm, so a typical recharge of 213 mm/y would be expected for a half-sealed area (and this does not take account of the fact that even sealed areas have direct recharge mechanisms via paved runoff soakaways, drainage via verges etc). Adding the losses of water mains, sewers and the additional seepage water deriving from garden irrigation the value is in a plausible range. This integrated approach shows that the groundwater recharge in the urban area has previously been greatly underestimated. In Figure 3.1.35 the spatial distributed groundwater recharge is displayed. Neighbourhoods, which are highly sealed reach only 110-200 mm. Neighbourhoods with a low proportion of impervious areas, are contributing with more than 550 mm per year to the groundwater recharge. Considering the total study area an average recharge of 332 mm for the year 1972 was modelled.

About 202 mm/y of drinking water are imported into the city from groundwater sources. This demonstrates that Rastatt would be able to secure its water supply by just using the aquifer beneath the city area. In practice, this is currently prohibited by water quality concerns and the 202 mm/y originate from sources outside the UVQ model domain, mostly upstream of the city.

The lateral inflow into the aquifer coming from the Black Forest Mountains is not included in the water balance but contributes significantly to the water balance as documented by the numerical groundwater model. Overall, the impact of the urban area of Rastatt on groundwater resources is rather low in terms of water quantity due to this lateral flow component and the agricultural and forest setting of the surroundings.

Sewer ex- and infiltration: The contribution of sewage exfiltration to the water balance is only 0.8 mm/y and seems to be rather low compared to the groundwater ingress which is with 3.67 mm/y almost 4 times higher. Figure 3.1.26 displays which pipes are exfiltrating and where infiltration is occurring. The amount of sewage exfiltration per asset ranges from 1 m³/y to almost 6,000 m³/y. So, the groundwater recharge through leaky sewers can increase significantly in the closer vicinity of sewer defects, whereas the effect on the study area is almost negligible. Groundwater ingress happens basically at the main collectors conveying the sewage out of the city. Their large diameters (1,200-2,000 mm) provide sectors where infiltration into the network can happen. The calculated groundwater ingress of 3.67 mm/y is very low compared to the estimation of 148 mm/y provided in Wolf et al. (2005) for the entire city area including suburbs. One reason is that some of the suburbs close to the River Rhine with exceptionally high water ingress are not included in the AISUWRS model domain. On the other hand, the infiltration modelling is a straightforward modelling exercise, which could not be compared to an actual flow measurement at the outlet of the sewer catchment. A significant uncertainty is therefore attached to this result and future studies are required to adjust the NEIMO parameters on infiltration to the local setting.

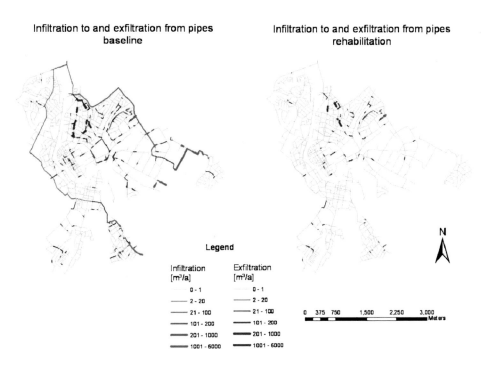

Figure 3.1.27. Left: ex- and infiltration within the baseline scenario. Right: reduced ex- and infiltration after rehabilitation of sewer defects with priority 1 and 2.

CCTV data versus generic defects: The description of the sewer condition according to the protocols from the CCTV inspections lead to significantly lower exfiltration and infiltration rates than the generic defect assumptions supplied by NEIMO. The exfiltration as well as the infiltration modelled with generic defects are more than 5 times higher (see Table 3.1.19). On the one hand, optical inspections are rather subjective and difficult as for example smaller joint displacements are generally coated by the colmation layer and are hard to identify (which leads to an underestimation of the defect frequency). On the other hand, the generic age/defect functions derived from Australian sewer systems might not be appropriate for German sewer systems.

Table 3.1.19. CCTV and generic comparison of exfiltration and infiltration rates listed for the respective event day.

Flow conditions	Dry		Medium		High		Storm		Total
	m³/d	[mm/y]	m³/d	[mm/y]	m³/d	[mm/y]	m³/d	[mm/y]	[mm/y]
Exfiltration									
Baseline CCTV	10.45	0.19	13.04	0.15	98.35	0.33	141.99	0.12	0.80
Baseline generic	86.29	1.58	112.73	1.32	466.21	1.58	709.38	0.60	5.08
Infiltration									
Baseline CCTV	110.55	2.02	108.57	1.27	90.54	0.31	84.69	0.07	3.68
Baseline generic	551.73	10.10	545.65	6.40	497.87	1.68	483.40	0.41	18.60

The steady state groundwater flow model using the assumptions from the baseline scenario produced the water balance for the extended model domain as displayed in Figure 3.1.29. Besides the urban groundwater recharge, the system is dominated by the influence of lateral groundwater flow and exchange with the surface waters (e.g. River Murg and River Rhine). The spreading of contaminants was demonstrated using the predicted boron loads to the groundwater (Figure 3.1.28).

Figure 3.1.28 and Figure 3.1.30 demonstrate that a large part of the aquifer is influenced by the boron emission of the baseline scenario. However, the absolute concentrations remain rather low with a maximum of 0.06 mg/l boron at a downstream observation point when the generic defects function is activated in NEIMO. If the condition information of the CCTV records is applied, the simulated concentrations in the groundwater are lower and reach a maximum of 0.029 mg/l boron. Overall it is remarkable that the simulated boron concentrations are usually below the measured concentrations in observation wells. Several reasons might be responsible for this phenomenon:

- Leaky private house connections were not incorporated into the AISUWRS model chain due to lack of data.
- Other sources than sewers contribute in a major fashion to the boron concentrations.
- Sewer condition is worse than expected from CCTV data.
- The mixing process simulated in the aquifer is overestimated by the groundwater model.

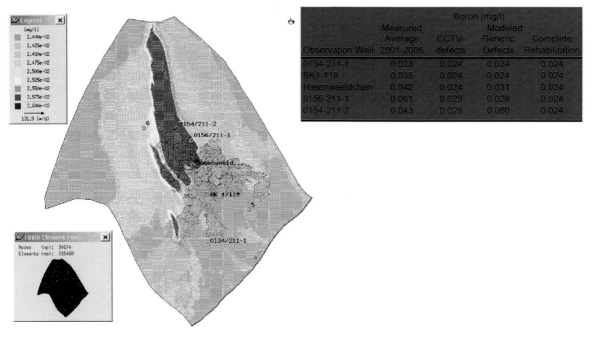

Figure 3.1.28. Spatial distribution of boron in groundwater as modelled for the baseline scenario.

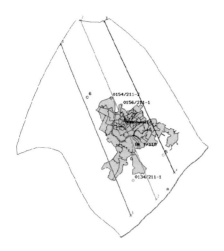 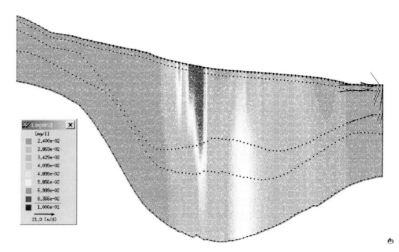

Figure 3.1.29. Profile lines selected for analysis in Feflow®.

Figure 3.1.30. Cross section showing boron concentrations along the red profile line in Fig 3.1.29.

3.1.8.2 Sewer Rehabilitation Scenario

The monitoring of the sewer network in Rastatt by the means of camera inspections allowed the implementation of CCTV data into the model chain. The recorded defects are classified into several categories according to a rehabilitation priority rating (ATV-M-149). The classes range from 1 to 7 (defect class 1=worst, immediate rehabilitation after the monitoring, to 7 which will be monitored again in the next period, no rehabilitation needed).

The only change in the input files of the model chain was performed in the CCTV file. All assets with rehabilitation priority 1 and 2 were removed from the NEIMO file, so these pipes are assumed to have no more defects. In Table 3.1.20 the number of the sewer assets represented in the rehabilitation scenario is listed. More than 55 % of the assets in Rastatt contain at least one defect. About 37 % of these assets were classified into rehabilitation priority 1 and 2. Consequently, even the rehabilitation of just the priority classes 1 and 2 requires a major economic investment. The typical length of one asset is 50 m.

Table 3.1.20. Assets represented in the model chain.

Scenario	Assets represented [nb]	Assets with defects [nb]	[%]
Baseline	3587	1989	55.45
Sewer rehabilitation	3587	661	18.43

Water balance: In the context of the total water balance (see Figure 3.1.31) the effect of the sewer rehabilitation plays a minor role as the exfiltration and infiltration were already represented with rather small quantities in the baseline scenario compared to the total water balance. However the changes are pronounced if the infiltration and exfiltration volumes are compared between the scenarios. The amount of exfiltrating sewage could be reduced by 40 % to 0.46 mm/y. In addition, the rehabilitation is very effective regarding the infiltration quantities and the reduction over the total study area reaches 93 %, corresponding to a total of 0.24 mm/y.

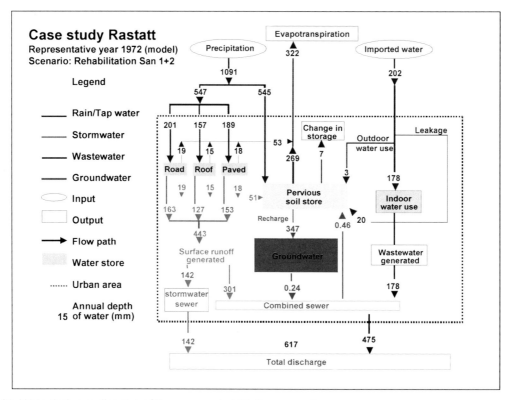

Figure 3.1.31. Water balance diagram of the sewer rehabilitation scenario.

Care must be taken in the assessment of the consequences of the reduced groundwater ingress. While the treatment costs in the WWTP are reduced, the function of the leaky sewer system as an artificial drainage system regulating groundwater levels must be considered. Complete renovation of the pipes can locally lead to rising groundwater level and therefore to flooding of cellars and basements.

Overall the sewer rehabilitation leads to slightly higher groundwater levels as 3.43 mm/y are not drained by the sewer system any more and reduction of recharge by sewage exfiltration is only 0.34 mm/y, resulting in a net surplus of 3.09 mm/y.

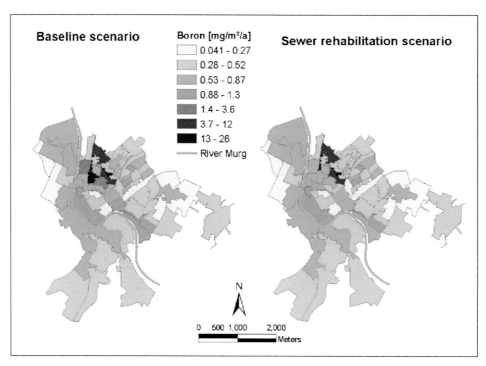

Figure 3.1.32. Impact of boron on the groundwater, comparison of the baseline and rehabilitation scenario.

3.1.8.3 Rainwater infiltration scenario

The infiltration scenario redirects surface runoff from pervious areas to gardens and open space with the intention to relieve the sewer network and to increase the urban groundwater recharge. Only areas (UVQ neighbourhoods) with less than 50 % sealed surfaces were taken into account as infiltration structures require space and as there is a minimum amount of unused open space required for social amenity purposes. Two different approaches to infiltrate surface runoff have been followed (see ATV-DVWK-A-138):
- Area infiltration
- Ditch/basin infiltration

The area infiltration means that the rainwater will be simply led to extensive used grasslands with no implementation of ponds to percolate into the ground. Its feasibility is determined by the ratio between effective paved area and the area demand for the infiltration volumes. If the space required for area infiltration is higher than 35 % of the public open space, ditch or basin infiltration will be applied. The ditches or basins were installed in the gardens on every residential property.

Water balance: In Figure 3.1.33 the total water balance of the infiltration scenario is displayed. The detouring of the surface runoff reduces the total discharge by 10.5 %. This results in an economical relief for the WWTP as less water needs to be treated. In addition, combined sewer overflows are reduced as a result of the lowered hydraulic overload of the sewer system. This applies to both controlled as well as uncontrolled sewer overflows (with the occasional flooding of a public park in Rastatt as an example).

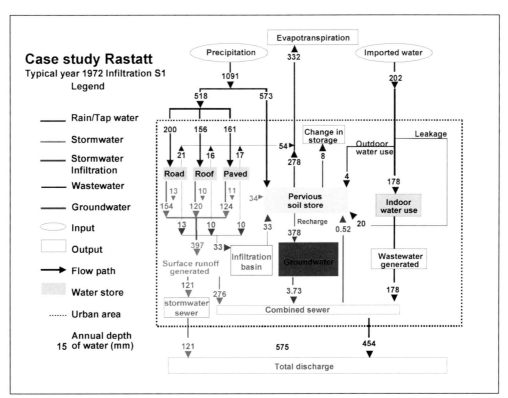

Figure 3.1.33. Water balance diagram of the rainwater infiltration scenario.

In Table 3.1.21 a comparison of the surface runoff quantities in the baseline and the infiltration scenario is displayed. It can be observed that even though extensive measures were performed in the study area the surface runoff could only be reduced by slightly more than 10 % in total. It has to be emphasised that this value is the average over the entire study area. Locally large portions of impervious areas had been uncoupled from the sewer system and are leading to a strong relief of the hydraulic burden of the sewer pipe. For the display of the strongest effects an analysis of individual neighbourhoods' water balances would be required.

The reduction effect is more pronounced during single rain events. The daily discharge arriving at the WWTP during a rain day will be reduced by more than 13 %. A hydrodynamic modelling would be helpful to see whether the pressure at bottlenecks in the sewer system of Rastatt could be reduced and the resulting flooding of adjacent cellars avoided. From the environmental point of view these measures can already lead to a significant reduction of the pollution impact on the receiving water bodies (i.e. River Murg, Federbach, protected areas), which are affected mostly during storm water events through the activation of the combined sewer overflows.

Table 3.1.21. Differences of the surface runoff between the baseline and the infiltration scenario.

Scenario	Yearly WWTP inflow [mm/y]	WWTP inflow on rain events [mm/d]
Baseline	443	17.08
Infiltration Scenario	397	14.80
Reduction	10.38%	13.35%

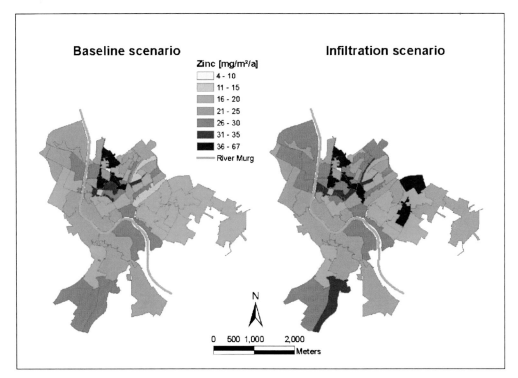

Figure 3.1.34. Comparison of the Zinc load in the baseline scenario and the infiltration scenario.

Contaminants: The infiltration measures redirect also heavy metals contained in the surface runoff (i.e. Pb, Zn, Cd, Cu) from the sewers to the soil in the model area. As the water is percolating through the active soil heavy metals are absorbed in the soil as long as the removal capacity is not exceeded. By the example of zinc the effects of runoff infiltration can be shown (see Figure 3.1.34). The study area-wide load increased by 12 % whereas the loads in neighbourhoods with infiltration measures are sometimes subject to an increase of more than 100 % from $20 g/m^2/y$ to $50 g/m^2/y$.

Groundwater recharge: A consequence of the surface runoff infiltration is a significant increase of groundwater recharge of approximately 30% (~60 mm/y) over the total study area, in which the increase could be well over 100 % in neighbourhoods with infiltration basins. Figure 3.1.35 illustrates a comparison of the groundwater recharge of the base line and the infiltration scenario.

Even though the groundwater recharge was increased, the effect on the groundwater level variations is negligible due to the strong lateral in- and outflow of the groundwater. Other case studies showed that infiltration measures or reduced industrial water pumping lead to an increase of the groundwater level and to a flooding of cellars in the affected area (see Goebel 2004). The groundwater flow system in Rastatt however is relatively insensitive to a rise in groundwater recharge in the urban focus area. This sensitivity was also tested by modifying a transient model, which exists for the model area (Kuehlers 2000). Groundwater recharge was increased by 35 % in the urban focus area. The increased recharge in this case only leads to a minor rise in groundwater levels in the range of millimetres at observation points inside the city (see example in Figure 3.1.36). The reasons for this lack of sensitivity are the high hydraulic conductivity of the study area, the high throughput of water due to the lateral flux from the Black Forest Mountains and the comparatively small area of increased recharge compared to the entire model domain. In addition, the numerous surface waters drain the groundwater body. These are all particular features of the hydrogeological setting of Rastatt; the effects on a city with similar urban infrastructure in a 'less-forgiving' geological setting, such as one with a thinner and less permeable aquifer would be much more marked.

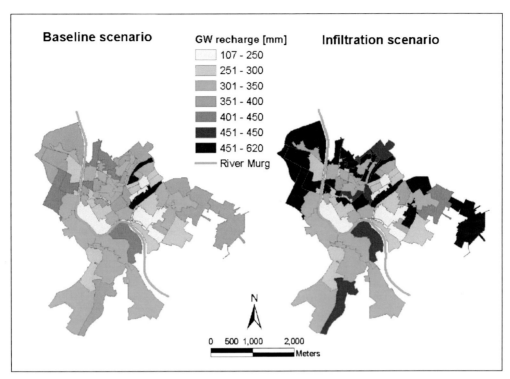

Figure 3.1.35. Comparison of the groundwater recharge of the baseline and the infiltration scenario S1.

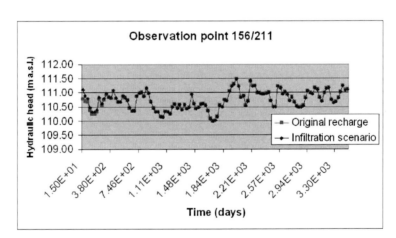

Figure 3.1.36. 30 year sensitivity testing with a transient model using a 35% increased recharge in the urban focus area.

The effect of the infiltration scenario on groundwater contamination could not be quantified in sufficient depth within this case study. The considered conservative marker species like boron and chloride are not subject to a major change as atmospheric deposition is not their main source and consequently the redirection of roof runoff does not pose a threat regarding these substances. A very relevant group of substances in the infiltration scenario are heavy metals. While loading changes to the combined soil-groundwater system from heavy metals are within the capability of the AISUWRS model chain, more research is needed to judge the adsorption and attenuation of heavy metals in the unsaturated zone, where pH and redox conditions are likely to mean that the majority of the load will be fixed. The relevance to groundwater is also extremely dependant on the technical implementation of the infiltration facilities. For example a regular replacement of the top soil layer in which most of the metals accumulate, or silt removal in the case of basins, would reduce the risk significantly.

3.1.8.4 Climate change scenario

The climate change scenario is based on predictions of the Danish Meteorological Institute (DMI) for the time period 2071-2100. The regional DMI climate model operates on a spatial resolution of 12 · 12 km. The simulation uses the A2 data scenario of the IPCC (2001) for the climate simulations. The simulation delivered data in daily time steps as needed by the UVQ model and was the climate model with finest resolution available for that area. Two timelines were made available by the DMI: 1961-1990 as control period and 2071-2100 as prediction period (forecasts for an earlier time period are accounted as not reliable (Prudence 2005)). The main deviations of the climate in 2071-2100 from the current climatic conditions are changing rainfall patterns. The area would receive higher amounts of precipitation in winter and early spring, the late summer would receive much less rainfall, and late spring, autumn and early winter precipitation values would remain approximately the same (Schaefer 2006). For the modelling of the climate change scenario, the year 2092 was chosen to be the typical year in the timeline from 2070-2100 with regard to groundwater recharge.

Water balance: The water balance in Figure 3.1.37 shows that the precipitation in the year 2092 is slightly decreased by approximately 10% and the groundwater recharge is reduced by approximately 15 % compared to the year 1972. These variations seem relatively small and a regional effect cannot be assessed by the use of the water balance diagram. For a sustainability assessment, a long-term model run, which would also account for the lateral inflow to the system, has to be set up and evaluated. However, for the year 2092, representing the long term average, the water balance is still positive and the groundwater recharge rate in the urban area exceeds the amount of imported water by 119 mm. Figure 3.1.38 illustrates the direct comparison of groundwater recharge in the study area Rastatt for the years 1960-1990 and 2070-2100. Note that the graph is based on data from the UVQ model only and does not contain the groundwater recharge that derives from water mains losses and sewer exfiltration.

The average groundwater recharge within the time period 1960-1990 is 339 mm/y with a spectrum from 95 mm/y up to 750 mm/y in the urban area. The average groundwater recharge in the period from 2070-2100 is only 294 mm/y. The range of recharge narrows to 90 mm/y (min) - 610 mm/y (max).

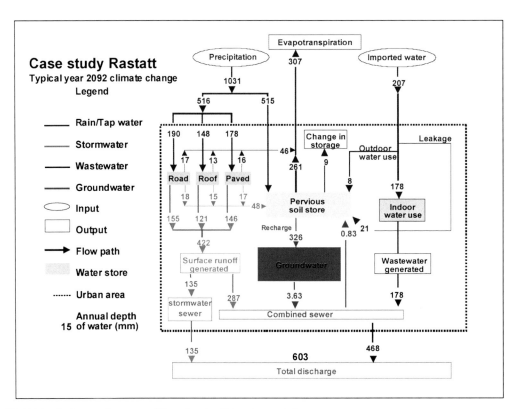

Figure 3.1.37. Water balance diagram of the climate change scenario.

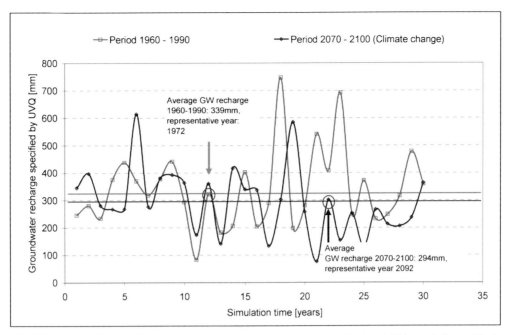

Figure 3.1.38. Groundwater recharge during the periods 1960 - 1990 and 2070 - 2100 modelled with the AISUWRS model array.

Exfiltration and infiltration: The effect of stronger rain events in future times increases the amount of sewage exfiltration only slightly from 0.80 mm/y to 0.83 mm/y. However, as the groundwater level is considered to be static, the portion of pipes, which are above the groundwater level is always constant. A fall in the groundwater level would lead to an increase of exfiltration and decrease of infiltration.

Reaction on the groundwater: The modelled impact on groundwater levels within the city area is low; however larger regional models are required to adequately describe the effects of this regional phenomenon which lead to changed boundary conditions. As the numerical groundwater model showed, the groundwater levels in the model domain are mainly controlled by the water levels of the River Rhine, the River Murg and the lateral groundwater inflow from the Black Forest and the northbound flow component in the Upper Rhine Graben. A groundwater model, which is able to depict the impact of climate change would need to include the entire upstream catchment (in this case more than one thousand km²).

The increased intensities of rainfall that the climate change model predicts lead to increased surface runoff and therefore to a reduced amount of water percolating into the soil and a higher hydraulic loading of the sewer network. As also the frequency of extreme rain events and combined sewer overflows is expected to rise, it is recommended to invest in decentralised infiltration measures as a relief to the sewer network.

3.1.8.5 Comparative scenario overview

In this section chapter the different scenarios are compared not only by the averages over the entire model domain but also by demonstrating the impact associated with the major development types. For the description of contaminant flow and loads, representative neighbourhoods (i.e. residential, industrial and low density areas (i.e. the area which has the lowest degree of sealing), see Figure 3.1.39) were selected. The evaluated water and contaminant balances are a compilation of UVQ and the NEIMO results. The results of the unsaturated zone modelling are not included. The system can be considered as a steady state approach, which passes the water and contaminant fluxes directly to the ground water model.

Boron, potassium, chloride and zinc were chosen as indicator substances for different urban contamination sources. All calculations assume conservative mass-transport without any sorption or decay. The red numbers in the following tables indicate a decrease in percent compared to the result of the baseline scenario. The black numbers indicate an augmentation of the respective parameter.

For the representation of residential areas 35 neighbourhoods were selected. 10 neighbourhoods of almost the same size were selected to represent industrial land use. To account for an almost "green" area the neighbourhood with the lowest degree of sealing in the study area was chosen.

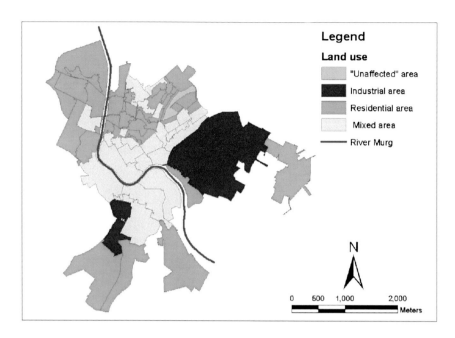

Figure 3.1.39. Study area Rastatt with residential, industrial and low density neighbourhoods represented in the comparative overview.

Boron (see Table 3.1.22): The only sources through which boron is introduced into the study area are tap water and the laundry as boron is a constituent of the washing powder. Other sources are not specified in UVQ due to the lack of published data from the literature. Consequently, the impact on the groundwater can only occur via water mains losses and even more importantly through sewage exfiltration. An additional influence is the irrigation of the gardens.

The sewer rehabilitation scenario therefore leads to the expected reduction of the total boron load to the aquifer of almost 30 %. The proportional amounts in the residential neighbourhoods are only slightly increased (32 % reduction) but can be explained by generally higher loads in the sewage. The effect in industrial areas is rather small with only 3 % reduction under the assumption that the industrial processes in these neighbourhoods do not produce boron emissions into the wastewater.

For the infiltration scenario a very low increase (<< than 1%) of the boron loads for the total study area is observed, The slight increase in the residential areas can be explained as most parts of sewer network act as combined sewers, so increased infiltration of the rainwater runoff component leads a lower dilution and to higher average concentrations in the sewage. Even though the exfiltration is lowered the load entering the groundwater is increased.

Table 3.1.22. Contaminant load of boron in the study area.

Scenario	Boron							
	Average		Residential		Industrial		Low density	
	Av [mg/m²/y]	change [%]	Av [mg/m²/y]	change [%]	Av [mg/m²/y]	change [%]	Av [mg/m²/y]	change [%]
Baseline	0.96	-	2.83	-	0.48	-	0.78	-
Sewer rehab	0.70	-27.24	1.90	-32.88	0.46	-3.04	0.64	-18.28
Rainwater infiltration	0.96	0.07	2.89	2.18	0.48	-0.32	0.64	-18.06

Potassium (Table 3.1.23): The input loads and concentrations could be defined for almost all indoor and outdoor sources. In contrast to boron, potassium loads were also specified for surface runoff. Consequently, the rainwater infiltration scenario causes an average rise in potassium loadings to the aquifer by c. 63 % compared to the baseline scenario. The effect is most pronounced (120 %) in the sparsely build-up areas (low density area) as large parts are used for the area or ditch infiltration. But also residential and industrial neighbourhoods show increased loadings around 60%.

As infiltrating surface runoff is the main source of potassium in the model chain, the rehabilitation of sewers contributes only little to the reduction of potassium input to the groundwater. The loadings can be decreased by 2 % on average, with up to 5 % in densely sewered residential neighbourhoods.

Accounting for an average groundwater recharge of 342 mm per year (=342 l/m²/y) and a load of 0.44 g/m²/y water, the concentration in the urban groundwater recharge amounts to 1.3 mg/l. This is obviously not enough to cause the monitored groundwater concentrations of 4.7 mg/l in urban background wells. The drinking water regulation limit of 12 mg/l is not approached.

Table 3.1.23. Contaminant load of potassium in the study area.

Scenario	Potassium							
	Average		Residential		Industrial		Low density	
	Av [g/m²/y]	change [%]	Av [g/m²/y]	change [%]	Av [g/m²/y]	change [%]	Av [g/m²/y]	change [%]
Baseline	0.17	-	0.19	-	0.18	-	0.20	-
Sewer Rehab	0.17	2.09	0.18	-5.35	0.18	-0.12	0.20	-1.10
Rainwater infiltration	0.28	63.14	0.30	58.64	0.26	48.87	0.44	120.72

Chloride (see Table 3.1.24): Chloride is considered as a good marker species due to its conservative behaviour and its relative abundance in the urban water cycle. Besides the indoor sources, loadings from roads and paved area could be implemented in the model approach. The load to the groundwater is relatively high compared to all other substances and results in a chloride concentration of the recharge water of 22.07 mg/l (7550 mg/m²/y with 342 l/m²/y). This corresponds reasonably well to the monitored chloride concentrations of 14.1 - 30.5 mg/l in the groundwater monitoring wells. This compares also well with the results from the Doncaster case study, where chloride was also found to be a suitable marker.

Due to the different sources the impact of individual technical measures on the chloride loadings is rather low. Sewer rehabilitation reduced the chloride loads by an average of 0.5 % (max. 1.5% for residential areas). The rainwater infiltration scenario results in an increase of 6 % chloride load on average. The most pronounced changes result from the installation of infiltration basins in the industrial areas, with their large pavements and parking lots. The salting of these areas contributes significantly to the chloride input. The effect in the low-density area can also be explained by a large portion of paved area.

Table 3.1.24. Contaminant load of chloride in the study area.

Scenario	Chloride							
	Average		Residential		Industrial		Low density	
	Av [g/m²/y]	change [%]	Av [g/m²/y]	change [%]	Av [g/m²/y]	change [%]	Av [g/m²/y]	change [%]
Baseline	7.55	-	8.97	-	6.31	-	3.19	-
Sewer Rehab	7.51	-0.50	8.84	-1.50	6.30	-0.04	3.18	-0.34
Rainwater infiltration	8.01	6.08	9.17	2.26	7.30	15.81	3.85	20.63

Zinc (see Table 3.1.25): Even though the drinking water regulation in Germany no longer lists a limiting value for zinc as a contaminant, it can be considered within the comparative overview to be representative of other heavy metals in the urban water cycle. For zinc it was possible to specify the input concentrations for both the indoor and outdoor sources. High loads of on average 22 mg/m²/y can be observed in the baseline and all other scenarios. The effect of the sewer renovation reduces the loads only by 2 % over the total study area. Generally the sewer rehabilitation scenario is not very effective with regard to the load reduction. The largest reductions occur in the residential areas (approximately 6 %). In the industrial and low density area the sewer rehabilitation shows only minor effect with less than 1.5 % load reduction.

A different picture is drawn by the infiltration measures, which increase the loads over the total study area by almost 13 %. A direct comparison of the typical residential and industrial areas shows that the load increases more in industrial areas (~25%) than in the residential areas with (7.4 %). As also noticed for chloride, the sparsely populated neighbourhoods and the industrial zones are subject to a more pronounced uncoupling of roof and paved areas. Consequently they are heavily affected by the infiltration facilities.

The limiting value of the WHO for zinc in drinking water is 5 mg/l. Considering an average recharge of 342 mm/y (see Table 3.1.26) and yearly input of 24.96 mg/m²/y the zinc concentration in the seepage water amounts to 0.073 mg/l. This suggests that the area impact is well below the limit. Nevertheless, there is certainly scope for higher concentrations to occur in the vicinity of the infiltration basins if no removal takes place.

Table 3.1.25. Contaminant load of zinc in the study area.

Scenario	Zinc							
	Average		Residential		Industrial		Low density	
	Av [mg/m²/y]	change [%]	Av [mg/m²/y]	change [%]	Av [mg/m²/y]	change [%]	Av [mg/m²/y]	change [%]
Baseline	22.10	-	27.80	-	18.52	-	12.03	-
Sewer Rehab	21.65	-2.03	26.24	-5.61	18.50	-0.15	11.86	-1.39
Rainwater infiltration	24.96	12.95	29.86	7.39	23.05	24.44	15.78	31.19

Groundwater recharge (see Table 3.1.26): The calculated scenarios exert only minor influence on the groundwater recharge predicted for the study area. Especially the sewer rehabilitation scenario affects the total recharge by less then 1 %. The actions performed in the infiltration scenario lead to an increase of the recharge of ~16 % over the total study area. An increase of 55 mm/y seems to be rather small for the measures set up in the model area. However, it should be noted that additional infiltration facilities were only assumed for roughly 50% of the neighbourhoods due to low amount of infiltration area available in the housing stock. In the residential areas a higher recharge of approximately 65 mm is predicted.

The communities recognised already the potential of rainwater infiltration and new buildings in industrial areas (which seal large areas) are already subject to uncoupling of roof areas and parking lots from the sewer system. Active infiltration via the natural topsoil is already performed, in particular in Baden-Württemberg. Actual examples can be found in the industrial areas (Industriegebiet Lochfeld) in the northeastern part of Rastatt.

Table 3.1.26. Variation of the groundwater recharge in the representative neighbourhoods resulting from the set up scenarios.

Scenario	Groundwater recharge			
	Average [mm]	Residential [mm]	Industrial [mm]	Low density [mm]
Baseline	342.1	352.5	311.5	523.7
Sewer Rehab	341.8	352.0	311.4	523.5
Rainwater infiltration	397.5	426.8	343.0	588.3

3.1.9 Conclusions

Within the experimental field investigations new hydro-chemical evidence for exfiltration from leaky sewers was found. The placement of observation wells in the direct vicinity of defective sewers using the geographically referenced sewer defect database provided opportunities for direct observation of sewage-influenced groundwater. Online probes, which monitored physico-chemical characteristics, detected daily variations in specific electrical conductivity of groundwater in the focus observation wells. The comparison with the city wide sampling campaign showed elevated concentrations of most parameters (including sodium, potassium, boron, phosphorous and ammonium) in wells close to defective sewers. The test site Kehler Strasse demonstrated that the colmation process requires several months to reach a pseudo-steady state under the transient conditions of a real sewer system. Typical exfiltration rates for the 2 cm wide transversal slit in a DN500 sewer amounted to 1-2 l/d, rising to approx. 10 l/d during rain events.

The simulations using the AISUWRS model chain quantified the loads to the aquifer for four scenarios; baseline, sewer rehabilitation, decentralised rainwater infiltration and climate change. In conjunction with the groundwater model it turned out that the loadings to the groundwater do not pose a threat for the marker substances considered (boron, chloride, zinc, potassium) if the loads are averaged over the individual neighbourhoods. Problems however may arise locally in the vicinity of large sewer defects.

The rehabilitation of sewers with class one and two priorities will reduce the loads to the aquifer by 30 % with regard to boron, and by smaller proportions for marker species which are also present as area sources (e.g. chloride).

The modelled rainwater infiltration scenario demonstrated that a large potential for the uncoupling of sealed surfaces from the sewer system exists. The dominantly sand and gravel sediments of the near subsurface allow quite simple technical measures to be used to promote the infiltration. The decentralised rainwater infiltration scenario results in 16 % higher groundwater recharge which is accompanied by increased loads to the soil/aquifer system of up to 31 % for zinc in individual neighbourhoods. The daily hydraulic load of the sewer system decreases by approximately 13 % over the entire network during rain events, with more pronounced changes in neighbourhoods with large infiltration facilities. Groundwater modelling has shown only changes of less than 1 cm in groundwater level arising from the additional infiltration of rainwater. This is in contrast to an example calculated by Goebel et al. (2004) but can be explained by the high hydraulic conductivities and the different aquifer setting. With regard to implementation challenges, the socio-economic studies have shown that rather high costs are associated with the proposed decentralized infiltration measures as they affect already existing housing stock. However, accounting for the hydrogeological setting in the urban area of Rastatt permitted realistic simulations of possible actions regarding on-site infiltration of rain water runoff from roofs and paved areas, and it was demonstrated that these could be performed without adverse effects due to the high hydraulic conductivity and the thickness of the aquifer. A reduction of the discharge quantities results in a lower pressure on the WWTP and the surface water bodies as well as in a reduced flooding of cellar basements.

The climate change scenario is characterised by a 10 % decrease in precipitation which results in a 15 % decrease in groundwater recharge. While the established groundwater model does not show any changes due to this reduction in the city area, a much larger model on catchment scale would be required to correctly predict the effects of climate change on groundwater levels.

3.2 A SANDSTONE AQUIFER: DONCASTER, UK

3.2.1 Background

3.2.1.1 Physical and geological setting

The Doncaster case study concerns the Sherwood Sandstone aquifer, which is the second most important in the UK after the Chalk. This regionally extensive sandstone formation is part of a more widespread Permo-Triassic Bunter and Lower Keuper red-bed sandstone sequence, which also forms productive aquifers elsewhere in north western Europe. In eastern England, it outcrops in a structurally controlled arc from south of Nottingham to the North Sea at Hartlepool, Co. Durham (Figure 3.2.1).

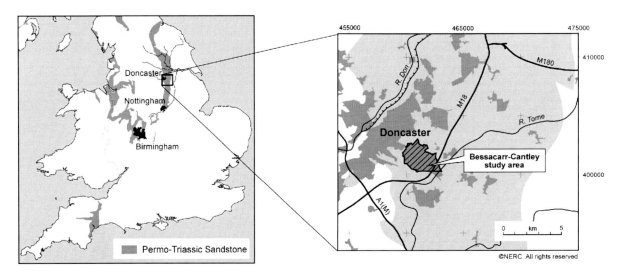

Figure 3.2.1. Location of Doncaster and Bessacarr-Cantley study area on Permo-Triassic sandstone aquifer.

The sandstones in the vicinity of the city of Doncaster, South Yorkshire have an outcrop width of about 16 km and dip gently towards the east-north-east at about 1.5°, being underlain by low permeability Permian marls and overlain to the east by Triassic mudstones (Figure 3.2.2).

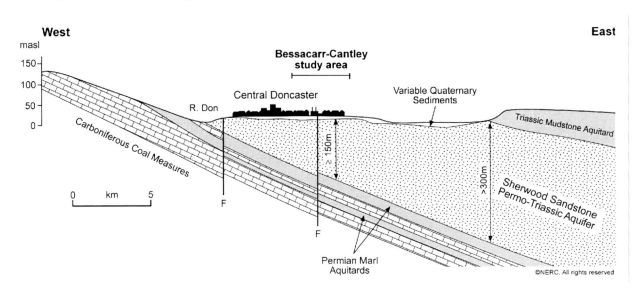

Figure 3.2.2. Sketch section across Sherwood Sandstone aquifer in vicinity of Doncaster.

They comprise a varied series of red and brown, friable to moderately cemented, fine to medium-grained sandstones, with sporadic thin layers and lenses of mudstone and mud-pellet conglomerates (Gaunt 1994). The formation has little topographic expression apart from isolated and subdued ridges on the western (basal) margin. One of these coincides with much of central Doncaster and another with the study area suburb of Bessacarr-Cantley (Figure 3.2.1). The aquifer is up to 300 m thick, increasing from its western edge to about 175 m to the east of Doncaster where the suburbs and nearby former coal mining villages are located. Quaternary superficial deposits ranging from glacial sand and gravel to peat and lacustrine silty clays overlie the sandstones in many places and these exert a major control on recharge processes, flow patterns and solute/contaminant transport (Smedley and Brewerton 1997). Bessacarr-Cantley is underlain either directly by the sandstone aquifer or by intervening permeable Quaternary sands and gravels of up to 8 m thickness (Figure 3.2.3).

3.2.1.2 Hydrogeological setting

Regional transmissivities of the unconfined Sherwood Sandstone aquifer, derived from pumping tests, lie in the range 100 – 700 m^2/d with a median of 207 m^2/d (Allen et al. 1997). Intergranular porosity measured from core samples is typically around 30%. Specific yield S_y values of around 0.1 are cited from both laboratory measurements and calibrated regional flow models but locally poor cementation of the upper part of the saturated aquifer implies that this is an underestimate for the South Yorkshire area, where S_y values are likely to be closer to 0.15-0.2. Permeability tests conducted by the study team using inflatable packers in a 59 m deep borehole (CWT) in Bessacarr gave a mean hydraulic conductivity of 5.25 m/d over the depth 27 – 59 m below ground level. Groundwater flow occurs from west to east but is controlled by the gradients induced by the water supply wells located in an arc east of the urban area.

Regionally, the Sherwood Sandstone aquifer is considered to be anisotropic as a result both of synsedimentary features (interbedded mud-rich horizons, presence of fining-upwards cycles, channelling) and post-diagenetic structural developments (bedding plane fractures, inclined joints, faults). Although fracture flow is widely recognised, intergranular flow tends to dominate in regional flow systems because fractures are often filled with sand (Allen et al, 1997). In addition there is evidence that locally the upper few tens of metres of the Sherwood Sandstone aquifer may be less indurated than equivalents elsewhere. The subdued, near sea-level elevations of most of the sandstone east of Doncaster resulted in wetlands until the early 20th century. This contrasts with further south along the strike of the formation, where the sandstone outcrop is sufficiently well cemented to form relatively high ground e.g. around Nottingham.

Although the lowest 40 m forms a discontinuous series of low ridges, elsewhere exposures and core recovery are poor (Gaunt 1994) and recent drilling experience in the Bessacarr area confirmed that above these basal beds much of the sandstone sequence is poorly cemented (Rueedi & Cronin 2003). Field sampling and data interpretation by the AISUWRS investigation team support the inference that bedding plane fractures and other features of a well-cemented sandstone sequence are infrequent throughout much of the uppermost 30 m of saturated aquifer and that intergranular flow is predominant. Below that, fractures in the more competent horizons alternating with the less competent sandy strata provide limited access for modern recharge to penetrate to significant depths.

The subdued topography around the study area makes assessment of the natural, pre-development flow system speculative. Brown and Rushton (1993) suggest that around Doncaster, groundwater would probably have drained from high recharge areas (drift-free or with permeable drift) in the centre and south west of the area, towards the east and north. If so, then the suburb of Bessacarr-Cantley would historically have been one of these zones, comprising a low eminence of about 20-25 m elevation draining outwards to the east, north and south towards encircling wetlands of about 5m elevation, underlain by a full aquifer with very shallow flow systems discharging to local watercourses. Part of this pattern persists below the study area to the present day, albeit under very different conditions of recharge and withdrawal.

3.2.1.3 Urban development setting

Doncaster, with a present-day population of about 200,000, is an old-established town on the River Don whose expansion accelerated during the early part of the 20th century into a traditional base of manufacturing, especially light and heavy engineering, glass making and textiles. It has also been closely associated with deep mining from the South Yorkshire coalfield, again from the early 1900s. For some years the town has been undergoing a transition and diversification away from this traditional manufacturing base into different economic sectors, with a notable increase in service industries. Doncaster is groundwater-dependent, and the city, nearby former mining villages and a large rural hinterland are supplied from a network of boreholes. The town's history of development has resulted in a dispersed urban setting, with an urban residential and industrial core but several major residential, commercial and industrial districts located peripherally and separated from the town by informal open space.

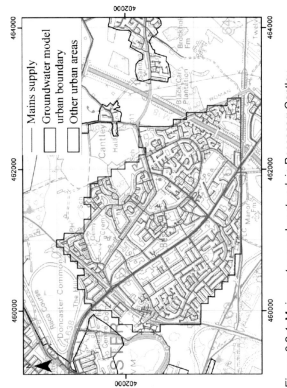

Figure 3.2.3. Quaternary and underlying solid geology of study area.

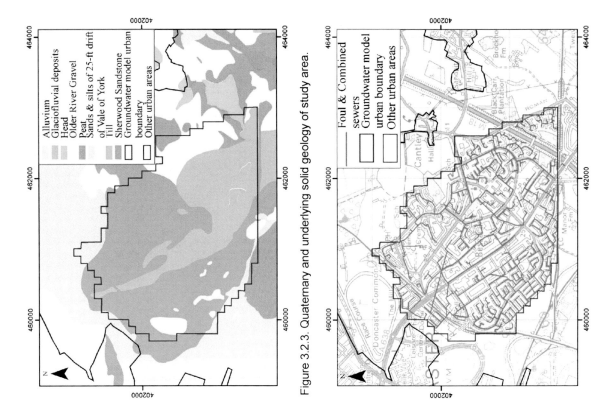

Figure 3.2.4. Mains water supply network in Bessacarr-Cantley.

Figure 3.2.5. Foul and combined sewer network in Bessacarr-Cantley.

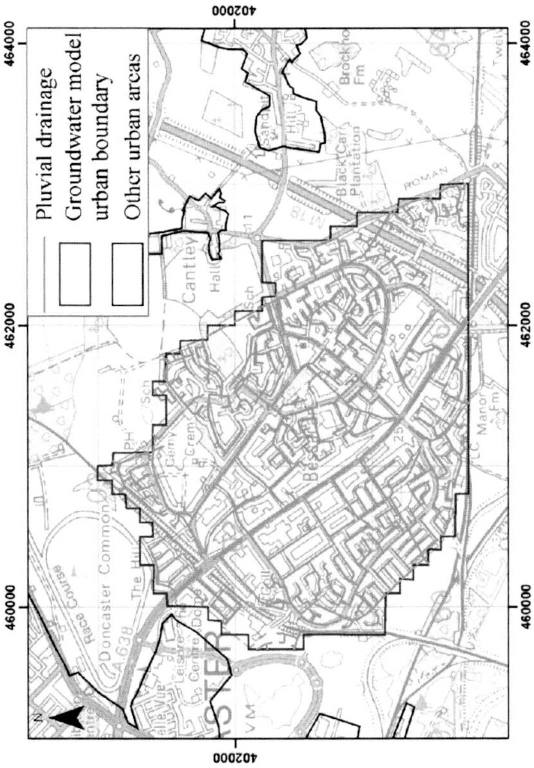

Figure 3.2.6. Pluvial (stormwater) piped drainage network in Bessacarr-Cantley.

The suburb of Bessacarr-Cantley is one of these districts, being located approximately 3 km southeast of the town centre. With a population of about 20,000 spread over an area of c.6 km², it has urbanised intermittently since the early 1920s and comprises a mix of residential properties and local services (schools with playing fields, retail, community buildings, public open space) (Figure 3.2.14). Town planning controls have kept the district geographically distinct, and both the urban footprint and its associated water infrastructure of mains supply, wastewater and pluvial drains are well-defined (Figures 3.2.4-6).

Bessacarr-Cantley was selected as the detailed study area for the following reasons:
- Nil or minimal and permeable Quaternary cover over the Sherwood Sandstone aquifer,
- Well-defined urban boundary and corresponding piped water infrastructure (piped water supply, sewers, surface water drainage, combined sewers),
- Mixture of land uses but predominantly residential and community facilities, with a range of housing types, developed over >80 years with the majority post-1930,
- Down-gradient of city and upgradient of water supply wellfield,
- Range of water table depths due to topography.

3.2.1.4 Urban water infrastructure

The piped water supply for the town of Doncaster, its suburbs and surrounding rural area is supplied by the Doncaster wellfield, a linked array of eleven pumping stations extending from just to the east of the town along a 15 km arc to the northeast and southeast. This wellfield supplies about 65 Ml/d to the Doncaster area, with water from four of the pumping stations being blended at the Nutwell Water Treatment Works to supply Bessacarr-Cantley. The average per person water usage is estimated at about 180 l/p/d including unaccounted-for water (Rueedi at al. 2005), with an average occupancy of approximately 2.5 persons per household.

The c. 8,000 households in Bessacarr-Cantley are served by a total length of water bearing pipe infrastructure of almost 220 km, via iron or plastic water mains and vitreous clay or concrete foul sewers and pluvial drainage systems (Table 3.2.1, Figs. 3.2.4-6). Only about 2 km of the 128 km of sewers (the oldest part) is a combined system, a feature common to most English cities. There are not thought to be any septic tanks in current use within the study area. Together with much of central Doncaster, the suburb's sewage is treated at a large downstream wastewater treatment woks located at Wheatley on the northern outskirts of Doncaster.

Table 3.2.1. Pipe infrastructure key statistics for Bessacarr-Cantley.

Pipe network type	Pipe asset count	Total length (km)	Materials comments
Mains supply	1135	91.6	84 % by length cast or ductile iron, 15 % PVC/PE
Sewer-foul & combined	1205	56.9	87 % by length vitrified clay, 12 % concrete
Drain-pluvial (stormwater)	1413	71.1	47 % by length vitrified clay, 53% concrete
Totals	3753	219.6	

3.2.1.5 Data sources

Available data from stakeholder and other organisations were required for two purposes:
- To provide background on the hydrogeology and the city setting (including the urban water infrastructure), initially to help identify a suitable detailed study area and then to inform decisions on scenario selection
- To populate the various models as much as possible with existing information, supplemented by data collected specifically by the project in order to constrain key variables.

The urban water model array has over 275 input fields, only a minority of which are provided as outputs from a preceding model (Table 3.2.2).

Table 3.2.2. Distribution of input parameters in the urban water model array, 2004*.

Model name	Short name	No of input fields	% total no. of input fields
Urban volume and quality	UVQ	116	43
Pipeline leakage	NEIMO	15	1
Unsaturated zone flow & transport model below leaking pipes	SLeakl	6	2
Unsaturated zone flow & transport model below open spaces	Posl	6	2
Saturated zone flow model	SFM	53	19
Saturated zone transport model	STM	63	23
Totals		278	100%

* Minor changes to input field totals for individual models occurred during development stage 2004-2005; the unsaturated zone model UL_FLOW has been employed in this case study

To assist organisation of the collection/collation of existing data and to identify where resources should be targeted for the acquisition of new field data, a simple information prioritisation ranking was devised (Table 3.2.3). This is a recommended procedure at project inception as different cities have different data availabilities. It could also with advantage be combined with a critical initial appraisal of sensitivity of individual models to particular parameters.

Table 3.2.3. Data collection and prioritisation.

Model criticality classification		Data deficiency importance
High	Key parameter needing city case-study-specific value; validity or confidence in the resultant model would be compromised without it	High; impact on model sensitivity likely to be major
Low	Secondary importance; useful to have case-study specific value, but generic, assumed or approximate value acceptable	Low
Data quality classification		Values sourced from?
A	Site-specific values available e.g. water levels, aquifer node hydraulic conductivity, pipe material class, soil distribution/thickness.	Available data from stakeholders and/or additional project field studies
B	Site-specific values unlikely but reasonable approximations need to be found from available data; field studies likely to be beyond scope/resources of the project	National, regional or company statistics e.g. sewer leakage rates, pipe failure rates, indoor water use.
C	Neither site-specific nor local-context values likely to be available i.e. every parameter not Class A or Class B.	Values would be defaults comprising working approximations, experimental or empirical estimates from technical literature searches, studies in analogous situations, laboratory studies.

A pilot exercise conducted in 2003 (Rueedi et al. 2004) for the Doncaster situation (Figure 3.2.7A) showed that UVQ is especially demanding in terms of number of parameters, but was the poorest served with site-specific values (Class A data, only 9% of the total). When collated, the results indicated that for the model array as a whole, it would only be practicable to populate about half of the input fields with site-specific (Class A) values (Figure 3.2.7B). The field data collection programme was directed to filling the most important gaps.

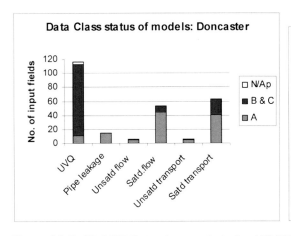
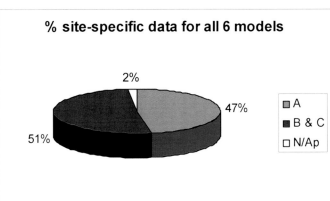

Figure 3.2.7. (A) 2003 Data class analysis for AISUWRS model input parameter fields, (B) Percentage of input fields in Doncaster case-study likely to be populated with site-specific data.

Data sources included the main stakeholder organisations (water utility, environmental regulator, municipal government) but specific data sets were required from numerous other agencies (Table 3.2.4).

Table 3.2.4. Key datasets and sources used in Doncaster case study.

Data Type	Source	Use
Topography (1:10 000, 1:50 000)	Ordnance Survey	P
Bedrock geology (1:50 000 & 1:250 000)	British Geological Survey	P
Superficial geology (1:50 000)	British Geological Survey	P
Soils Data	National Soil Resources Institute (NSRI)	P
Urban land use (town planning scale)	Doncaster Metropolitan Borough Council	P
Sewer infrastructure (Pipe infrastructure, soakaways, Overflows, Outfalls, Waste Water Treatment Works, CCTV)	Yorkshire Water	P
Mains Pipe infrastructure (Pipe infrastructure, night time flow records, water abstraction and raw water quality data for public supply wells	Yorkshire Water	P
Water levels, regional model parameters (groundwater flow), scenario modelling data	Environment Agency	P
NextMap© digital terrain model	Intermap Technologies Inc	P
NERC digital terrain model	Centre for Ecology and Hydrology (Wallingford)	P
Meteorological data (precipitation, evaporation)	Meteorological Office	P
Regional and multilevel monitoring database	Project Specific Data collected by Robens Centre & BGS	P
Land cover map, 2000	Centre for Ecology and Hydrology (Monks Wood)	S
Aquifer vulnerability and groundwater source protection zones	Environment Agency	S

P: required for model setup and calibration   S: required for background study, site and study area selection

3.2.2 Field investigation programme

3.2.2.1 Objectives

The field investigations complemented the collection and analysis of existing data, particularly by obtaining detailed information about groundwater flow conditions and the urban water mass balance fluxes in and around the study area. This better understanding was used to calibrate the models and to provide a solid basis for the alternative water use scenarios.

3.2.2.2 Activities

The field investigation programme comprised:
- Site selection. Drilling, installation and subsequent monitoring of 5 multilevel borehole sites located in Bessacarr-Cantley. (Figure 3.2.8). The 5 sites sampled up to 7 different depth intervals each in the shallow aquifer down to about 60 m. below ground level (Figure 3.2.9). Their design permitted pumped water samples to be taken and also continuous water level and electrical conductivity measurement using programmable in-well data loggers (Rueedi and Cronin 2003). Multilevel boreholes are particularly useful for detailed groundwater quality studies because contaminant concentrations in bedded deposits can vary markedly in the vertical direction. In some situations, the zone contributing to the well contamination may occupy only a small part of the total aquifer thickness. This zone could otherwise go undetected, or could mistakenly be assumed to represent conditions over the entire aquifer depth.
- Sampling of groundwater in Bessacarr (the 5 multilevel sites) and selected existing private wells in the surrounding periurban and rural areas (5 boreholes, reduced from an initial network of 12).
- Assessment of water quality in the pipe network, comprising sampling at 3 sewer inspection manholes and 2 stormwater outfalls plus the collation of mains water quality data provided by Yorkshire Water.

A wide range of microbiological and hydrochemical parameters were sampled and analysed (Table 3.2.5) in order to characterise shallow groundwater beneath the study area, periurban groundwater in the surrounding rural area and urban groundwater in the city proper. These data provided background concentrations and contaminant loads used to calibrate the UVQ and the groundwater models (Section 3.2.8).

150 Urban Water Resources Toolbox

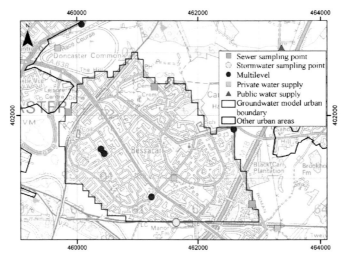

Figure 3.2.8. Field sampling network in Bessacarr-Cantley including multilevel and regional wells and sewer monitoring points.

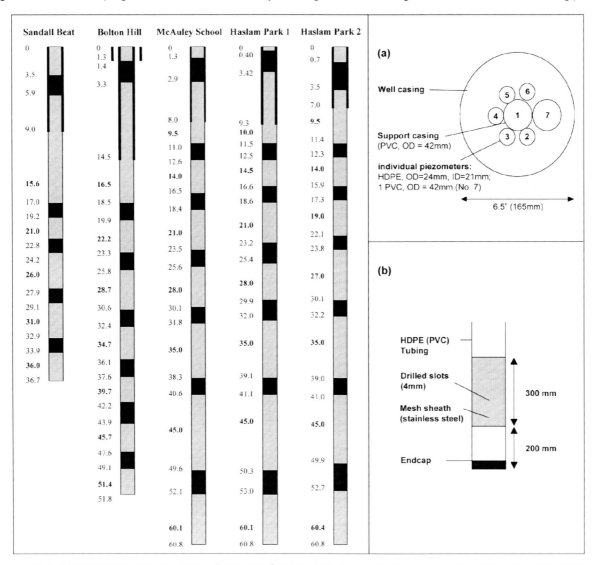

Figure 3.2.9. Multilevel research boreholes, Bessacarr-Cantley; thick lines indicate steel casing, dark areas the location of bentonite seals, numbers in bold the mid-depths of 0.3 m long sampling port a) Arrangement of plastic pipes inside the open borehole. b) details of screened section at lowest part of each sampling tube.

Table 3.2.5. Determinands and analytical methods; AISUWRS field sampling programme.

Characterisation group	Determinand	Method	Sampling Objective
Microbiological	Faecal coliforms, total coliforms, faecal streptococci, sulphite reducing clostridia	Filtration, culture and enumeration	WW
	coliphage, and enteric virus	Assay and culture	WW
Field hydrochemical	pH, temp, Eh, DO_2, SEC	Field meters in in-line cell	GP
Major and minor constituents (laboratory determinations)	HCO_3	Field alkalinity titration	GP, UR
	Na, K, Ca, Mg, SO_4, Si, Al, B, Ba, Be, Cd, Co, Cr, Cu, Fe_{tot}, La, Li, Mn, Ni, Mo, Pb, P_{tot}, Sc, Sr, V, Y, Zn, Zr, (As, Se)	ICP-AES	UR
	Cl, TON, NO_3-N, NO_2-N, NH_4-N	Automated colorimetry	UR
	DOC	Carbon analyser	
Stable isotopes and residence times	CFCs	GC	RT
	SF_6	GC	RT
	$\delta^{13}C$	Mass spectrometry	UR

| WW | Wastewater indicators | GP | General physicochemical parameters |
| UR | Potential indicators of urban recharage | RT | Residence time indicators |

3.2.2.3 Results

The following summary is an interpretation of results drawn jointly from the available data analyses and from the field investigations. More details are provided in 8 papers published in project field investigations final report CR/05/028N (Morris et al. 2005) and available to download from the AISUWRS website http://www.urbanwater.de/ (see numbered reference set at end of the chapter).

Objective 1. Describe vertical variations in lithology, structure and vertical hydraulic gradients in the aquifer.
- New information from both drilling logs and cores from the 5 dedicated multilevel piezometers showed that the local sandstone aquifer is unstable or weakly cemented to depths up to 30m. This contrasts with other regions where the Triassic sandstone is moderately-to-well cemented (e.g. Birmingham, Nottingham), and fracture-flow acknowledged to be important.
- Vertical hydraulic gradients in the multilevel wells were found to be small ranging from 0 to 0.0018. This could be a consequence of the small topographic differences in the research area where the highest points are only a few tens of meters above sea level (Figure 3.2.10).

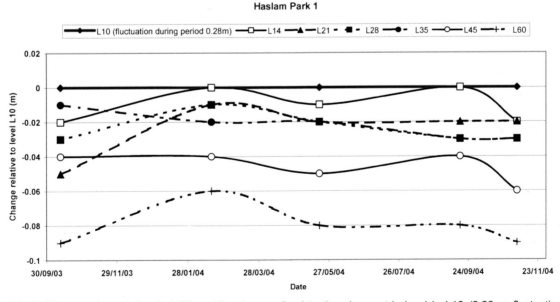

Figure 3.2.10. Piezometric variation in HP2 multilevel normalised to the piezometric level in L10 (0.28 m fluctuation over period). L10, L14 etc are sampling port depths (in m. below ground level). Note synchronicity of variation at all ports down to 60 mbgl and small magnitude of relative head difference (from Morris et al. 2006).

- Packer testing of the Cantley Water Tower observation well showed little difference in the bulk permeability of the near water table zone and that between 40-60m depth, both giving values of about 5 m/d.
- Surprisingly, seasonal changes in water level were well propagated through the aquifer, with deep access ports not only responding to rainfall recharge events but also showing negligible time-delay compared with ports located close to the water table. It would, however, appear that while the upper 50m of saturated aquifer shows good hydraulic continuity (probably as a result of linked fracture systems in the harder more well-cemented horizons) the strong stratification evidenced by the vertical distribution of water quality indicators suggest that only relatively small volumes of water are involved in vertical movement.
- It is assumed that the vertical hydraulic gradients strongly depend on the long-term pumping regime at the public water supply wells located down-stream of the focus study area. Overall, the downward gradients at the multilevel wells indicate that, as expected, Bessacarr-Cantley is located in the recharge area of the aquifer.

Objective 2. Detail the distribution and persistence of standard sewage indicators and sewage-derived viruses and their seasonal fluctuations.
- Measurements taken during this project have confirmed the general sanitary engineering observation that daily variations in sewage volumes in a suburban area are significant. The varying temporal contributions of grey water and toilet flush water were deduced using scenario comparisons (Rueedi et al 2005). The observed trend follows a strong peak in the morning with another increase in the afternoon, whereas sewage volumes during the night are low. Chemical, isotopic and microbial results have shown large daily variations in sewage indicator concentrations. One of the major drivers for these variations is toilet waste as it correlates with the largest variations in measured flows and contains large amounts of nutrients (nitrate and phosphate) and most faecal micro-organisms. In fact, detailed temporal measurements of micro-organisms (e.g. total coliforms, faecal coliforms, sulphite reducing clostridia (SRC), enteric viruses) have proven to be a useful method to infer the differences in domestic wastewater input required by the UVQ model (toilet, bathroom, laundry, kitchen).
- Concentrations of microbial sewage indicators (bacterial as well as viral indicators) and pathogenic viruses in sewage were found to vary over several orders of magnitude, both on a daily and an annual time-scale (Rueedi et al. 2005). This has to be considered when using them to assess quantitatively the influence of sewage on microbial groundwater quality. Chemical parameters were found to vary significantly on a daily time-scale in the sewer but the quarterly sampling suggests the variation may be less on an annual time-scale (Table 3.2.6).

Table 3.2.6. Orders of magnitude of indicator bacteria numbers at various stages of urban cycle (Cronin et al. 2005)

	Human faeces*	Sewer sampling (Doncaster) cfu or pfu/100ml	Groundwater sampling (Doncaster) cfu or pfu/100ml	Orders of magnitude decrease
Faecal coliforms	$10^6 - 10^9$	$10^5 - 10^7$	$<10^0 - 10^1$	4 to 7
Total coliforms	$10^7 - 10^9$	$10^6 - 10^8$	$<10^0 - 10^2$	4 to 8
Faecal Strep	$10^5 - 10^8$	$10^5 - 10^7$	$<10^0 - 10^2$	3 to 7
SRC	$10^3 - 10^{10}$	$10^3 - 10^6$	$<10^0 - 10^2$	1 to 6
Enteric viruses	up to 10^{12}	$<10 - 10^2$	$<10^{-2}$	2 to >5

*Number of faecal indicators commonly found in human faeces expressed as cells per gram of faeces (wet weight) (from Gleeson & Gray, 1997).

- The largest frequency of positive detects of the bacterial faecal indicators, and to a lesser extent the enteric viruses, were found in the shallow intervals (0-30m) of the multilevel wells (Figure 3.2.11) where the largest sewage contributions to recharge were estimated from the major and minor ion mass balance calculations (Rueedi et al, in review) It should be noted though that the magnitude of these microbial faecal indicators was low and the groundwater is not grossly contaminated. Interestingly, indicator micro-organisms as well as enteric viruses were also found at depths of up to 60mbgl. This suggests deep penetration of modern (<50 year old) water. The occurrence of faecal indicators in Doncaster corresponds with the profiles of the groundwater dating tracers CFC-11, CFC-12 and SF6 (Morris et al. In press) and is consistent with the synchronous seasonal changes noted by pressure transducer readings in the multilevels down to 60 m depth. Similar findings came from depth-specific groundwater sampling campaigns in Nottingham and Birmingham, where indicator micro-organisms and pathogens were detected to similar depths (Powell et al. 2003)

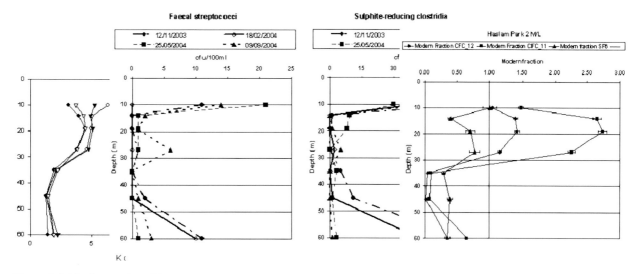

Figure 3.2.11. Composite of hydrochemical (K), microbiological (FS, SRC) and residence time indicator (CFC-11, CFC-12, SF6) depth profiles for multilevel HP2, showing clear influence of urban recharge to approximately 30m depth.

- Table 3.2.7 indicates a high number of positive detects of faecal indicators in the aquifer and a broad correspondence between regional and multilevel wells, although counts were quite low throughout the sampling campaigns. Faecal indicator microorganisms, sulphite reducing clostridia (SRC) as well as faecal streptococci were detected to depths of 60 mbgl. SRC spores are known to be long-lived and therefore they are expected to survive in groundwater over a timescale from months to years. However, faecal streptococci, with a half-life of typically 10s to 100s of days, are less persistent. Their occurrence is therefore more surprising. The conceptual picture of groundwater flow in this area needs to recognise the presence of very recent recharge at significant depths. It is concluded that flow is likely to be a product of slower (matrix–flow) and faster (fracture-flow) components. This conceptualisation has risk implications for sewage-derived contaminants such as viruses because it implies a more extensive penetration of pathogens than would otherwise be suspected from the evidence of groundwater dating tracers and intergranular hydraulic flow calculations (Morris et al, in press).

Table 3.2.7. Faecal indicator sampling results summary. Results are cumulative of all sampling campaigns in Doncaster (July, November 2003; February, May, September, November 2004) (from Cronin et al. 2005).

Faecal indicator % positives →	Depth-specific multilevel intervals	Regional wells	Sewers	~Cost £ Sterling
N samples	154	45	43	
Field TTC %[1]	18	11	100	<0.3
E. coli %	18	16	100	~10
Total coliforms %	34	24	100	~10
Faecal Strep. %	40	24	100	~15
SRC %	44	47	100	~15
Coliphage %	1	7	100	~25
N samples	60	3	17	
Enteric virus[2] %	12	0	100	>100

[1] Analyses of thermotolerant coliforms were undertaken in the field using a portable DelAgua testing kit as well as samples being sent for laboratory filtration and confirmation (shown in the next row named E. coli), [2] Combination of results from two methods.

Objective 3: Assess relative magnitude of sources of urban groundwater recharge and their effects on the quality and availability of water for public and private supply:
- Mains leakage was estimated from Yorkshire Water hourly night time flow records 1998-2003 (see 3.2.7.1) Leakage rates of the 6 leakage control zones ranged from 1.1 to 5.3 l/property/h, with a 5-year average of 1.82± 0.53 l/property/h. This value corresponds to an average leakage rate of 9.7% of all imported water or 22±5 mm/y equivalent recharge. Irrigation values averaged 7300 l/property/y corresponding to 4.9 % of total supplied water and 10 ± 3mm/y equivalent recharge, calculated by subtracting winter water usage from observed demands during spring and summer. Losses beyond the household connection, i.e. pipeline leaks on properties, wastage inside the house due to faulty ball cocks etc. were not included.

- Current leakage rates from the sewerage system were deduced using mass balance calculations to be about 20 to 45 mm/a equivalent recharge corresponding to a total leakage of 7-15% of annual sewage throughput (Rueedi et al. submitted). Such values are of similar order to previously reported sewer volume losses (typically 3 to 5%).
- The use of pollution indicators proved more difficult than anticipated because species that have been demonstrated as useful elsewhere, such as chloride, nitrate and sulphate, show little more variation than that encountered in neighbouring rural catchments (Table 3.2.8). Although the range of indicators used was confined to inorganic and microbiological markers, it suggests that in hydrochemical and microbiological terms, the adverse effect of urban recharge on the measured quality parameters of the underlying groundwater in the Bessacarr-Cantley area has so far been limited. This is ascribed to the combined effects of a non-industrial prior land-use history, locally high storage capacity in the friable upper aquifer and particularly the availability and ready infiltration of dilution from precipitation entering urban green space areas.
- A complicating feature is that nearby rural wells are not free of anthropogenic influences, and these may mask the influence of sewage-derived recharge. Private boreholes, usually withdrawing groundwater from shallow depths are influenced by urban recharge in and near the urban areas and by agricultural practices over the last 50 + years. Public supply wells, located farther outside the urban area and withdrawing water from depths below 30 mbgl are less influenced.

Table 3.2.8. Comparison of concentration ranges of potential sewer leakage indicators with those for other parts of urban water infrastructure in Bessacarr-Cantley.

Marker species	Concentration range mg/l					
	Bessacarr-Cantley study area				Rural/periurban	
	Waste water n=29	Mains supply n=30-479*	M/levels 0-30m n=75	M/levels 30-60m n=65	8 private wells n=30	3 public supplies n=30-410
Cl^-	60 – 90	26 – 41	10 – 170	15 – 110	15 – 90	20–80
SO_4^-	60 – 100	27 – 46	30 – 140	20 – 160	20 – 350	30–80
HCO_3^-	400 – 575	180 – 240	90 – 300	35 – 275	100 – 550	100–220
K^-	17.5 – 22.5	2 – 3	1.5 – 13	1.5 – 6.5	2 – 28	2.5–3.5
B	0.15 – 0.5	BDL (0.05)	0.04 – 0.14	0.01 – .09	0.025 – 0.1	<0.1
NH_4-N^-	25 – 75	<.02	-	-	<0.01 – 0.5	<0.02
TON	<2	0.5 – 10	2.5 – 13.5	5 – 17	<0.1 – 30	5–16
DOC	30 – 110	N/A	1 – 5	0.7 – 2	1.5 – 7	N/A
Data source	FS	YW	FS	FS	FS	FS

*	Depending on parameter measured
BDL	Below detection level
Wastewater	3 sites: Burnham Close, Everingham Rd, Warning Tongue Lane
Mains supply	Nutwell combined raw (blend of Armthorpe, Boston Park, Nutwell, Thornham Pumping Station waters)
Multilevels:	5 sites: Haslam Park 1 & 2, Bolton Hill, McAuley School, Sandall Beat
Private wells	8 sites (Beechtree Nurseries, Doncaster Racecourse, Gatewood Grange, Misson Quarry, Warning Tongue Lane, , Elmstone, Crowtree and Lings Farms)
Public supplies	3 sites (Nutwell, Rossington Bridge and Armthorpe pumping stations, various boreholes)
FS	Data collected by AISUWRS project team Jun 2003 – Nov 2004
YW	Data from Yorkshire Water raw water quality surveillance archive Jan 1999 – Mar 2004

- Comparison of multilevel and regional well analyses confirms the aquifer system's geochemistry is complex even before urban recharge is taken into account. Wide local variations are apparent; with groundwater in the south being calcite dominated and almost saturated with oxygen while that to the north tends to be reducing and usually contains higher concentrations of iron and manganese. These variations are probably controlled by the permeability of overlying superficial deposits in the recharge zones.
- Monitoring of the multilevels provided detailed depth profiles of various groundwater parameters. No significant seasonal variations were detected for either the chemical or the microbial parameters measured in the groundwater. However, they had quite site-specific profiles which were neither consistent between different indicators at the same site, nor between different sites for the same indicator (Figs. 3.2.12, 3.2.13). Both conservative and non-conservative indicators were affected. This prevented simple comparisons but detailed analysis showed broad similarities that enabled an approximate distinction of likely contamination sources (Figure 3.2.13).

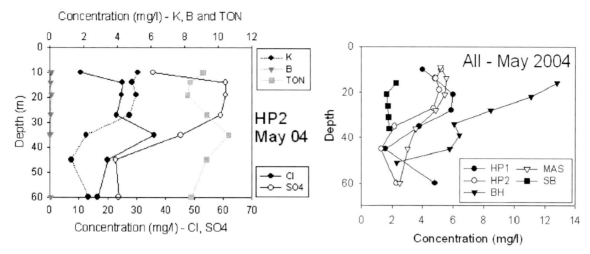

Figure 3.2.12. Contaminant indicator behaviour 2; distinctive depth patterns between conservative indicators at same site (example from HP2) and between sites (potassium profiles for all 5 multilevels).

- Nitrate levels were not found to be significantly or consistently elevated in the upper aquifer, and ammonium concentrations were universally low (which is consistent with the aerobic condition of the aquifer throughout). In fact, nitrate concentrations were sometimes lower in the shallow levels than at depth. This suggests that either a) nitrate is being stored in the unsaturated zone or b) the observed sewage contamination originates mainly form sewage overflows and nitrogen was consumed during passage through the soil zone or c) that it is being converted to gaseous nitrogen (and lost to the system) in complex bacterially-mediated nitrification/denitrification reactions via nitrite in the vicinity of leaks or during passage through the underlying unsaturated soil zone.
- The urban tracers proving most useful to quantitatively assess urban recharge were potassium, sodium, boron, and SF6. The most useful qualitative tracers were the alkalinity $\delta 13C$ ratios, the faecal indicators sulphite reducing clostridia (SRC) and faecal streptococci, as well as CFCs. These demonstrated clear contaminant influence on profiles 0-30m, with some influence detectable to at least 60m (Figure 3.2.11), but also showed that the rural background is often as high as the urban loading.

3.2.3 AISUWRS model setup for Bessacarr-Cantley study area

3.2.3.1 UVQ

UVQ requires a wide range of input data to physically characterise each neighbourhood (Table 3.2.9)

Table 3.2.9. Date sources used to populate the UVQ for the Bessacarr-Cantley study area.

Source	Data Type
Meteorological Office	Precipitation and potential evapotranspiration
Yorkshire Water	Hourly night time water supply (1998-2003)
Yorkshire Water	Daily water supply (1997-2004)
Yorkshire Water	Sewage flow volumes for dry and wet days in July and August 1993
Yorkshire Water	Foul sewage and storm water network
Doncaster Metropolitan Borough Council (DMBC)	Land use map and photogrammetry for town planning
DMBC	Population statistics
Ordnance Survey	Land use details
Field measurements	High resolution sewage flows in one neighbourhood over one day
Field measurements	Sewage and storm water quality (seasonal and one day high resolution)

As a basic driver UVQ needs daily precipitation and potential evapotranspiration (PET) records. The UK Meteorological Office provided weekly data for 1970 to 2003 inclusive and daily records for 2003 to 2004. Additionally, daily precipitation data were available from 1970 to 1999. 1970 was used as substitute for the missing years 2000 to 2002 where no daily data were available. Daily PET data would also have been useful but they are not considered crucial for the calculations because the daily variations are usually not as large as for precipitation amounts. The large amount of information needed to populate the UVQ model required a systematic step-by-step set-up mode.

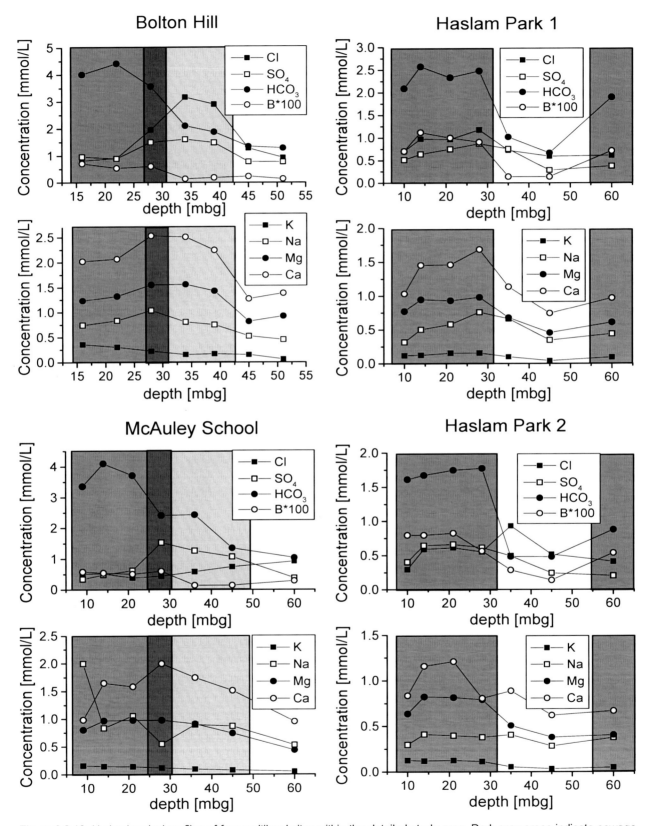

Figure 3.2.13. Hydrochemical profiles of four multilevel sites within the detailed study area. Dark grey areas indicate sewage contamination influence; light grey areas show other influences (possibly stormwater). (Rueedi et al. submitted).

Step 1

Each UVQ neighbourhood is required to be homogeneous in terms of water use, surface coverage/characteristics and soil properties. Furthermore, each neighbourhood must only have one outflow per drainage system (foul sewage and storm water). Therefore, the first step of setting up UVQ required detailed analysis of both land use and OS maps and pipe network drainage systems (Figures 3.2.14 A, B).

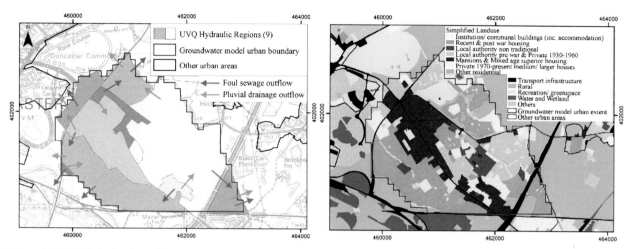

Figure 3.2.14. A) dedication of UVQ regions based on pipeline network information. Black arrows indicate foul sewage outflow and blue arrows show stormwater outflows. B) Simplified land use map of focus study area.

As the separation of the two drainage networks already led to 9 regions (UVQ regions in Figure 3.2.14A) the land use map had to be significantly simplified to keep the final number of neighbourhoods as low as possible without losing important information. It was therefore decided to distinguish principal land use types in terms of water usage types (residential, schools and commercial/health care) and also to split residential areas into two groups to account for the older central part of the suburb, which is dominated by large houses with large gardens, whereas the newer parts (post 1950) are small houses or terrace houses with small gardens. As a result, 20 neighbourhoods were distinguished (Figure 3.2.15).

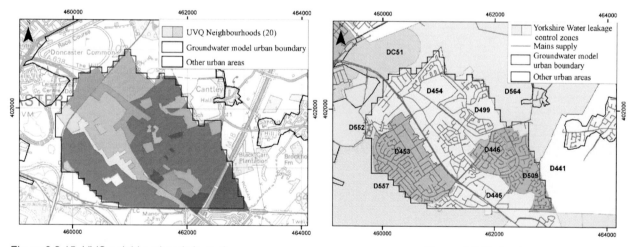

Figure 3.2.15. UVQ neighbourhoods in study area. study area.

Figure 3.2.16. Leakage control zones in study area.

Step 2

In a second step water supply volumes had to be determined for each neighbourhood and split among the different usage types distinguished in UVQ (e.g. toilet and kitchen use). This scheduling step included assigning mains leakage and water usage for greenspace irrigation.

The water supply network is managed and controlled over six leakage control zones (LCZ) (Figure 3.2.16). These zones do not match with the UVQ neighbourhoods previously selected. Therefore, all water supply data provided was split among the neighbourhoods proportional to the estimated number of inhabitants. The night time usage data were analysed to find the amount of leakage leaving the system prior to consumption. Measured volumes (Gross leakage) were corrected using national standard values to represent night time residential and commercial actual night-time usage. The average leakage per property (Net leakage) is used to calculate the percentage of leakage used in each neighbourhood. Table 3.2.10 shows, besides the leakage rates, the number of properties in each LCZ and the number of days reported.

Table 3.2.10. Summary of Leakage measurements from Yorkshire Water for all LCZs.

LCZ No.	Gross Leakage [L/property/h]		Net Leakage [L/property/h]		No. properties	No. days reported
	mean	2003	mean	2003		
D445	7.2	6.04	5.3	4.20	785	1717
D446	3.2	3.32	1.5	1.65	941	1932
D453	3.0	3.24	1.1	1.40	2266	1485
D454	4.0	4.12	2.1	2.23	2080	1705
D499	2.6	2.59	1.2	1.25	1132	1941
D509	2.7	2.69	1.2	1.18	1119	1996

The daily water demand records from Yorkshire Water were used to estimate the total water usage in each LCZ. To split the total usage into household use and garden use the recorded values were separated as indicated in Figure 3.2.17, where a baseline usage (for households) was defined (see red line). Water needs higher than the baseline, as usually observed in spring and summer, are assumed to be used for garden irrigation. Table 3.2.11 summarizes the results for LCZ D499.

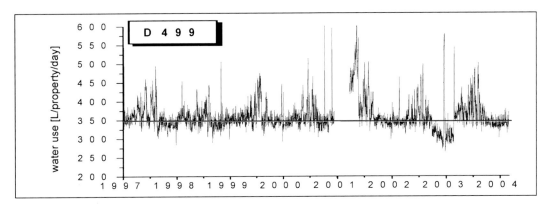

Figure 3.2.17. Average daily water demand per property for leakage control zone D446.

Table 3.2.11. Deducing garden watering use; example using data for leakage control zone D499.

LCZ # D499	Net Leakage %	Average water use L/prop/day	baseline L/prop/day	garden use L/prop/day	leakage corrected L/prop/day	total garden use L/prop/y
1997	8.0	361.9	350	11.9	10.9	3996
1998	8.1	354.5	350	4.5	4.1	1510
1999	8.0	361.5	350	11.5	10.6	3862
2000	8.1	357.7	350	7.7	7.1	2583
(2001)	(7.3)	(392.8)	(350)	(42.8)	(39.7)	(14482)
2002	8.2	351.3	350	1.3	1.2	436
2003	8.0	359.1	350	9.1	8.4	3056
Average	8.1	357.7	350	7.7	7.1	2574

Step 3

The soil parameters play an important role in UVQ because they define the infiltration capacity on unpaved areas both under normal conditions and for onsite options of alternative sanitation (e.g. rain water harvesting or grey water infiltration). However, this process is only represented with a very simple approach that tries to reproduce the balance between infiltration and evaporation. Therefore, the parameters involved were tested for their sensitivity to changes and subsequent effects on the water balance. Results of this sensitivity analysis can be found in Rueedi and Cronin (2005).

To calibrate the soil parameters lysimeter outflows could be used. However, no such measurements were available in this study area. Therefore, an alternative way was applied which basically relies on the assumption that gardens are only irrigated when it is needed and that the irrigation volumes exactly cover the actual demand. Under these assumptions the measured water supply volumes from Yorkshire Water can be used to calibrate local soil parameters. Ideally, observed patterns of water usage in gardens (total use minus baseline use) should be obtained, permitting the optimal choice of soil parameters. It was found that the patterns could be reproduced on a monthly basis but that the model fails to model daily usage properly. Figure 3.2.18 shows the results for neighbourhood 4.

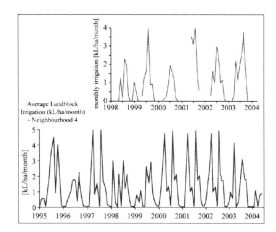

Figure 3.2.18. Comparison between observed (top) and modelled (bottom) water volumes used for garden irrigation.

Step 4

The drainage system in the study area is mostly separated, with just a few of the oldest assets being combined sewer in the oldest part of Bessacarr. Furthermore, there is always some stormwater draining through the foul sewage system (mostly through wrong connections), and vice versa. To estimate the volumes or ratios of stormwater drainage through the foul sewage system, results from a past engineering planning study (DMBC, 1994) were used. This study measured a few rain events to assess current capacities and plan future construction activities. The measurements were made on both the mostly separated system (H7, H8, H9) and the fully separated part of the system (H13, H14, H17) and rainfall amounts were monitored continuously (R02). Figure 3.2.19 displays a dry day (top) and the rain event from July 16, 1993 (bottom) for both outfalls where sewage flow volumes plotted on the event day were corrected for dry weather flow. It is quite obvious that both systems are influenced by storm events even though the influence on the partly separated sewage system is significantly greater (Table 3.2.12).

Table 3.2.12. Results of sewage flow monitoring for monitoring points H09 and H14.

Event	Precipitation [mm]	R02 Flow Surplus [L]	Estimated Total runoff [L]	H09 (mostly separate) %	Flow Surplus [L]	H14 (fully separate) Estimated Total runoff [L]	%
1/7/1993	19.2	227109	5450800	4.2	70170	6198000	1.1
16/7/1993	58.8	428849	18186000	2.4	205223	20470500	1.0
4/8/1993	39.6	360136	12011000	3.0	63392	13550500	0.5
5/8/1993	58.8	623211	18186000	3.4	162536	20470500	0.8
Average				3.5			0.85

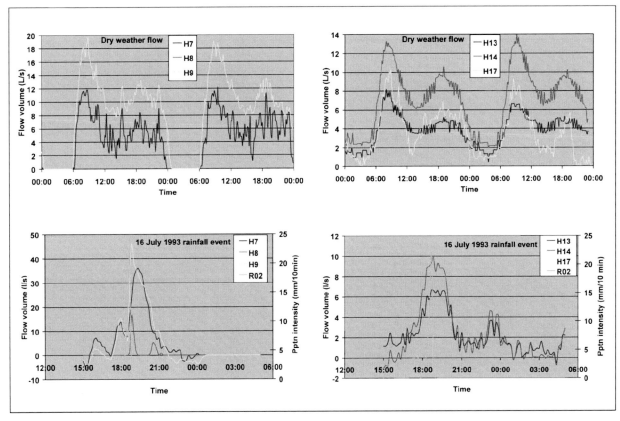

Figure 3.2.19. Comparison of dry weather and rainfall event sewer flows for partially separated and fully separated reaches of drainage system in Bessacarr-Cantley.

Step 5

UVQ requires both total water supply volumes and also water usage allocated volumetrically to toilet, kitchen, laundry and bathroom use. Table 3.2.13 shows from a comparison of usage between different studies/standards that bathroom use seems quite consistent but that toilet flush volumes depends a lot on the source. The values used in UVQ are approximately an average of published proportions. For business water usage absolute usage volumes per unit were taken from Adams 1999 and the split to the different water usages estimated.

Table 3.2.13. Distribution of water usage in households and in small businesses.

In-house usage					
	(US EPA, 1980)	(Almeida et al., 1999)	(ABA, 2004)	(Heaney et al., 2002)	Adopted for UVQ runs
	%	%	%	%	%
kitchen	10.6	13	23.2	19.9	15
bathroom	23.1	40	35.7	20.5	30
toilet	40.7	30.8	26.8	30.2	35
laundry	25.2	16.2	14.3	25.9	20
Commercial/community usage (after Adams, 1999)					
	Schools	Hotels	Hospitals	Offices	
	%	%	%	%	
kitchen	5.0	30.0	10.0	5.0	
bathroom	30.0	30.0	55.0	30.0	
toilet	65.0	30.0	30.0	60.0	
laundry	0.0	10.0	5.0	0.0	
total [l/p/d]	57	182	625	49	

Step 6

The final step to calibrate the UVQ model was to determine and adapt contaminant loads and/or concentrations. Both the foul sewage system and the storm water system were analysed for hydrochemistry (major and minor), indicator micro-organisms (e.g. total coliforms, sulphite reducing clostridia) and enteric viruses. As none of the different sources of foul sewage and storm water were analysed separately first guesses had to be taken from literature (where available) and adapted to the observed values from the field campaigns. Tables 3.2.14 A and B show guess values and values finally used to obtain optimal agreement with the field measurements. It can be seen that no guess values were found for a number of contaminants, particularly regarding the loads for storm water sources.

Table 3.2.14A. List of all contaminants used by UVQ. The top part contains guessed source loadings per household per day and used values for calibrating the model. The lower part contains results of the calibration compared to the field measurements.

Contaminant markers →		B	Cl	K	N	P	Zn	E.coli*106
Water use category ↓		Grey water [mg/capita/day]						
Kitchen	guess	?	?	30	180	24	4.3	10
15%	calibrated	0.1	50	40	50	30	0.4	20
Bathroom	guess	0.1	13.5	3	396	18	90	150
30%	calibrated	0.1	50	0	100	30	4	1000
Toilet	guess	?	?	3000	12750	820	11	300
35%	calibrated	0.1	5000	2000	2500	1200	1	2000
Laundry	guess	20	40	10	295	51	5.8	150
20%	calibrated	50	50	10	70	100	0.6	500
Sampling site & neighbourhood		Foul Sewage [mg/L]						
BC	calibrated	0.360	71.0	17.0	38.0	9.72	0.063	2.17
(NH17)	measured	0.384	70.0	19.1	36.7	10.5	0.059	1.95
EVR	calibrated	0.370	79.0	20.0	42.0	12.0	0.071	2.56
(NH20)	measured	0.352	83.4	21.0	46.0	9.91	0.07	2.90
WTL	calibrated	0.374	70.0	17.0	38.0	9.58	0.062	2.50
(NH18)	measured	0.44	69.7	18.8	38.3	10.35	0.078	1.98

Table 3.2.14B. List of all contaminants used by UVQ. The top part contains guessed source concentrations for calibrating the model. The lower part contains results of the calibration compared to the field measurements.

Contaminant markers →		B	Cl	K	N	P	Zn	E.coli*106
Origin ↓		Runoff sources [mg/L]						
Road	guess	?	400	?	0.7	0.23	0.3	?
	calibrated	0.07	300	2.5	1.5	0.2	0.03	1000
Pavement	guess	?	?	?	2.5	0.17	0.1	?
	calibrated	0.07	40	2.5	1.5	0.2	0	1000
Roof	guess	?	?	?	?	?	?	?
	calibrated	0.07	0	2.5	1.5	0.2	0.03	1000
Roof first flush	guess	?	?	?	8.6	0.44	0.2	?
	calibrated	0.14	0	5	3	0.4	0.06	1000
Rain	guess	?	2.1	0.08	2.5	?	0	0
	calibrated	0	2.1	0.08	2.5	0	0	0
Sampling site & neighbourhood		Stormwater [mg/L]						
HC	calibrated	0.070	106.1	2.5	1.33	0.20	0.028	1000
(NH17)	measured	0.071	109.0	2.37	1.33	0.23	0.034	1100

Mains leakage was calculated from minimum night time flow measurements from 6 leakage control zones (LCZs), less a UK industry-standard value for night-time domestic and commercial usage. The resulting net leakage per property value for each LCZ was distributed among the corresponding neighbourhoods. The average of 10% of total imported water is low for the UK as a whole.

3.2.3.2 NEIMO

Raw data required for this model came from the datasets listed in Table 3.2.15. These were used with ArcView to construct the 3 input files required for each pipe network type used in the model[1] (see Chapter 2 for model description):

- Pipes.csv
- Connections.csv
- Uvq_nbhoods.csv
- Successor.csv

These are combined within the Decision Support System.with a fourth file (Uvq_nbhoods.csv) containing volumetric and contaminant load fields for the 20 neighbourhoods in the study area and which is generated by UVQ.

Table 3.2.15. Data sources used to generate input files for NEIMO.

Source	Data type
Yorkshire Water (YW)	Pipe network database; sewer, combined sewer, pluvial drainage systems
Doncaster Metropolitan Borough Council (DMBC)	Town planning/land use database, photogrammetry
National Soil Resources Institute (NSRI)	Soil properties database
AISUWRS project field investigations	UVQ outputs; volume of wastewater generated and contaminant loadings
Ordnance Survey	Land use detail

Pipes.csv:

Table 3.2.16. Summary statistics Bessacarr-Cantley drainage pipe networks.

Drainage function	Asset Count	Total network length (km)
Foul sewer	1170	54.7
Combined sewer	35	2.1
Surface water/pluvial/stormwater	1413	71.1
Miscellaneous (transition, overflow, unspecified)	9	0.2
Totals	2627	128.1
Project area	6.3 km^2	

This file provides information about the pipe assets in the sewer network serving Bessacarr-Cantley (Table 3.2.16) and was mainly populated from the YW databases. These datasets provided most of the required information per asset (unique asset reference, gravity flow or pressure pipe type, nominal diameter, length, pipe material), but some fields had to be derived indirectly:

- GRADE; pipe slope was not recorded, and was assumed as 1% following general engineering practice
- The date of installation field CDATE was very incomplete and was inferred with the help of the DMBC land use database by referencing the asset to the typical age of the housing type in the immediate vicinity (Figure 3.2.20A).
- MAT; the pipe material was logged but the type of joint material was not available. Joint type has changed with time; pre-1960 mortar joints were used, 1960-1970 was a transitional period then post-1970 rubber ring joints became the norm. The pipe age inferred for the CDATE field was aggregated into pre- and post-1970 subsets in order to deduce the joint type according to installation date. There were also some material types that the model does not yet cater for such as brick or ductile iron; these were assigned a default concrete material type. Table 3.2.17 shows the classification scheme adopted to populate the MAT field (Figure 3.2.20B).

---

[1] The file asset_nodes.csv is required only if running MapObjects ©, which was not used for this case-study

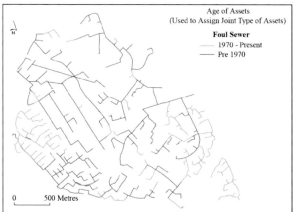
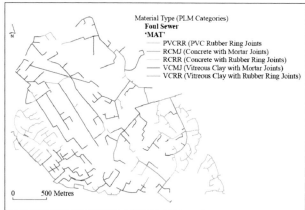

Figure 3.2.20. (A) Division of foul sewer network using GIS to assign pipe age (CDATE) and (B) joint type in MAT field.

Table 3.2.17. Pipe classification scheme to deduce joint types, Bessacarr-Cantley.

Pipe Material	NEIMO joint category inferred from pipe date, deduced from housing age	
	Pre-1960 and transitional period 1960-1970	Post-1970
Vitrified clay	VCMJ	VCRR
Concrete	RCMJ	RCRR
Asbestos cement	RCMJ	RCRR
Cast Iron	RCMJ	RCRR
PVC	PVCRR	PVCRR
Brick	RCMJ	RCRR
Ductile Iron	RCMJ	RCRR

- GROUND; a spatial query using the soil properties database permitted the first-pass division of the study area into sand- or clay –dominant matrix, with assets in each zone correspondingly referenced (Figure 3.2.21)
- NBH_ID; the neighbourhood reference number was inserted from UVQ output files. Where an asset crossed more than one neighbourhood it was assigned manually to only one according to relative length.

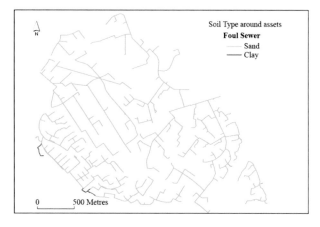

Figure 3.2.21. Assigning a soil properties theme to pipe network to determine reactive or non-reactive soil conditions.

Connections.csv

This file distributes the number of connections across neighbourhoods using the asset ID and the number of properties connected to that asset. A GIS spatial query assigned a UVQ neighbourhood ID to all assets completely within a neighbourhood polygon. The c. 5-10% of assets spanning multiple neighbourhoods were assigned manually according to relative length. Property connections were not recorded and so were assigned rationally to the smaller pipe diameters as an average number per neighbourhood:

Connections per asset = $\frac{\text{Total connections in UVQ neighbourhood}}{\text{<300 mm diameter assets count}}$

The 20 neighbourhoods in the study area comprised 34 polygons, as several UVQ neighbourhoods involved multiple areas (Figure 3.2.15). Some of these were too small to contain assets which could be allocated a connections value (Figure 3.2.22). This can occur as an artefact of the UVQ delineation or if the neighbourhood contains just a few large buildings e.g. a school. In order to represent these areas either a >300mm asset was used or a nearby asset from another neighbourhood was reassigned with the connections value and neighbourhood identity of the empty one.

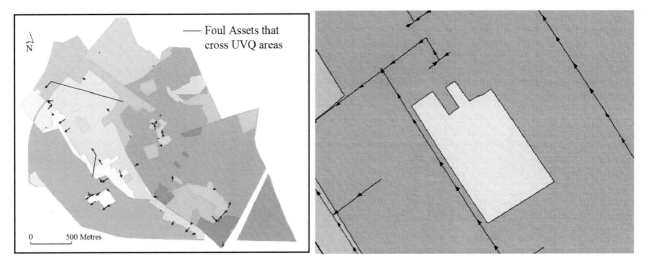

Figure 3.2.22. UVQ neighbourhoods with multiple areas and/or no assets to assign connections to.

Successor.csv

This file gives the upstream/downstream relationship between pipe assets and requires that the network conform exactly to a tree structure, with numerous house connections at the upstream extremity and a single outlet pipe at the downstream end (see Section 2.2). A number of manual corrections were required to generate a modified asset file because the sewer network in Bessacarr-Cantley, which has evolved over >80 years, does not fit the tree-structure assumption. As this is probably a common occurrence in European cities, and the compilation of this file gave the most tedious problems, the main data correction stages that were required are listed and illustrated below:

- Some assets had no upstream and others no downstream entry (Figure 3.2.23). There were various reasons for this occurrence:

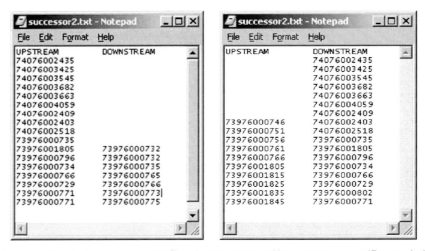

Figure 3.2.23. Incomplete entries in successor parent file; absent upstream/downstream asset IDs needed for input file to run.

The largest group, with no upstream predecessor, comprised entries for the first (most upstream) asset in a pipe run (Figure 3.2.24). This top tier had to be removed (a requirement only for the successor file, not the other model input files). A second group of entries that needed to be manually entered were connections from foul sewer to combined (and vice versa), pressure pipe connections to gravity mains and other special asset categories needed to complete a pipe run (also illustrated in Figure 3.2.24). A third group with missing downstream asset ID entries that had to be entered manually by inspection were those assets that exit the edge of the study area and then re-enter (Figure 3.2.25A). This is a product of the neighbourhood definition exercise, which will in most cases be similarly derived from a land use dataset i.e. a different source to that providing the pipe network. A final small group comprised inventory omissions; in this case the asset was manually matched with the next succeeding pipe ID (Figure 3.2.25B).

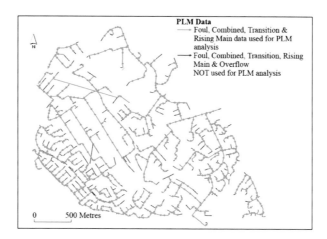

Figure 3.2.24. Successor file compilation: removal of top tier rows with no upstream predecessor.

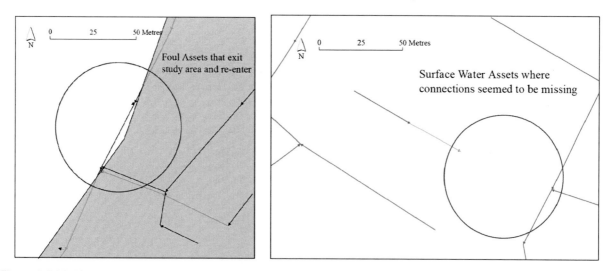

Figure 3.2.25. Manual entries for missing downstream entries due to (A) assets exiting and re-entering study area and (B) inventory omissions.

- The sewer and drainage systems in Bessacarr-Cantley also had five exit points, which by creating five separate hydraulic systems would have unnecessarily multiplied the number of model runs in the model chain. An acceptable simplification was achieved by creating a virtual asset and ID as the downstream entry for each exit pipe asset; this virtual asset represented the trunk exit main of the system in question and in effect linked the 9 UVQ regions illustrated in Figure 3.2.14A so that they became a single foul sewer (and corresponding pluvial drainage) network that conforms to the required tree structure.
- There were also a few instances where one asset flowed into two (as can happen where relief drains are installed either to provide extra capacity or to relieve storm surcharging in combined systems). In these instances one entry was manually deleted to preserve the tree structure (Figure 3.2.26).

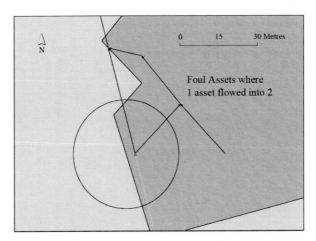

Figure 3.2.26. Manual deletion of one successor file entry was necessary where one pipe asset flowed into two.

Calibration and results of NEIMO application to study area

As there were no field leakage measurements available to calibrate the NEIMO model to actual observations, the model was only adapted to obtain the recharge concentrations observed in shallow groundwater i.e. leakage was adjusted to provide enough contaminant load to give observed shallow groundwater concentrations. This may be a way of calibrating the model in areas where no detailed observations of pipeline leakage are available, as is the case in most urban areas.

Foul sewer leakage is calculated by NEIMO and includes losses from the combined sewer system (Figure 3.2.27). The latter is a minor component in this area, comprising <4 % of the foul sewer network length of 56.9 km. Storm drain leakage is also calculated by NEIMO as a separate exercise on the pluvial drainpipe network. In Doncaster, the NEIMO used generic defect files to calculate leakage volumes from information on pipe material, age, etc.

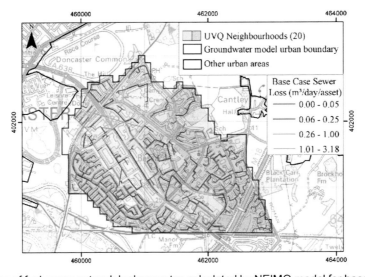

Figure 3.2.27. Visualisation of foul sewer network leakage rates calculated by NEIMO model for base case scenario.

The first results suggested that most leakage (97%) occurs through joints and only about 3% is through cracks. However, results from Rastatt suggest that the two types of leakage contribute similarly to the overall leakage volume. It was found that, even using the upper range of recommended conductivities (0.0001 m/s) of the colmation layer, insufficient leakage volumes were generated to reproduce observations of shallow groundwater contamination indicator concentrations i.e. modelled leakage was much too low, by a factor of at least 10. The crack leakage appears to be underestimated and so the colmation layer hydraulic conductivity was increased iteratively until acceptable rates were produced (K_cracks in Table 3.2.18). An acceptable result provided 35% of leakage through cracks and 65% occurred through joints. The total leakage volume is about 10% of the annual flow or related to the length of pipes: 1910 m^3/km/year or about 0.06 L/km/sec. The resulting volumes for leakage therefore have significant uncertainty, as they could only be deduced as a product of mass balance estimates without the benefit of volumetric field leakage measurements as a crosscheck of results. Nevertheless, the exercise was able to incorporate leakage in a semi-quantitative sense and thereby to successfully demonstrate model linkage

In reality the leakage (or, more precisely, the extra loadings) could originate from other locations in the wastewater removal system such as from house-to-street connections or from sewage overflows after sever storm events, neither of which are considered in the model chain. Another possibility is that current set of generic curves systematically underestimates the crack losses. The volumes displayed in Figure 3.2.27 and the leakage parameters displayed in Table 3.2.18, therefore, should be seen only as indicative values because they incorporate other leakage sources/pathways.

Table 3.2.18. List of model parameters used in NEIMO.

DeltaL A [m]	0.010	Mannings N	0.013
DeltaL B [m]	0.010	K_cracks [m/s]	0.001
DeltaL C [m]	0.008	K_joints [m/s]	0.0001
DeltaL D [m]	0.006		

3.2.3.3 UL_FLOW

The unsaturated flow model (UL_FLOW) was applied in the case study to assess both average and minimal residence times of water in the unsaturated zone. The latter could be used as an indicator for the local risk of contaminating groundwater. Note that this model only considers spatial recharge occurring from open space, septic tanks and rain - or grey water infiltration structures. The model requires soil properties and depth of the unsaturated zone for each neighbourhood, and in effect the soil zone thickness is equated to that of the unsaturated zone. Soil properties were taken from the local soil properties data provided by the National Soil Resources Institute NSRI and include spatial information on different lithologies, soil textures and corresponding parameters of water transport, as listed in Table 3.2.19.

Table 3.2.19. List of soil properties and textures of local soil types 551, 712 and 821 in the study area.

Depth	Description*	Texture**	Van Genuchten α	Van Genuchten n	Residual porosity	Total porosity	K_{sat} [m/s]
soil 551							
0 - 0.25	A	MSL	0.0911	1.3487	0.04	0.53	4.52E-05
0.25 - 0.55	Bw1	LS	0.237	1.4396	0.04	0.459	7.17E-05
0.55 - 1	BC	S	0.125	1.4438	0.03	0.478	5.64E-05
>1	C	S	0.158	1.574	0.03	0.463	8.13E-05
soil 712							
0 - 0.15	A	C	0.0432	1.2173	0.06	0.592	1.14E-05
0.15 - 0.5	Bg1	C	0.0386	1.2092	0.06	0.497	2.80E-06
0.5 - 0.7	Bg2	ZC	0.0344	1.2048	0.08	0.454	6.90E-08
0.7 - 1.3	BC	C	0.0352	1.2045	0.08	0.443	9.20E-08
>1.3	C	ZC	0.0349	1.2061	0.08	0.41	5.70E-08
soil 821							
0 - 0.15	A	MSL	0.0844	1.3234	0.06	0.574	4.64E-05
0.15 - 0.45	Bg1	LS	0.1001	1.3733	0.05	0.509	6.87E-05
0.45 - 0.8	Bg2	LS	0.1166	1.422	0.03	0.41	3.44E-05
0.8 - 1	BC	S	0.1291	1.455	0.03	0.456	5.32E-05
>1	C	S	0.1489	1.5181	0.03	0.445	6.42E-05

* Soil horizons used in standard soil descriptions
** Soil texture; C Clay, MSL medium sand loam, LS Loamy sand, S sand, ZC silty clay

The depth of the unsaturated zone was calculated by GIS analysis using the ground elevation from the NextMap© digital terrain dataset and the water table surface. The latter was locally refined using data from the project's own research boreholes supplemented by Environment Agency regional observations and produced from the ground water flow model. Figure 3.2.28 shows the resulting depths of the unsaturated zone in Bessacarr-Cantly. A statistical analysis of the results provided average, minimal and maximal depths of the unsaturated zone in each neighbourhood (Table 3.2.20).

Figure 3.2.28. Map displaying depths of the unsaturated zone in the study area. Black lines denote neighbourhood boundaries.

Table 3.2.20. Statistical description of depth distribution and corresponding soil types in all 20 neighbourhoods.

UVQ	Mean	St. dev.	Min.	Max.	Soil type	UVQ	Mean	St. dev.	Min.	Max.	Soil type
1	1.28	0.35	0.69	2.04	soil 712	11	13.92	2.54	6.44	17.83	soil 551
2	13.32	2.14	5.84	17.19	soil 551	12	9.61	4.2	3.38	16.68	soil 551
3	3.24	0.91	1.51	4.7	soil 551	13	10.21	2.04	6.15	12.85	soil 551
4	14.54	1.83	3.59	16.95	soil 551	14	4.01	0.45	3.28	5.33	soil 551
5	6.96	2.67	2.6	13.06	soil 551	15	9.61	3.25	3.44	16.95	soil 551
6	3.76	0.66	1.47	4.87	soil 821	16	8.82	3.89	1.77	18	soil 551
7	5.89	1.03	4.31	7.94	soil 551	17	3.91	2.07	0.52	12.49	soil 551
8	12.88	1.11	10.86	14.36	soil 551	18	7.35	1.61	4.26	11.65	soil 551
9	2.83	0.55	2.05	3.92	soil 821	19	4.31	1.85	1.78	9.16	soil 551
10	9.28	2.41	5.15	14.87	soil 551	20	4.6	0.9	3.29	7.53	soil 551

As the model assumes a constant depth of each UVQ neighbourhood it could be considerably improved by splitting each neighbourhood into sub-areas with smaller depth ranges. For example, a GIS overlay of UVQ neighbourhoods with one-meter contours of depth map would provide areas with depth ranges of only one meter and would, therefore, more accurate predictions of risks to groundwater.

3.2.3.4 Saturated zone flow and contaminant transport models

The saturated zone flow and contaminant transport models are the final stage in the model chain and for the Doncaster case study MODFLOW and the solute transport module MT3D were employed.

The groundwater flow model:

The groundwater flow model was derived from the regional water resources management model established in 1993 by I T Brown and K R Rushton from the University of Birmingham (Brown and Rushton 1993), extended and slightly modified in 1997 by M Shepley of the Environment Agency (Shepley 2000). The original model is regarded as a well-calibrated regional groundwater model that adequately represents aquifer conditions in the Doncaster area. The following stages were undertaken to provide a suitable platform for the modelling of groundwater flow and solute transport in the detailed study area and especially to facilitate the running and comparison of different urban water scenarios:

The original 2-D model was translated into an equivalent 2-D MODFLOW transient model. This process is described in Neumann and Hughes (2003) and included re-discretization from the original nodal mode to a block-centred grid of 1 km by 1 km (Figure 3.2.29A). The original model simulates the period 1970-1997 and the translated MODFLOW version simulates the same period of time using monthly stress periods divided into four time steps. The hydrologically/hydrogeologically -defined boundaries of the 1020 km² regional MODFLOW model and other features were retained. Groundwater hydrographs generated to compare the MODFLOW modelled output with the original model data were generally in very good agreement, and the resultant model was accepted as a faithful translation of the original.

(1) In order to provide higher resolution in the focus study area, a 9.25 km² sub-regional area corresponding to Bessacarr-Cantley was re-discretized in the MODFLOW 2-D transient model as a 100 m by 100 m grid. (Fig 3.2.29B). The urban boundary was also approximated within this fine-mesh grid so that any part of the detailed study area could be referenced to a particular groundwater flow grid cell address (Figure 3.2.29C). New, more detailed water level data collected during the field study programme enabled re-calibration of water levels in this area.

(2) A steady-state version of the transient model was created; this model used 27-year averages of rainfall recharge, urban recharge, pumping volumes, water levels in the Quaternary cover and regional boundary inflows and outflows derived from the original simulation time period 1970-1997. An acceptable approximation of head distribution in the study area was achieved (Figure 3.2.30). A steady state version by its very nature is an average and cannot represent seasonal or other time-dependent changes in input or output but it has the advantage of avoiding the additional complexity of transient changes (e.g. seasonal or annual changes in abstraction or rainfall recharge) whose effects would also have to be taken into account when comparing different scenarios. Steady-state simulations were also quite suited to scenario modelling because input parameters could be quickly changed, runs could be performed relatively rapidly and results readily compared.

(3) As a final stage the steady-state two-dimensional version was converted into a three-dimensional version by the division of the aquifer into four layers (Figure 3.2.29D). The main reason for this step was to facilitate meaningful modelling of contaminant transport. The final version used for the scenario modelling was therefore a 3-D steady state regional model with extra sub-regional discretization in the vicinity of the detailed study area.

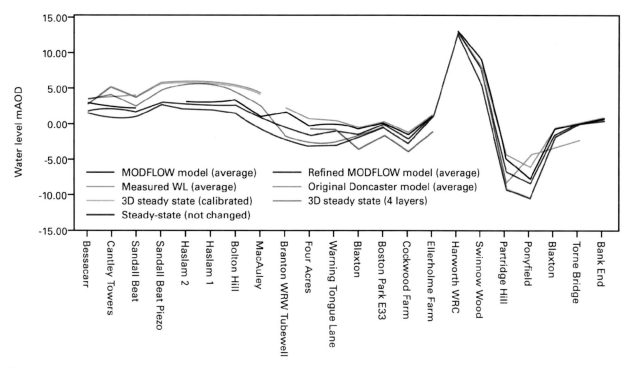

Figure 3.2.30. Comparison of heads at reference observation wells distributed across the Doncaster regional model for the various stages in the evolution of the final model used for scenario comparison (3-D steady state).

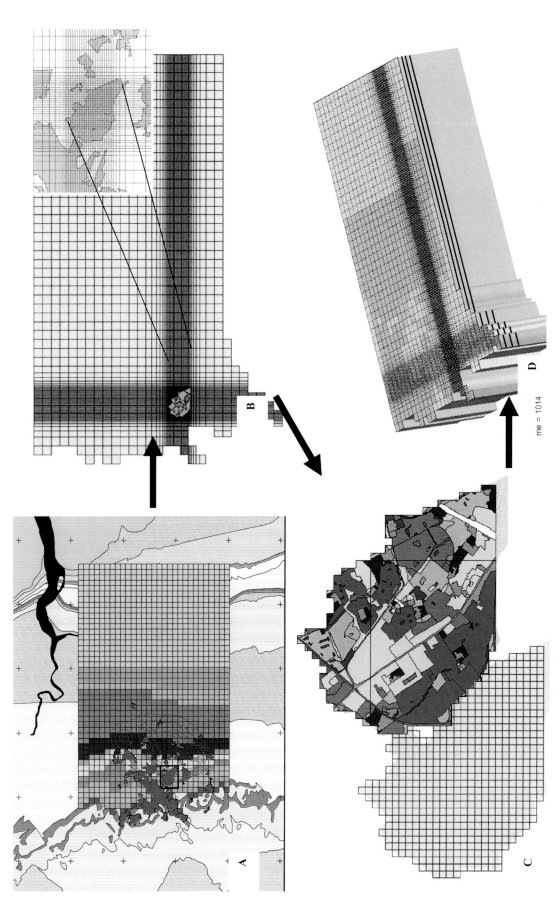

Figure 3.2.29. Evolution of groundwater flow model for Doncaster case-study: (A) the original 2-D transient regional water resources model (B) translation into MODFLOW and refinement, including discretization to 100 m cells around study area (C) conversion to 2-D steady –state (D) separation into layers to provide 3-D steady state version.

For comparison of alternative water use scenarios, a groundwater flow base-case was created. This is a steady-state simulation that uses the annual averages of recharge and pumped abstraction and river loss/gain for the transient model calibration period of 1970-1997 throughout the regional model except in the nodes within the study area boundary. Head difference contours were produced using the contouring package Surfer™ via a spreadsheet file that compared the exported head distribution output files of the particular scenario with the base-case values.

The groundwater contaminant transport model:

The MT3D code was used to model saturated zone contaminant transport in the Doncaster case study. MT3D employs heads, fluxes across cell interfaces in all directions, and locations and flow rates of various sinks and sources as solved by a calibrated numerical groundwater flow model, so it can borrow the same numerical grid structure used in the MODFLOW model, with which it readily interacts. MODFLOW output files, provided via the Groundwater Vistas™ (GV) pre- and post-processor software, were employed for the solute transport modelling, which used MT3D executable files.

So, for the Doncaster case study, the solute transport numerical model emulates the flow model and is therefore a 3-D, four-layer regional model that includes the extra sub-regional discretization in the vicinity of the Bessacarr-Cantley study area. Although the groundwater flow model is run in steady state mode and produces constant head values, the simulation of contaminant movement is transient and produces contaminant concentration values for the model cells at specified time steps. The length of each individual time step is dictated by the conditions that satisfy the stability of the explicit part of the solute transport numerical solution. This associates a concentration with each of the moving particles and at the end of each time increment, the contaminant concentration at a cell is evaluated from the concentrations of moving particles which happen to be located within the cell. Aquifer porosity affects the real velocity of the groundwater flow and consequently the movement of the contaminant and the location of particles. In this model, the aquifer porosity values are assumed to be identical to the specific yield values used in the flow model.

Two representative contaminant indicators were used to demonstrate urban recharge quality effects. Potassium and chloride are common urban recharge markers and the field monitoring had shown both to be elevated above rural background in shallow ground water beneath the study area (see Table 3.2.8). Both are conservative contaminants, i.e. they do not decay. However, it is assumed that the movement of these contaminants through the porous medium is affected by hydrodynamic dispersion. The coefficient of hydrodynamic dispersion is defined as the sum of the coefficient of diffusion and the coefficient of mechanical dispersion. Bulk diffusion can be estimated reasonably well for granular media; however, a realistic quantification of the mechanical dispersion coefficient and its relative contribution is difficult (Domenico and Schwartz, 1998). Values of dispersivity are observed to be scale-dependent (Zheng and Bennett, 1995). Values of 75 m, 25 m and 2.5 m were selected as regional values for the longitudinal, transversal and vertical dispersivities respectively. A diffusion coefficient of 8.64×10^{-6} m2/d (Freeze and Cherry 1979) is also used in the solute transport simulations.

The model boundary is defined by the impermeable boundary specified in the flow model where both the dispersive and advective fluxes are equal to zero. MT3D allows the inclusion of a specified-concentration boundary which acts as either a source providing solute mass to or a sink taking solute mass out of the simulated domain. This useful feature allowed cells representing urban areas (Doncaster and other adjacent urban areas) in the regional model that are nearby but outside the sub-regional discretized zone to be assigned a loading too. The average of all-recharge value for all Bessacarr-Cantley neighbourhoods for the given scenario was used in these specified-concentration boundaries. To set the initial conditions of the contaminant concentrations a background concentration of the contaminant, selected from observed field values, was set everywhere in the model domain. A numerical simulation for a period of time of 70 years was then undertaken and the produced concentration results were used as initial conditions for subsequent simulations.

3.2.4 Scenario modelling results and related topics

The Decision Support System permits the previously described models to be linked in a chain to produce a simulation of the actual situation. This status quo condition became the base case, which is derived from the actual situation as follows:

- The UVQ model output run represents the reference year 1997.
- The NEIMO model input and output represent the same reference year 1997 and are daily runs incorporating a distribution of daily pluvial drainage flows during the year at dry flow, average, high flow and flood condition days at the ratio 269:78:18:1, according to the statistical distribution in the total observation period (1970-2004).
- No unsaturated zone model was required as the groundwater model was run to steady-state conditions.
- The groundwater flow model was calibrated in transient mode for the years 1970-1997, then a steady state version used to represent equilibrium conditions (see previous section).

- The groundwater solute transport model was run in transient mode to represent the evolution of contaminant distribution. The first 70-year stage represents the evolution of aquifer from the late 1920s as it responded to urbanisation, while the second 100-year stage represents a prognosis for the future with a continuation of the urban recharge loadings under present conditions i.e. the contaminant concentrations were applied for 170 years starting from pre-urbanisation rural background conditions in 1927. Although this is a simplification, the vast majority of the housing stock in this previously rural district post-dates the mid-1920s.

The base case is the first of the five scenarios illustrated in the following pages and was the basis against which four different water and contaminant balance scenarios were compared. These are representative scenarios devised to qualitatively illustrate the effect on the underlying aquifer of changes that are outside the immediate control of local stakeholders, such as climate change or national trends of changing demand (exogenous effects) and those brought about by engineering or regulatory interventions, such as pipe renovation or better use of roof runoff (Figure 3.2.31).

Action Scenario	Scenario description	Exogenous effects			
		Climate change		Demand change	
		2020 Moderate (likely case)	2080 (Worst case)	Alpha 2025 (likely case)	Gamma 2025 (best case)
0 Actual situation; no additional action planned	1 Base case; Status quo used for calibration and comparison with other scenarios	2A	2B	3A	3B
A Sewer rehabilitation B Reuse of water (by 100% of households)	A All sewers in pre-1945 areas renovated				
	B1 Roof runoff for garden watering				
	B2 Roof runoff for garden watering & toilet flush				
	B3 Grey water from bath/laundry for toilet flush				
Implication of scenario →	↑Existing climate and wellfield pumping rates	↑Slightly lower recharge	↑Much lower recharge	↑Slightly higher water demand	↑Much lower water demand

Figure 3.2.31. Scenario modelling for the Doncaster case study. Grey areas show the scenarios discussed below.

For the base case, the predicted water table contour map is shown, and to facilitate comparison, the equivalent map for each of the other four scenarios is compared to this head distribution as water table difference contours at completion of each flow model run. For these latter four scenarios the contaminant transport model is also run with an initial 70 year actual situation stage and then the corresponding scenario condition for the next 100 years, to represent the effect of change from the present day.

Each scenario set thus comprises a water balance diagram, a water table difference contour map (water table elevation map for the base case scenario), trend graphs for potassium and chloride contaminant indicators for a central location in Bessacaar-Cantley (Haslam Park) and for the closest downgradient borehole of the public supply wellfield (Nutwell 2), and a comments section.

Case study Doncaster
Base case

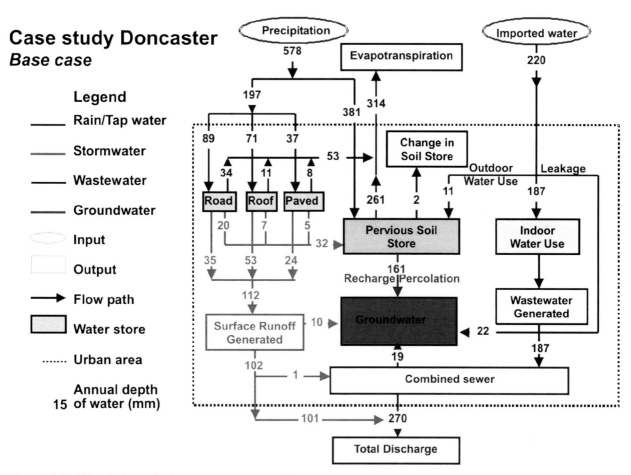

Figure 3.2.32. Water balance for Doncaster case-study: the base case (all values in mm/a equivalent recharge).

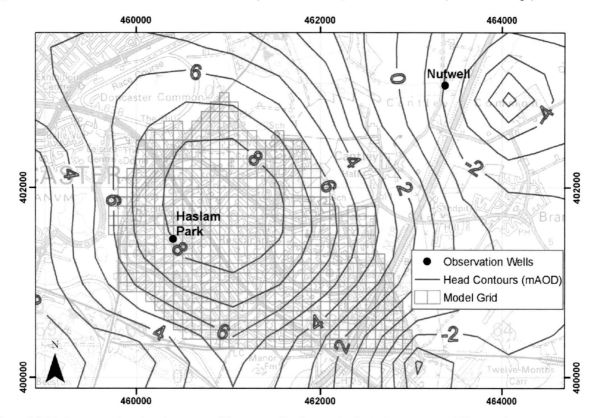

Figure 3.2.33. Base case head contour map of Bessacarr-Cantley; nodes in study area are at 100m spacing

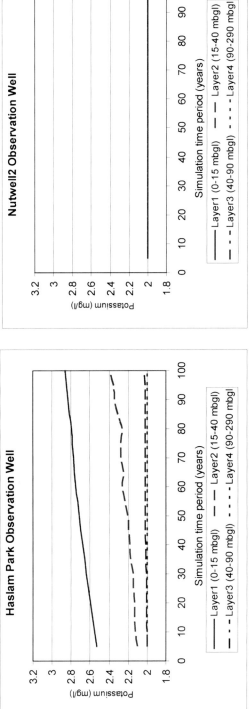

Figure 3.2.34. Potassium trends at sample location within urban area.

Figure 3.2.35. Potassium trends at down gradient location near public supply boreholes.

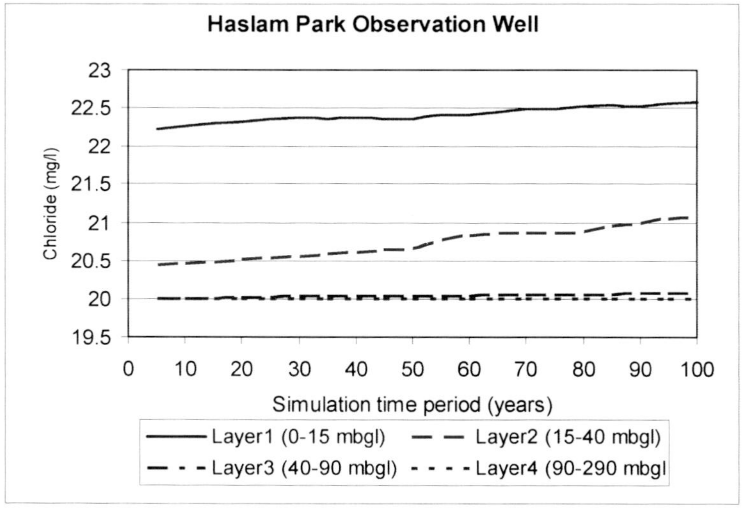

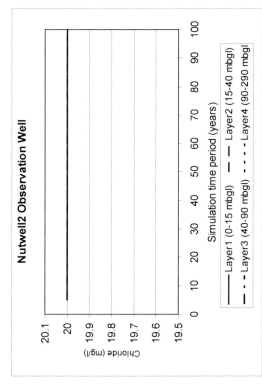

Figure 3.2.36. Chloride trends at sample location within urban area.

Figure 3.2.37. Chloride trends at down gradient location near public supply boreholes.

Comments; the Base Case (1)

- Local head contours (Figure 3.2.33) are strongly influenced by the effects of pumping from the nearest of the 11 public supply boreholes in the Doncaster wellfield (which comprise >90% of pumped withdrawals from the aquifer). The composite cones of drawdown of two of these of 11 pumping stations (Nutwell/Thornham and Rossington Bridge) can be seen to the NE and SE of the study area. A further influence is the presence or absence of superficial deposits, the less permeable of which curtail natural recharge from rainfall. The aquifer forms a small area of raised topography and is either at outcrop or beneath thin but permeable cover beneath almost all of the study area. The interaction of these two contrasting effects is thought to contribute to the continued presence of a groundwater mound, which is about 3m above the equivalent area to the northwest.
- MT3D simulations in transient mode are employed using the steady state model for head distribution with a run time of 100 years into the future. These 100-year simulations of the conservative indicators of potassium (K) and chloride (Cl) illustrate the probable future effect of urban recharge in the study area if conditions are unchanged. For these simulations the K and Cl concentrations were set at observed average rural background values in the surrounding rural area (2.0 mg/l and 20.0 mg/l respectively). For the base case average recharge concentrations in the focus study area ranged from 2.66 to 5.43 mg/l for K and 13.5 to 57.3 mg/l for Cl across the neighbourhoods. These were applied as recharge concentrations.
- The influence of other urban areas in the model was also taken into account by assigning the average contaminant value from the focus study area to all other dominantly urban nodes in Doncaster (4.36 mg/l for K, 32.1 mg/l for Cl).
- Groundwater flow in the detailed study area occurs both laterally and vertically, being strongly driven by the abstraction wells which are located downstream and penetrate up to 150m into the saturated aquifer. Laterally the peak concentrations of the contaminant plumes for the indicators progress at rates of about 40-100 m/a. The rate of vertical penetration from the topmost layer is estimated to be less than 3 m/a and can be observed in the Haslam Park graphs (Figs. 3.2.34, 3.2.36) as the additional effect on indicator concentration below the uppermost layer, at the top of which recharge is applied. Note that both upper layers (1 and 2) for both indicators start the simulation period with a slow upward trend. This is because they are still responding to the loading applied over the previous 70 years when urbanisation was under way. Layer 3 starts to respond after about 15 years (equivalent to 70+15=85 years of urban recharge application), equivalent to a downward vertical contaminant movement of the around 0.75 m/a.
- The Cl and K start-concentration values of c. 22.2 mg/l and 2.53 mg/l respectively for Layer 1 are comparable to, but somewhat lower than, the observed field measurement values from the corresponding upper 15-20m zone of the two Haslam Park multilevels. Sample analysis results from the latter averaged 25.9 mg/l and 4.9 mg/l respectively. This is considered an acceptable result for calibration purposes because the corresponding 100-year concentration contours indicate that higher values of 24-30 mg/l Cl and 3-3.5 mg/l K do indeed occur in the central part of the study area, just east of the chosen observation point.
- The implications for groundwater quality management is shown in both indicator cases (Figs. 3.2.34-3.2.37); the rising trend is subdued and limited to the uppermost few tens of metres of saturated aquifer within the urban area (Haslam Park), while no trend can be detected even after 100 years at the surrogate wellfield receptor (Nutwell 2).

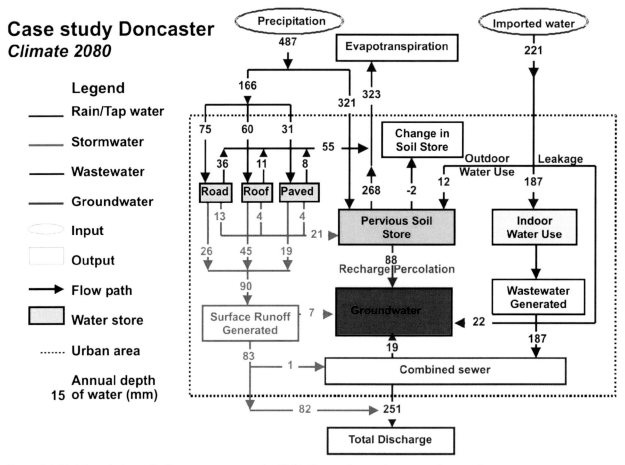

Figure 3.2.38. Water balance for Doncaster case study: 2080 climate change (worst case) scenario.

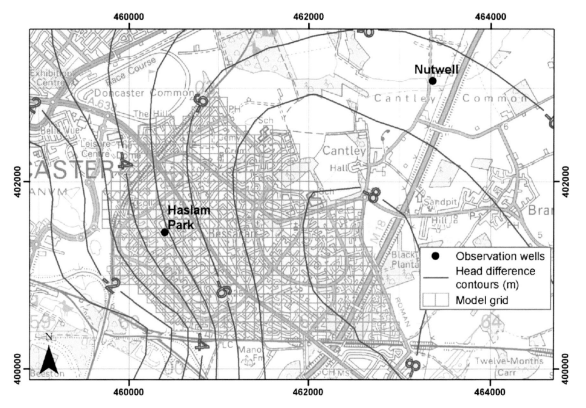

Figure 3.2.39. Climate change (2080, worst case) contour map of head difference from base case, Bessacarr-Cantley.

Urban Water Resources Toolbox 177

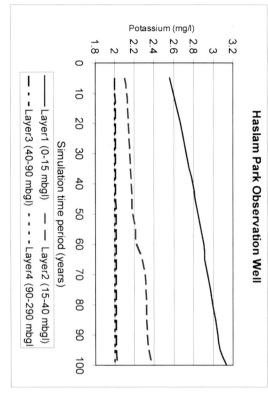

Figure 3.2.40. Potassium trends at sample location within urban area.

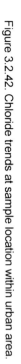

Figure 3.2.42. Chloride trends at sample location within urban area.

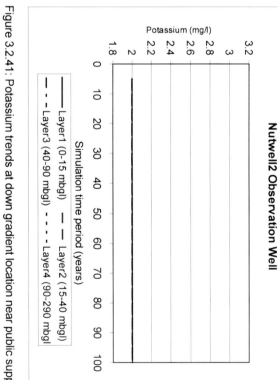

Figure 3.2.41: Potassium trends at down gradient location near public supply boreholes.

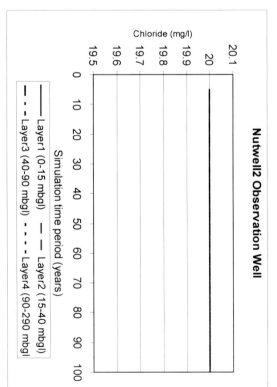

Figure 3.2.43. Chloride trends at down gradient location near public supply boreholes.

Comments; the 2080 climate change (worst case) scenario (2B)

This climate change scenario is one of two that were modelled by the project team for Doncaster. It uses data for the UVQ climate file from the UK Climate Impact Programme UKCIP (precipitation) and from the UK Centre for Ecology and Hydrology CEH (potential evapotranspiration PE). The illustrated scenario is for the High Emissions (worst-case) scenario using predicted climate data for the period 2071-2100. This scenario predicts slightly wetter and warmer winters and hotter and much drier summers.

- The changes in amount and seasonal distribution of rainfall and PE to the input climate files can be observed in the water balance (Figure 3.2.38), where compared with the base case total rainfall is reduced by over 15% and PE by 3%. Although these are modest changes, they produce larger and more long-lasting soil moisture deficits, which strongly reduces groundwater recharge (55% of the base case value).
- Note that for the climate scenarios only the rainfall magnitude is amended in the climate files; the scenarios do not take into account changes in event frequencies and intensities. Both are factors that could strongly affect the magnitude of groundwater recharge. Furthermore, the UVQ model was not validated against unsaturated zone measurements (e.g. using a lysimeter). As a consequence, though it accurately schedules the destinations of mains water and precipitation through the urban settings, it does not necessarily represent the reality of annual recharge distribution. Such uncertainties need to be considered before basing decisions on the results.
- Figure 3.2:39 shows contours of head difference between this scenario and the base case. The strong reduction in recharge through gardens and green space causes widespread falls in head which are most marked along the eastern edge of the study area.
- The result is to depress water levels towards sea level across much of the area, with below sea level values in the east. The easterly location of maximum fall reflects the effect of abstraction from the nearby wellfield public supply boreholes of Nutwell, Thurnham and Rossington Bridge as their composite cones of depression interacting with the original groundwater mound centred on Bessacarr-Cantley.
- The reduced availability of dilution from the pluvial recharge component will result in higher concentrations of K and Cl in total recharge (the main source of these contaminant markers is from sewer leakage). This can be identified in the higher average K and Cl recharge concentrations generated by the mass balance calculation of UVQ components (Table 3.2:21) and observed as an increased rate of rise in shallow groundwater concentrations for both markers below the urban area of 10-23% compared to the base case simulation (Figs 3.2:40 and 3.2:42).
-

Table 3.2:21. Comparison of neighbourhood output contaminant concentrations for K and Cl in UVQ, climate change scenario.

Scenario	Average concentration in mg/l for all neighbourhoods*	
	Potassium (mg/l)	Chloride (mg/l)
Base case	4.37	32.1
Climate change 2080	5.93	40.6

*n=20

- As both markers start from concentrations that are very low compared with drinking water guidelines, there would only be a public health significance by analogy, if similarly persistent contaminants were present at concentrations approaching maximum admissible limits and the change in rate of contaminant accretion in the upper aquifer provoked exceedance.
- Only the upper two layers show the effect of contaminant loading, and the rate of rise is not marked, while effects of penetration below the upper few tens of saturated aquifer is negligible. This is not an artefact of the layering arrangement adopted in the model as isotropy has been assumed (Kh/Kv ratio of 1), but instead demonstrates both the weak vertical head gradients that exist (a feature confirmed by field observations of heads in the 5 project multilevels) and the strong dilution effect of porewater in this high-porosity sandstone aquifer. As vertical dispersion coefficients were considered to be higher than normally assumed (due to predominantly vertical flow) a wider range of horizontal dispersion rates would produce a stronger effect on the deeper aquifer layers.
- At the wellfield indicator location (Nutwell, Figs 3.2:41 and 3.2:43) no effect is observed in any layer.
- In summary, the observations suggest for this scenario suggest that although the water levels are markedly depressed, there is not a particularly adverse effect on water quality, either beneath the study area or downgradient at the wellfield.

A sandstone aquifer: Doncaster, UK

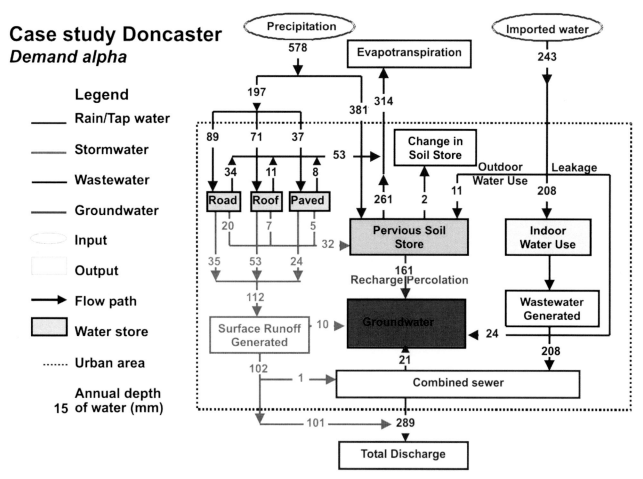

Figure 3.2:44. Water balance for Doncaster case study: the 2025 demand increase (alpha) scenario.

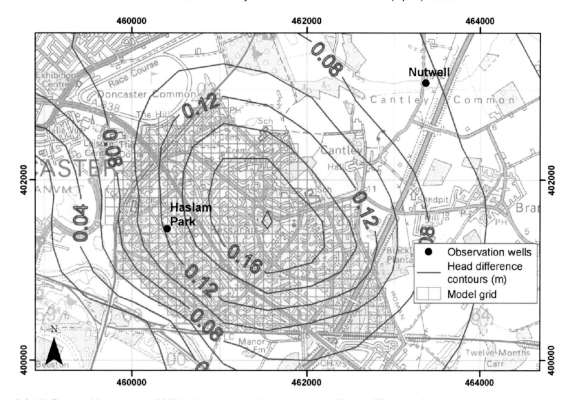

Figure 3.2:45. Demand increase to 2025 (alpha scenario) contour map of head difference from base case, Bessacarr-Cantley.

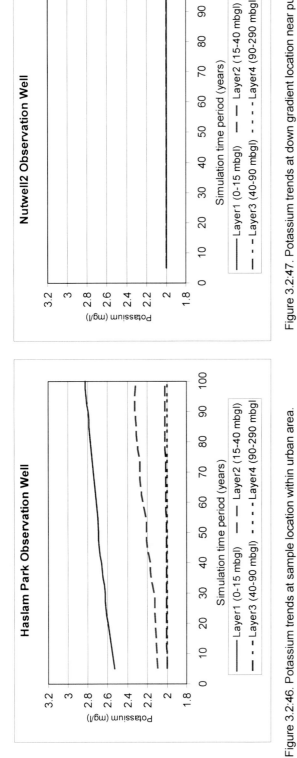

Figure 3.2:46. Potassium trends at sample location within urban area.

Figure 3.2:47. Potassium trends at down gradient location near public supply boreholes.

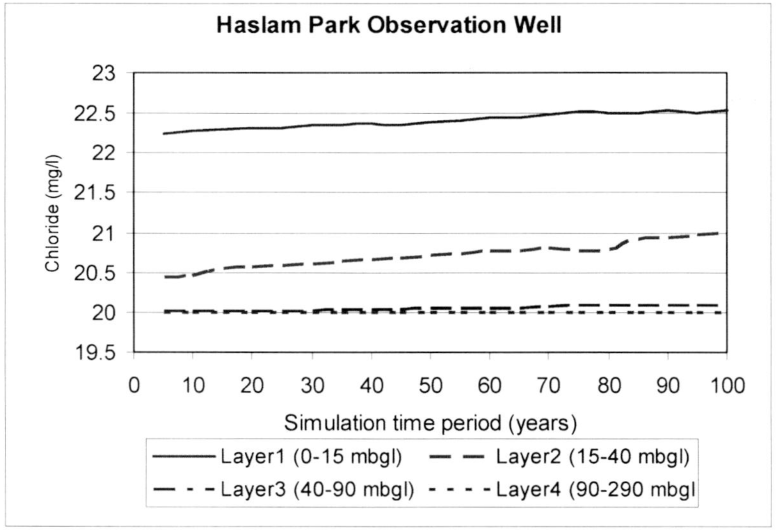

Figure 3.2:48. Chloride trends at sample location within urban area.

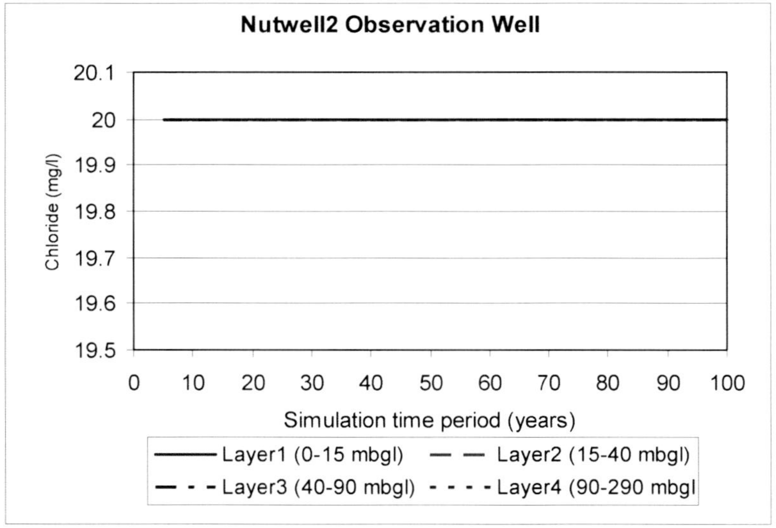

Figure 3.2:49. Chloride trends at down gradient location near public supply boreholes.

Comments; the demand increase (alpha) scenario (3A)

This demand change scenario is one of the two that were modelled by the project team for Doncaster. These were drawn from four future water demand scenarios described in the Environment Agency national strategy document 'Water Resources for the future' (EA 2001). All four water resource-planning scenarios factor in drivers of demand (cost, household usage, leakage, economic and industry trends, irrigation needs) to assess future water demand. These are based on four models of possible or desired societal change used for national foresight planning (Table 3.2.22). The alpha scenario is illustrated here.

Table 3.2.22. Societal change and consequent water demand scenarios used for national planning in UK (modified from EA 2001).

National foresight planning scenarios →	Provincial enterprise	World markets	Local stewardship	Global sustainability
EA water demand scenarios →	Alpha	Beta	Delta	Gamma

Household water consumption patterns for each scenario were incorporated in UVQ as changes in water usage. Compared with the base case, scenarios alpha and beta increased water consumption by 14% and 9% respectively, while scenarios delta and gamma both reduced it by c.25%. Scenarios alpha and gamma were modelled and the highest water demand scenario (alpha) is illustrated in this example.

- The increased household demand is illustrated in the water balance (Figure 3.2.44), with an 11% increase in wastewater generated. However, while the impact on wastewater drainage and treatment facilities would be marked, the effect on the underlying aquifer is slight, because the extra sewer and combined sewer leakage arising from the increased flows only adds an extra 2% to total groundwater recharge.
- Groundwater head change in the underlying aquifer is negligible at <0.2m (Figure 3.2.45). In reality, a possible consequence of the higher water consumption envisaged in the alpha scenario, with its consequent increased mains water demand, might be increased abstraction from the supply wells in the vicinity. For simplicity's sake this eventuality was not incorporated in the scenario, but depending on the location and extent of increased pumping, groundwater heads on the northeast or southeast of the study area would fall.
- Similarly, for the urban observation well (Figs 3.2.46, 3.2.48) there is no discernible effect on urban contaminant markers concentrations or penetration compared with the base case because, because as Table 3.2.23 shows, K and Cl concentration maxima and minima for the 20 study area neighbourhoods is virtually unchanged from the base case version.
- As with the base case, no effect is observed in any layer at the wellfield surrogate observation well.(Figs 3.2.47, 3.2.49)

Table 3.2.23. Comparison of UVQ neighbourhood output contaminant concentrations for K and Cl, alpha scenario.

Scenario	Potassium (mg/l)		Chloride (mg/l)	
	Max*	Min*	Max*	Min*
Base case	5.43	2.66	57.3	13.5
Alpha	5.32	2.66	56.7	13.4

*n=20

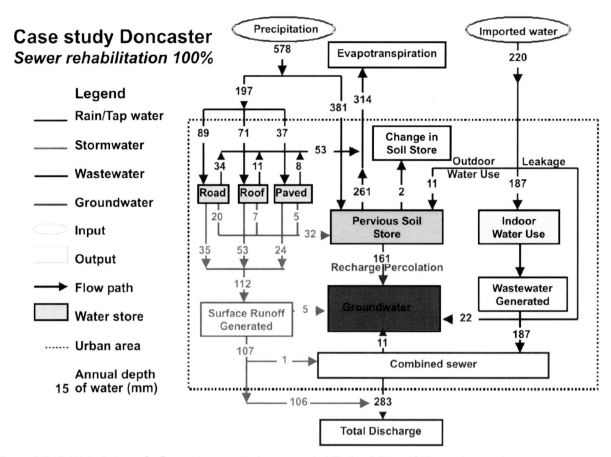

Figure 3.2.50. Water balance for Doncaster case study: sewer rehabilitation (all pre-1945 areas) scenario.

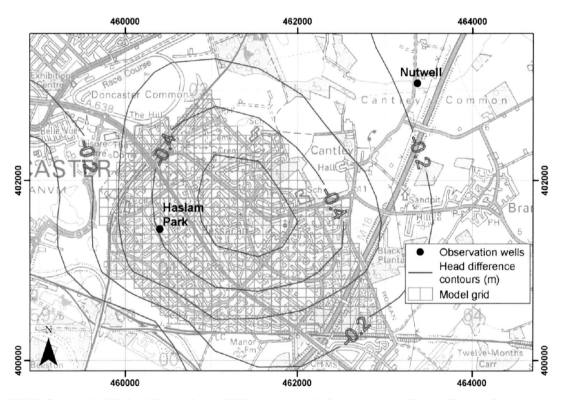

Figure 3.2.51. Sewer rehabilitation (all pipes in pre-1945 areas renovated) contour map of head difference from base case.

A sandstone aquifer: Doncaster, UK

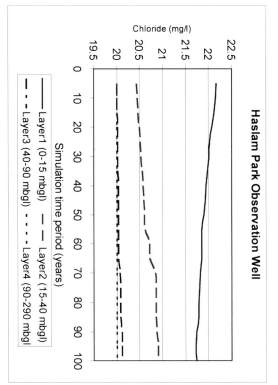

Figure 3.2.52. Potassium trends at sample location within urban area.

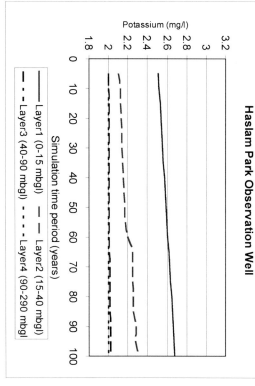

Figure 3.2.53. Potassium trends at down gradient location near public supply boreholes.

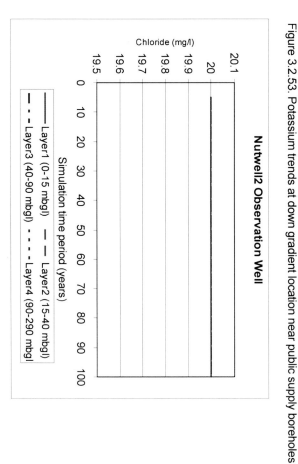

Figure 3.2.54. Chloride trends at sample location within urban area.

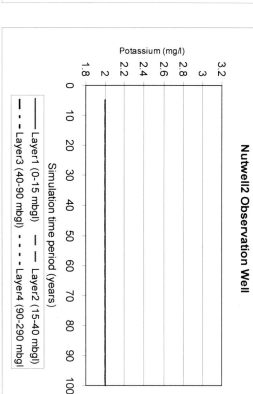

Figure 3.2.55. Chloride trends at down gradient location near public supply boreholes.

Comments; the sewer rehabilitation (all pre-1945 pipe arrays) scenario (A)

This scenario illustrates an engineering intervention in the urban water system by modelling a simple pipe renovation option. In this scenario, the inputs to and outputs from UVQ are unchanged from the base case and instead the input to the NEIMO is amended.

For this demonstration scenario the input files were amended to simulate the renovation (replacement or relining) of all pre-1945 sewers, irrespective of material, diameter or location. About 36% (19.4 km) of the total foul and combined sewer system and 33% (25 km) of the stormwater drain network were affected (Figure 3.2.56). As about half of the pipe assets do not have an installation date recorded, the pre-1945 pipe subset was identified by assuming that all undated pipes were contemporaneous with the date of urbanisation. Then all pipe assets of that age in districts with a predominantly pre-1945 housing stock were renovated. For simplicity of approach no phasing was adopted and it was assumed that all renovation was completed prior to the simulation.

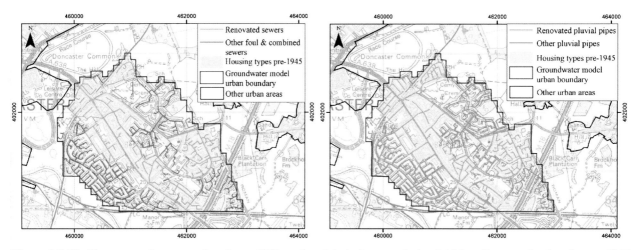

Figure 3.2.56. Pipe renovation scenario; all pre-1945 and undated pipe assets in districts with a predominantly pre-1945 housing stock were renovated.

- The water balance (Figure 3.2.50) shows that the contribution to total recharge of these two pipe networks is reduced by 45% as leakage from old pluvial and wastewater pipe network assets is curtailed.
- The impact on total recharge is much more modest, because the two largest components (garden/green space infiltration and mains leakage) remain unchanged, and the total only reduces by 6%. In consequence, the effect on groundwater contours is muted; the mound observed in the base case is slightly subdued as heads fall across the area by up to 0.5 m. (Figure 3.2.51).
- As expected, the reduction in wastewater leakage has a beneficial effect on net recharge water quality (Table 3.2.24) and this in turn brings an observable benefit to shallow aquifer water quality beneath the study area; chloride concentrations fall (in effect staying in the background concentration range) and the rate of rise in potassium concentration is slowed compared to the base case (Figs 3.2.52, 3.2.54). The slightly different behaviour of the two tracers originates from their different loads in the two sewer pipe networks compared to the other recharge sources. Chloride, for example, mainly originates from the stormwater (road runoff) and also, to a lesser degree, from the wastewater system, whereas potassium is predominantly introduced through the foul sewage route.
- There is no observable effect on the wellfield surrogate observation well (Figs. 3.2.53, 3.2.55)

Table 3.2.24. Comparison of UVQ neighbourhood output contaminant concentrations for K and Cl, sewer pipe renovation scenario.

Scenario	Potassium (mg/l)		Chloride (mg/l)	
	Max*	Min*	Max*	Min*
Base case	5.43	2.66	57.3	13.5
Sewer renovation	4.52	2.83	53.0	11.0

*n=20

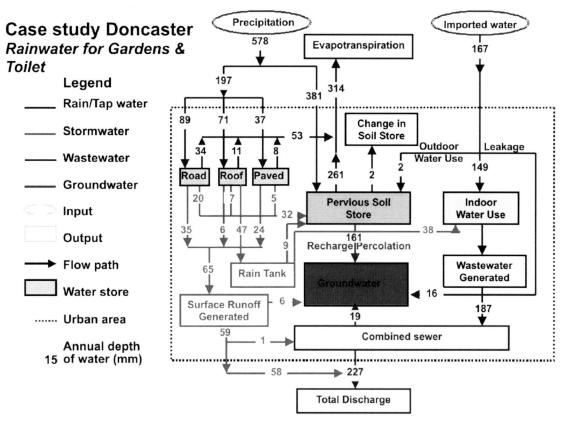

Figure 3.2.57. Water balance for Doncaster case study: re-use (roof runoff for garden watering + toilet flush) scenario.

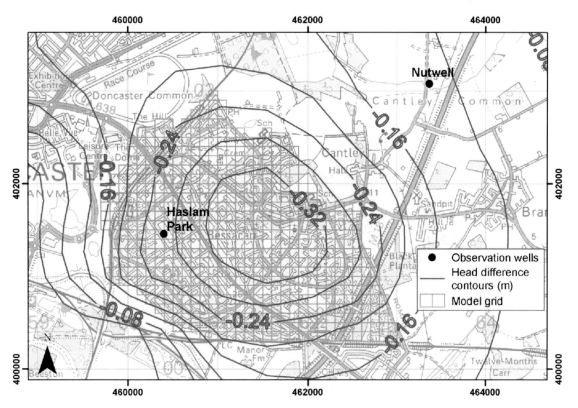

Figure 3.2.58. Water reuse (roof runoff for garden watering + toilet flush) contour map of head difference from base case.

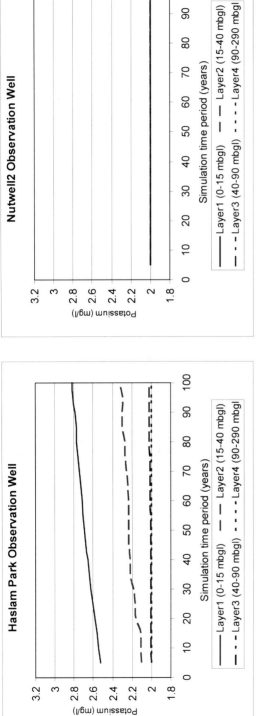

Figure 3.2.59. Potassium trends at sample location within urban area.

Figure 3.2.60. Potassium trends at down gradient location near public supply boreholes.

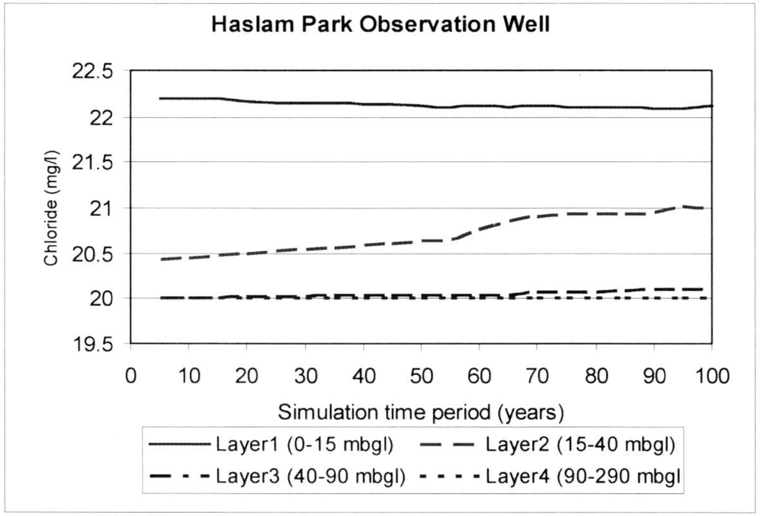

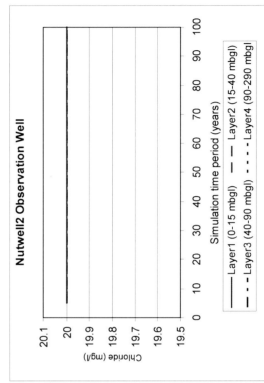

Figure 3.2.61. Chloride trends at sample location within urban area.

Figure 3.2.62. Chloride trends at down gradient location near public supply boreholes.

Comments; the water reuse scenario (roof runoff for garden watering + toilet flush) (B2)

This scenario shows an example of how the effects of different options for proactive water re-use can be explored. Four different options involving re-use were assessed. Three options used rainwater from roof runoff stored in tanks as a mains water substitute, one for garden irrigation only, one for garden irrigation and toilet flushing and one for all household uses. The fourth option had a mains water source but re-used grey water from bath and laundry for toilet flushing. These scenarios have been the subject of an economic assessment (see Section 5).

This subset of scenarios was employed to demonstrate the potential of the model chain to optimise alternative water use strategies in economic terms. For this purpose, the UVQ model was run for a series of different rain tank sizes to produce the necessary database for economic assessment. The output of UVQ included parameters important for a cost-benefit analysis (reduction of mains water consumption, number of system failures/overflows and system reliability). The scenario presented here, namely a 6000 l raintank for garden irrigation and toilet flushes was found to be the optimal strategy for Doncaster. However, the calculated amortisation time of >50 years shows its limited potential for implementation at current water pricing tariffs.

The option employing roof runoff via a tank and pump for toilet flushing and garden watering is illustrated here.

- The water balance (Figure 3.2.57) shows that although there is a reduction in mains water demand (24%) this is substituted by stored rainwater from roof runoff. Thus foul sewer leakage is unchanged but there is a reduction in pluvial drain leakage. As the mains leakage calculation in UVQ is expressed as a percentage of total supply, the leakage value is also correspondingly reduced. This is an artefact of the calculation method but even with this falsely-derived extra reduction, the total groundwater recharge reduction is still <5%.
- The impact on head contours (Figure 3.2.58) is correspondingly small, with heads reduced by 0.35 m or less across the area. In reality, there would probably be a counterbalancing rebound as abstraction from local supply wells fell in response to reduced mains demand.
- The reuse strategy does appear to have a beneficial effect, because the trend in chloride increase in the shallow aquifer is halted while the trend of increase for potassium is slightly reduced (Figs. 3.2.59, 3.2.61). This is because the main effect of this scenario, besides the reduced mains water consumption, is the significant reduction in stormwater flow and leakage, where stormwater is the main source of chloride in groundwater. Recharge bulk concentrations from the UVQ neighbourhoods are more variable, but the mean values are not greatly changed (Table 3.2.25).
- There is no observable effect on the wellfield surrogate observation well (Figs. 3.2.60, 3.2.62).

Table 3.2.25. Comparison of UVQ neighbourhood output contaminant concentrations for K and Cl, water re-use scenario.

Scenario	Potassium (mg/l)			Chloride (mg/l)		
	Max*	Min*	Average*	Max*	Min*	Average*
Base case	5.43	2.66	4.37	57.3	13.5	32.1
Water re-use	5.55	2.63	4.41	97.7	13.1	31.2

*n=20

3.2.4.1 Related scenario modelling results: using UL_FLOW to assess likely unsaturated zone residence times

The UL_FLOW model was applied to the worst-case climate scenario (2080) where both mean and minimal depths of each neighbourhood were tested for their sensitivity to the results. It was found that, as expected from the recharge rates, mean residence times for the climate scenario are about twice as much as for the base case (Table 3.2.26). However, the results for minimal residence time were striking in that minimal residence times hardly change for the worst-case climate scenario even through the average recharge is only half as much as in the base case. This is a case of compensating effects; the shortest residence times occur in winter where recharge rates are slightly increased whereas the major decrease in average recharge occurs during summer.

An additional case was modelled to show the potential of the model to estimate optimal sizes of infiltration fields (leach fields) for household grey water where all three cases assumed mean depths of the unsaturated zone. It was found that, particularly for the low-lying areas, grey water leach fields without any additional technical measures (e.g. pre-treatment) had to be quite large to maintain minimum acceptable soil residence times as a groundwater quality safeguard (Table 3.2.26). $10m^2$ leach fields, for example, would lead to minimal residence times below 100 days for most neighbourhoods whereas $50m^2$ leach fields suggest residence time between 100 and 400 days, a time that seems more acceptable to minimise microbial risk to groundwater.

Table 3.2.26. Minimal (min.) and mean residence times (in days) of different scenarios for all 20 UVQ neighbourhoods (NH).

NH	Base case mean depth		Base case min. depth		Climate 2080 mean depth		Climate 2080 min. depth		$10m^2$ leach-fields		$20m^2$ leach-fields		$50m^2$ leach-fields	
	min. days	mean days	min. days	mean days	min. days	mean days	min. days	mean days	min. days	mean days	min. days	mean days	min. days	mean days
1	39.6	937.2	21.1	536.8	37.9	1819	21.1	652.4	14.3	30.5	20.3	60.5	26.7	131.5
2	175	3600	515	1751	1802	4911	545.7	2599	108.9	215.0	188.9	208.8	357.5	461.9
3	112.4	801.6	28.2	589.9	99.6	1281	24.4	1004						
4	1999	3951	225.3	1266	2087	5431	216	1978	109.1	217.5	190.3	211.8	366.1	434.1
5	658.3	1896	124.8	982.4	668.1	2758	107.1	1611	55.2	63.9	91.2	121.7	173.7	274.3
6	219.8	1236	42.5	811.1	204.4	1933	35.3	1418						
7	520	1785	298.6	1424	552.2	2640	305.7	2170						
8	941.5	2039	751.4	1764	946.9	2913	744.7	2540						
9	114.4	904.4	69.4	770.1	98.6	1457	57.9	1275						
10	1113	2668	422.1	1634	1105	3764	461.8	2474	91.3	104.1	155.9	198.0	283.0	416.3
11	1718	3532	553.6	1808	1923	5153	634.1	2786	115.9	225.3	199.6	220.5	374.3	483.4
12	460.5	1512	75.8	749.8	996.2	3375	157.2	1745						
13	2373	4329	1256	2729	772.7	2661	336.6	1830						
14	252.7	1307	181.3	1160	249	2020	166.1	1847						
15	1056	2460	192.5	1136	1057	3486	178.1	1801	85.6	98.6	145.5	159.9	268.4	375.6
16	1038	2529	76.1	985.9	1027	3635	62.47	1705	80.9	92.8	137.6	177.3	253.8	380.6
17	245.9	1251	4.4	639.9	240.2	1931	4.4	1329	25.4	38.0	42.4	69.5	80.08	163.9
18	710.5	1903	275.3	1249	711.7	2741	272.6	1923	57.9	67.2	95.9	128.1	183.8	266.0
19	297.4	1407	74.6	948.1	303.8	2135	61.08	1628	38.3	54.1	60.2	92.1	107.6	216.1
20	321.8	1420	183.4	1151	334.9	2149	168.3	1816	40.8	46.5	64.4	98.6	116.1	231.3

3.2.5 Conclusions and observations on outcomes

It should be stressed that while the results of the scenario runs are quantitative in a mechanistic way, insofar as the models employed in this case study have individually been successfully operated and their outputs linked to produce numerically consistent products, the outcomes should only be viewed qualitatively. It was not within the resources available to this project to assess the effects of uncertainty on the final outputs from the model train on what is very much a prototype application. Nevertheless, uncertainty must be an issue in sequentially applied models with many input fields, some complexly derived, that combine a necessary reliance on available data with only a limited facility to apply dedicated data collection and interpretation for calibration and validation purposes.

With these caveats about the likely strong effects of uncertainty on results, there are some general observations that can be made about the outcomes of the scenario modelling conducted for the Doncaster case study:

(1) The UVQ model has provided crucial information on the urban water and contaminant flows. Outputs of UVQ were used both as stand alone results to assess economic feasibility of alternative scenarios and within the DSS to provide input flows and concentrations for the subsequent models.

(2) Water usage distributions inside the household (toilet, bath, laundry and kitchen) may be highly variable in a study area, depending on the size of a neighbourhood, and is only approximately known even on a large scale. Consequently, model results for alternative scenarios considering separation of these water usages (e.g. grey water for toilet flush or rainwater for toilet flush) are similarly uncertain.

(3) The actual water quality of the different usage types in a household is even more uncertain and may vary largely from study area to study area. Reference values were found for a number of pollutants but for some contaminants hardly anything is known.

(4) The degree of surface sealing of an urban area is often not very well known and the degree of connection of areas such as roadside verges, pavements, informal open areas etc with the drainage network(s) is even more uncertain. Even if stormwater volumes were monitored to calibrate the model and constrain the assumptions, it would be difficult to really validate the numerous assumptions. These uncertainties will have a direct impact on scenarios considering rainwater-harvesting strategies (e.g. rain water for garden irrigation or decentralised rainwater infiltration).

(5) The water quality of surface runoff from the different sealed areas is not very well known even though abundant reference values are available for the contaminants with the highest potential for contamination (e.g. heavy metals). These references can help to constrain the assumptions made and to decrease the uncertainties of the model results.

(6) As discussed in Rueedi and Cronin (2005) the representation of the recharge and evapotranspiration process in UVQ is quite crude. Even if the total recharge were known with a high degree of certainty – which is usually not the case – the actual reaction to a rain event may make a big difference to the effects and results of alternative scenarios (e.g. climate change or rainwater infiltration). It may be possible to obtain reasonable representation of the recharge process with the simplified approach used in UVQ but it would need additional work to calibrate the parameters and provide recommendations for future users (e.g. effects of different soil types, different vegetation, etc.).

(7) Once an acceptable steady-state version of the transient groundwater flow model (with an expanded-grid sub-region within a coarser regional grid) had been established, running scenarios to generate head contour maps was rapid and straightforward. All scenarios were PC-run and took had less than 1hour each.

(8) The 2080 High Emissions (worst-case) scenario led to the most striking water level declines across the study area, due to the major reduction in recharge through gardens and green space. This has the effect of reducing groundwater heads close to or below sea level, with correspondingly lower levels further east in the wellfield area. The effect on groundwater heads is much more marked than in that of the other scenarios, all of which show water table rises or declines of no more than ± 0.5 m compared with the base. The pipe renovation scenario, for instance, resulted in only about 10% equivalent decline, even though in engineering terms it was quite radical, with about a third of the pipe infrastructure reclassified as rehabilitated. This underlines the relatively small contribution of sewer leakage volumetrically to the water balance compared with on-site rainfall recharge.

(9) The effect on contaminant indicator trends in terms both of total load and of absolute concentrations is very subdued in all of the scenarios (Figure 3.2.63). This is partly due to the relatively modest applied load, and partly to dilution effects, both from precipitation-derived recharge and from pore water with a low background concentration (effective porosity 15%). This suggests that as a receptor, the aquifer system beneath the study area may be relatively resilient against many dissolved–phase contaminants; background concentrations are low

in terms of potability norms, applied loads are modest, there is available dilution and the system is hydraulically well-damped.

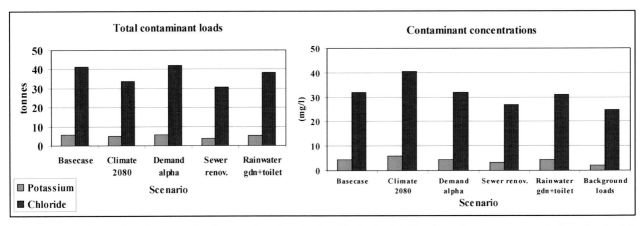

Figure 3.2.63. Comparison between the 5 scenarios of contaminant indicator total loads and average concentrations for the 20 study area neighbourhoods.

(10) The usefulness of the model train in scheduling contaminant loads from different parts of the pipe infrastructure is well illustrated in Figs 3.2.64 A and B. These show the concentrations of recharge from different neighbourhoods for two contaminant indicators. While broadly similar in pattern, there are differences because the source of potassium is exclusively from foul sewer system leakage (left-hand figure) whereas chloride (right-hand figure) comes from both foul sewer and pluvial system (de-icing salt in winter).

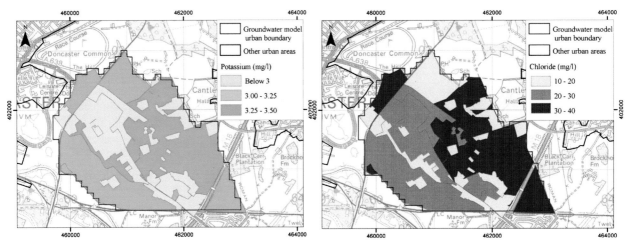

Figure 3.2.64. Comparison of urban contaminant loads (expressed as net concentrations in mg/l) generated by different neighbourhoods for (A) potassium and (B) chloride, example from pipe rehabilitation scenario.

(11) The observation that concentrations in the wellfield are not going to be affected for the next hundred years is mainly a consequence of the large residence times of groundwater flowing from the urban areas towards the wellfield. Abstraction wells in the close vicinity of the urban areas are expected to be more vulnerable to urban recharge.

3.3 A LAYERED AQUIFER SYSTEM: LJUBLJANA, SLOVENIA

3.3.1 Background, physical and geological setting

The Ljubljana basin that encloses the city of Ljubljana is a large tectonic depression comprising a flat area of approximately 300 masl elevation, set amongst hilly and mountainous surroundings of up to 1300 masl. Its northern part constitutes Ljubljansko polje ("Ljubljana field"), from where most of the city's water is abstracted and it is this area which is discussed in the Ljubljana case study. Its southern part constitutes Ljubljansko barje ("Ljubljana moor"), a marshy area subject to flooding stretching to the south and southwest of the city (Figure 3.3.1).

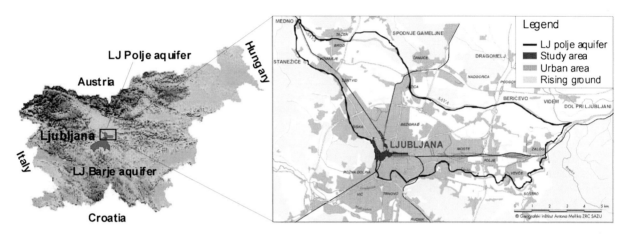

Figure 3.3.1. Location of Ljubljana and study area on Ljubljansko polje aquifer.

The Ljubljansko polje, part of the Ljubljana basin tectonic depression, is drained by the River Sava (of which the Ljubljanica River is a tributary) and contains a thickness of up to 120 m of Pleistocene and Holocene fluvial sediments. These comprise a phreatic aquifer composed of partly conglomeratic Pleistocene gravel, deposited on a basement of Palaeozoic clastics. These sediments may be loosely consolidated and pervious, or tight and impervious. No Holocene lacustrine sediments are present here. Low permeability clay layers immediately to the north and to the east of the Permo-Carboniferous outcrop of the hill of Rožnik in the west of the city give rise to perched aquifer conditions in a limited area (Figure 3.3.2).

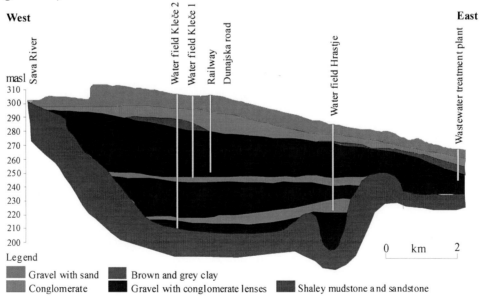

Figure 3.3.2. Ljubljansko polje cross-section (Sava River to wastewater treatment plant).

3.3.2 Hydrogeological setting

The phreatic groundwater in Ljubljansko polje is receiving recharge from rainfall, from the river Sava in its upper reach, and via lateral inflow from the Ljubljansko barje multi-aquifer system. The aquifer structure may be complicated as a result of the presence of conglomerate layers, so that locally perched groundwater may exist, but it is relatively simple compared to the multi-aquifer system existing to the south, within the Ljubljansko barje Quaternary sediments.

Aquifer parameters are summarised in Table 3.3.1 below. It should be noted that the Ljubljansko polje is one aquifer hydrodynamically, but with heterogeneous hydraulic conductivity. The upper layers consist mostly of gravels; the lower layers consist mostly of conglomerates.

Table 3.3.1. Aquifer parameters for Ljubljansko polje - T -values are calculated from hydraulic conductivity values deduced from limited past pumping tests of the aquifer (Prestor, 1998).

Aquifer Unit	Transmissivity (m2/d)	Storativity	Porosity
Ljubljansko Polje	Ranges from 43.2 m^2/d (conglomerates) to 4752 m^2/d (gravel)	$10^{-2} - 10^{-1}$	5 – 15 % (effective)

The combination of the humidity of the local climate, the snow-rain flow regime of the River Sava (as the main source of groundwater recharge), a large regional groundwater gradient and the high porosity of the sediments together create fast flow and high regeneration of the dynamic reserve of the Ljubljansko polje groundwater. The mean annual rainfall over Ljubljana is approximately 1400 mm/yr, with less than 50% lost to evapotranspiration. The rest goes towards aquifer recharge (Brečko, 1999) either falling directly on the area to provide immediate infiltration or as runoff from the fractured sandstone and shale hills in Ljubljana. The aquifer is also replenished from influent recharge from the River Sava (2 – 3 m^3/s) and from the barje multi-aquifer system (0.3 m^3/s) (IRGO, 2002). See Figure 3.3.3 below for a general conceptual schematic of groundwater recharge dynamics.

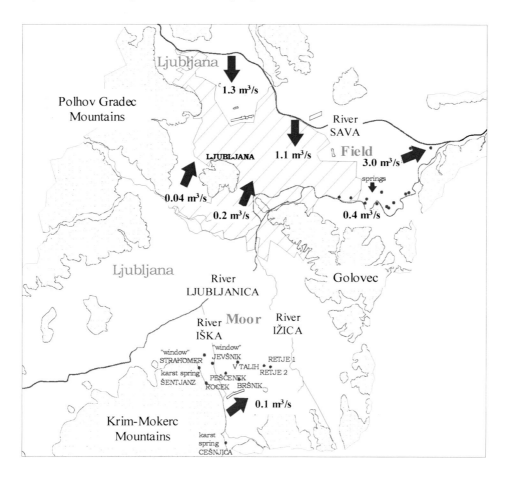

Figure 3.3.3. Conceptual map of major water fluxes in Ljubljansko polje (Jamnik et al., 2001).

3.3.3 Urban development setting

Ljubljana, the capital and largest city of the Republic of Slovenia is an important political, cultural and economic centre with a population of 270000 inhabitants. It is situated in the heart of Slovenia, 298 m above sea level, in a broad basin between the Alps and the Adriatic Sea. Ljubljana represents an important crossroads between the Mediterranean, and Central and South-eastern Europe.

Settlement began in the Ljubljana basin in prehistoric times, with the ancient "Amber Route". The development of the modern city began in the second half of the 19th century when the city underwent widespread political, economic, town-planning and architectural revival. Further major renovation followed later, after the last major earthquake of 1895. The most intensive urban land use development occurred after the Second World War through the growth of residential and industrial areas in the inner city of Ljubljana. Private housing development intensified from the 1960s with the availability of state subsidies. Suburbs expanded as unplanned dispersed housing developed in the 1980s with an insufficient provision of communal infrastructure (i.e. water supply and sewerage system) in smaller settlements, but with increasing daily commuting and transport congestion. From 1987 onwards the population declined in Ljubljana, with agglomeration occurring as (sub)urbanisation shifted to de-suburbanisation - a trend which was further reinforced in the 1990s (URBS PANDEN, 2003).

The city became the capital of the Republic of Slovenia when the country gained independence in 1991. Today, Ljubljana stands at the intersection of the currently expanding national highway system and at the junction of rail routes to Austria, Italy, and ports along the Adriatic Sea as well as eastwards. The City Municipality covers 275 km^2 and now has eight surrounding communes (Brezovica, Dobrova - Polhov Gradec, Dol pri Ljubljani, Ig, Medvode, Škofljica, Velike Lašče and Vodice), collectively forming the Ljubljana agglomeration of over 903 km^2 area.

The detailed study area within the city of Ljubljana was selected about 1 km north of the town centre. It spreads over an area of approx. 76 ha with a population of about 2300 and comprises residential properties, commercial/business premises and industrial areas. The area was chosen because of its position overlying the Ljubljansko polje aquifer, its well-defined water infrastructure, the mixture of land uses, its location up-gradient of the Hrastje wellfield and the presence of already-existing observation wells in the general vicinity.

3.3.4 Urban water infrastructure

All of the water supply for Ljubljana (102.4 ML/d in 2001) is abstracted from groundwater. The Vodovod-Kanalizacija Public Utility (VO-KA) provides, manages and maintains all water supply, sewerage, wastewater treatment and drainage services in the city.

Groundwater is pumped from four wellfields on Ljubljansko polje (Kleče, Šentvid, Hrastje and Jarški Prod) and one on Ljubljansko barje (Brest). There are a total of 39 active water pumping stations, reaching depths of 40 m to 100 m, equipped with submersible deep well pumps with yields of 15 to 100 L/s. The average per person water usage for the whole of Ljubljana is estimated at about 150 L/c/d. The Kleče wellfield supplies the water for the study area (c. 2.7 ML/d), one exception being the Union brewery with its own production wells (c. 1.700 ML/d). The average per person water usage for the study area for residential properties is estimated at about 180 L/c/d. Considering all land uses (i.e. including industrial sites, swimming pool etc) the water usage would be 800 L/c/d.

The town of Ljubljana has a total length of water-bearing pipe infrastructure of c. 1300 km; 69% of it is sewerage system, 31% represents water mains (Table 3.3.2). Within the sewerage system, there are 22.8% foul sewers, 24.7% pluvial drainage systems (stormwater), 52.3% combined system and 0.2% overflow channels. Concrete pipes prevail in the sewerage system, but cast iron is the most abundant among mains water pipes. Ljubljana has c. 12000 active septic tanks, with 10 of them within the study area. In the study area 14.115 km of combined system is present, with 92% of the pipe assets of concrete and 8% of PVC. The sewage of the whole town of Ljubljana is treated at a central wastewater treatment works located on the eastern outskirts of Ljubljana. There are 8.377 km of water mains in use within the study area, mostly made of cast iron.

Table 3.3.2. Pipe infrastructure key statistics for the town of Ljubljana.

Material	Shape	Foul Sewers (km)	Drain-Pluvial (km)	Combined System (km)	Overflow Channels (km)	Total Length Sewage system (km)	Water Pipes (km)
Asbestos cement	circular	16.9	3.8	2.0		22.8	54.1
Steel	circular	1.2	0.1	0.2		1.5	7.4
Masonry/stone	rectangular			0.03		0.01	
Cast iron	circular	0.3	0.1	0.4		0.7	156.0
Ductile iron	circular	4.2	21.6	2.4		28.3	44.9
Masonry/brick	rectangular/oval			2.2		2.2	
Polyethylene	circular	4.5		1.3		5.7	78.8
PVC	circular	50.9	13.9	10.4		75.2	77.2
Concrete	circular	125.4	185.8	416.7	1.8	729.7	
INSITUFORM*	circular			20.7		20.7	
Reinforced polyester	circular	6.1	1.1	24.1		31.3	4.0
Total		209.5	226.4	480.43	1.8	918.11	422.4

*pipe relined with hose lining system: onsite impregnation and installation of unsaturated polyester and epoxy "sock" inside damaged channel

3.3.5 Data sources

Data requirements for each of the AISUWRS models (Urban Volume and Quality – UVQ, Pipeline Leakage Model – NEIMO, unsaturated flow and transport models – POSI & SLeakI, saturated flow and transport model) necessitated substantial additional investigation and data acquisition. A summary of data requirements, sources and provision is given in Table 3.3.3.

Table 3.3.3: Data requirements for the AISUWRS models.

Model	Data	Source	Format	Comments
UVQ	Urban cadastre – land uses, block sizes, roof/paved/open areas etc.	SMA*, digital ortho-photos (DOFs), raster map analysis	GIS / St	No previous cadastre available with required detail. Best available data in DOF and city plan rasters. Spectral analysis of aerial photos using IDRISI and Arc View GIS software used for definition between surface types.
	Household occupancy	Ministry	GIS-v	
	Water consumption rates per type of use (kitchen, bathroom etc.)	VO-KA**, literature, other case-studies	St	Average personal consumption rate available for study area (but no spatial distribution). Division between use-types can only be made by literature/other case-study comparison.
	Contaminant data (water supply, storm and wastewater, all sources)	ARSO***, VO-KA, literature, other case-studies	St	Quality analysis for pumped (drinking) water. Stormwater data lacking N, P and SS concentrations. Complete quality wastewater analysis from 4 sampling points within study area. Contaminant loads for different usage within households obtained from literature, other case studies and from wastewater sampling results.
	Sewer and drainage distribution maps	VO-KA	GIS-v	Primary sewers all Ljubljana, secondary for selected study area only.
	Water supply records	VO-KA	St	
	Wastewater flows (combined system)	VO-KA	St	24-hour measurements at two locations, data also obtained from calculation from the model of the whole sewage system.
NEIMO	Detailed asset data	VO-KA	GIS-v	Primary sewers all Ljubljana, secondary for selected study area only. Some data missing. Joint types not available, but inferred from installation date/pipe material. No burial depths (only z coordinates) – estimated.
	Soil types	Faculty for Agriculture	GIS-v	
	CCTV records	VO-KA	Text files	Sample records received and translated to English. Translation of coding system to Australian system

POSI, SLeaki				determined. Representative sample set (not all, due to translation and interpretation requirements) not acquired. Generic CCTV file is in use for NEIMO.
	Water supply leakage data	VO-KA	St	Estimated gross losses from three large areas and point data for known damages (1992-2003) only.
	Soil types		GIS-v	Soil map. Soil cover relatively thin and homogeneous. Unsaturated rock zone more important.
Visual MOD-FLOW	Unsat. zone profiles	Existing IRGO data, new mapping	Sp	Unsaturated zone is deep (up to 30m), and hence more significant than soil cover. Situation is complex, with very heterogeneous sediment distribution in Ljubljansko polje. Profiles being derived/improved from existing and new geophysical borehole data and provided in new hydrogeological maps.
	Unsat. zone hydraulic profiles	Existing data, new mapping	Sp	Limited data available (spatially heterogeneous). Incorporated in new mapping.
	Borehole data	All local borehole owners	St/Sp	Records from IRGO, VO-KA, ARSO and other borehole owners/drillers supplemented and collated in hydrogeological map.
	Observation Wells	IRGO, VO-KA, ARSO	St	Piezometric monitoring. Limited pumping test data.
	Production Wells	VO-KA	St	Daily abstraction values. No data for private abstractions (except for Union Brewery).
	Recharge rates	Existing data, ARSO, field studies	St	Estimates available from previous modelling and field work, supplemented with new studies (lysimeter, mapping).
	Surface water bodies	ARSO, field studies	Sp/St	Included in hydrogeological mapping.
	Material parameters for selected contaminants	Literature	St	Sorption and decay (with respect to different aquifer materials) dispersion and advection coefficients required.
	Concentration inputs		St	No measurements. Estimates and model predictions only.
	Measured concentrations	Field studies, VO-KA, ARSO	St	AISUWRS field studies supplemented with all data from VO-KA monitoring and available ARSO monitoring.

GIS - GIS datasets   St – Statistical datasets   Sp – Spatial data

*SMA – The Surveying and Mapping Authority of the Republic of Slovenia
**VO-KA – The Vodovod-Kanalizacija Public Water Utility
***ARSO – Environmental Agency of the Republic of Slovenia

3.3.6 Field investigation programme

3.3.6.1 Objectives

Field investigations aim to improve the understanding of the whole urban water cycle and provide the necessary knowledge and data for developing systems to assess the impacts and sustainability of urban water systems. The objectives of the field investigation programme in Ljubljana were:
- To acquire all data required to populate and practically apply the set of AISUWRS models, which will be used to assess current and potential impacts of urban wastewater on local groundwater.
- To conduct geological and hydrogeological mapping over the whole Ljubljana city area to bring together and improve currently uncoordinated data and support vulnerability and risk assessments in relation to the impact of the urban environment on the underlying aquifer.
- To address important scientific research problems, including: the identification of recharge and water budget parameters, their effects on groundwater quality and understanding of the dynamics of the urban aquifer in Ljubljana.

3.3.6.2 Activities

AISUWRS field investigations in Ljubljana included:
- Geological and hydrogeological mapping;
- Groundwater quality sampling (Figure 3.3.4) for investigation of the influence of urban contaminants on the local aquifer and to provide background concentrations and contaminant loads for the UVQ and groundwater model;
- Vertical profiling of physical water quality parameters in boreholes (Figure 3.3.4) to improve understanding of local vertical heterogeneity in the aquifer profile;
- Piezometric monitoring for understanding and modelling aquifer recharge, flow and storage dynamics;
- Construction and operation of an urban lysimeter for analysis of urban infiltration and recharge (Figure 3.3.4);
- Sewage quality measurements and 24-hour sewage flow measurement for the calibration of the UVQ model and for comparison with the groundwater quality data;
- Research, organisation, and pre-processing of other data for the fulfilment of modelling requirements within the AISUWRS project (urban water supply and drainage infrastructure, land-use data, population data, existing geological and soils data, etc.).

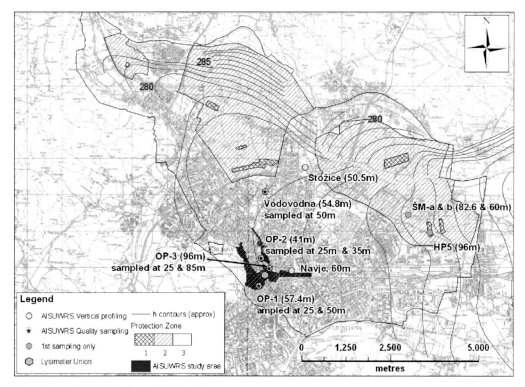

Figure 3.3.4. Locations of AISUWRS monitoring sites.

Groundwater was sampled and analysed for basic chemical, microbiological and isotope parameters four times from summer 2003 to autumn 2004 at selected locations (Figure 3.3.4). Where possible, samples were taken using Bimbar inflatable double packers to allow sampling at specific depths and efficient purging of the sampling zone of the borehole.

Parameters for analysis were selected to provide an overview of the groundwater chemistry with selected focus on potential markers of urban sewage pollution (B, Cl^-, NO_3^-, SO_4^{2-}, E. coli, ^{15}N). Samples were analysed for:
- Major ion chemistry (NO_3^-, SO_4^{2-}, Cl^-, HCO_3^{2-}, Ca^{2+}, Mg^{2+}, Na^+, K^+) plus TOC, NH_4^+, NO_2^- and PO_4^{3-}. [VO-KA water analysis laboratory, Ljubljana];
- ICP Mass Spectrometry for major elements and metals (including Al, As, B, Ba, Be, Cd, Co, Cr, Cu, Fe_{total}, Mn, Mo, Ni, Pb, P_{total}, Sc, Sn, Sr, V, Zn). [ACME Analytical Laboratories Ltd, Vancouver, Canada];
- Microbiology:- *Escherichia coli* (E. coli) and total coliform bacteria (Membrane filtration method, SIS EN ISO 9308-1), total microorganisms – number of colonies at 37°C and 22°C (SIST EN ISO 7899-2), Enterococci (SIS EN ISO 7899-2) and *Pseudomonas aeruginosa* (SIS EN ISO 12780). [IVZ RS (Institute for public health, Republic of Slovenia), Ljubljana];
- Isotopes: H/D, $^{16}O/^{18}O$, $^{15}N/^{14}N$. [Joanneum Research, Graz, Austria].

Vertical profiling of boreholes was carried out at six locations, up to six times from summer 2003 to winter 2004, using a Mini Sonde multi-parameter probe to obtain groundwater physical parameters: temperature (T), acidity (pH), electrical conductivity (SEC), and dissolved oxygen (DO) over depth.

3.3.6.3 Results

The products of collating old and new data from geological and hydrogeological field studies were a new geological and a new hydrogeological map of the Ljubljansko polje (Ljubljana field) and Ljubljanasko barje (Ljubljana moor) areas at 1:25 000 scale. These maps aided the development of the groundwater flow and transport models.

No significant differences in the composition of groundwater in respect of depth or season were observed at the selected sampling sites. Instead, the differences in the concentrations are more noticeable between selected locations (Figure 3.3.5). It is common for all selected sites that the major cations, Ca and Mg, were released into the groundwater from a proportion of matrix dissolution of limestone and dolomite. But a proportion of the Ca is also most likely a contribution resulting from road salting, as Na is displaced from the surfaces of smectite-type clays by ion exchange.

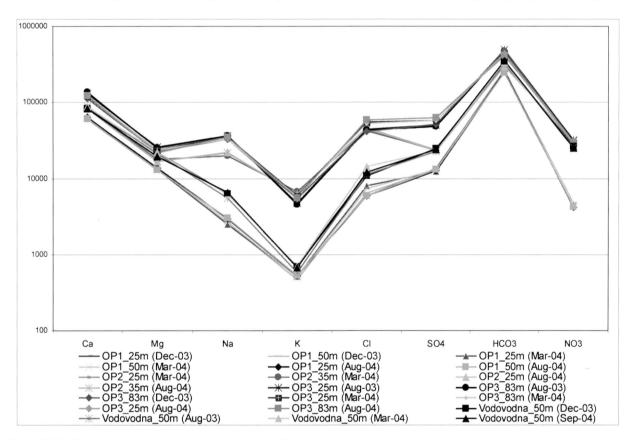

Figure 3.3.5. Major cations and anions at selected sampling sites.

The groundwater of the southwestern part of Ljubljansko polje aquifer, where the research area is situated, is relatively uncontaminated with some exceptions (two observation wells, OP2 and OP3, respectively), where an anthropogenic impact on groundwater can be observed. Groundwater is mostly influenced by saline infiltrate from road salting, and probably also by localised pipeline leakage of the sewerage system, evidenced from the microbiological results and boron content. Indicators of faecal contamination (*Escherichia coli*, total coliforms, enterococci and *Pseudomonas aeroginosa*) have been found in some groundwater samples in upper levels of observation wells. On three out of five sampling sites, microbiological parameters exceeded the maximum admissible concentrations (MAC) in drinking water defined by Slovenian national regulations for drinking water (URL RS, 2004). Concentrations of boron are up to 163 µg/L and are also mostly found in the upper parts of the observation wells. Isotopic ratios of $\delta^{15}N$, $\delta^{18}O$ and δD indicate a complex source of N compounds. Nitrate concentrations did not exceed the MAC in groundwater (50 mg/L as NO_3^- (URL RS, 2005 or the MAC in drinking water (also 50 mg/L as NO_3^- (URL RS, 2004) at any observation wells. OP2 is the only location with a significant amount of phosphate (PO_4^{3-} was found slightly in excess of the groundwater standard (0.2 mg/L), 0.215 mg/L in March 2004), indicating an influx from septic tank outlets, leakage from the sewerage system or fertiliser application. Phosphate is accompanied by potassium, which is also a typical constituent both of agricultural sewage and urban pollution. Total chromium concentrations exceeded the MAC in groundwater (30 µg/L (URL RS, 2005)) at Vodovodna observation well significantly (57.8 µg/L in August 2004, 60.7 µg/L in December 2003, 71.3 µg/L in March 2004 and 53.0 µg/L in August 2004). The source is from historical Cr(VI) pollution from nearby industry. Transport of Cr(VI) is in the direction towards the south eastern part of the Hrastje wellfield. Chromium levels decrease to significantly below admissible standards before it reaches the wellfields at the edge of the city.

Sewage water quality data were used for the calibration of the UVQ model and for comparison with groundwater quality data. Generally, the average concentrations of wastewater for Cl^-, B, K^+, PO_4^{3-}, NO_2^-, Fe and NH_4^+ are higher than are observed in groundwater. Exceptions are groundwater samples from two observation wells (OP2 and OP3) with higher concentration of B and Cl^- in comparison with residential/commercial wastewater. Sr, Ba and Al are also higher in groundwater of these two wells in comparison with wastewaters. The wastewater in Ljubljana contains high numbers of the microorganisms listed above, but an effective removal is taking place in transit through a relatively thick unsaturated zone. The two observation wells that continuously stand out with relatively high concentration means of chemical and microbiological parameters are situated near railway tracks and in a slightly excavated area with an unpaved car park, respectively.

There is general agreement that urban groundwater pollution occurs locally through areas where the unsaturated zone is thinner and more permeable, and probably also by downhole movement from upper polluted horizons in unpumped boreholes whose design interconnects different aquifer horizons (e.g. observation wells), either within the casing or between borehole walls and its casing (the probable situation at the observation wells OP2 and OP3).

The monitoring results suggest that the situation is under control for now, but we should be aware that increased quantities of salt water from road salting causes gradual desorption of inorganic pollutants (e.g. heavy metals) and organic pollutants (e.g. PAHs) from the sediment in the unsaturated zone and they could reach the aquifer in the future.

It should be pointed out that, during the AISUWRS project, water from pumping wells tapping the Ljubljansko polje aquifer - in pumping stations Kleče and Hrastje, as well as water from pumping wells of Union brewery- indicated no presence of microorganisms and all measured chemical parameters were below maximum admissible concentrations (MAC) in drinking water defined by Slovenian national regulations for drinking water (URL RS, 2004).

A seasonal distribution of major ions and selected trace elements were also determined during the sampling campaign. Some data obtained within the AISUWRS project were collated with historical data collected by the VO-KA water utility to assist this study. The seasonal variations of the anions and cations are related to the influx of fresh water from rainfall. The highest concentrations are found after longer or stronger rain events (Figure 3.3.6).

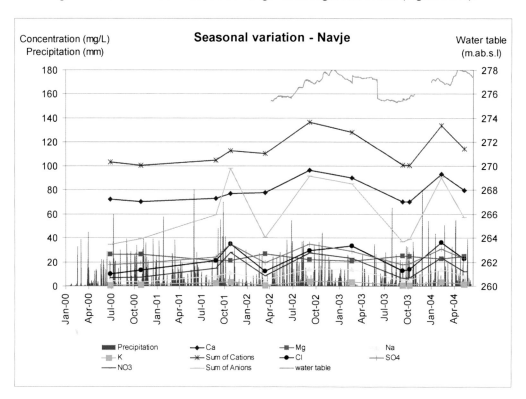

Figure 3.3.6. Seasonal variation of major ions at Navje observation well during the 5-year period. The well head of Navje observation well is at 298 m.ab.s.l, the well is 60 m deep.

Vertical profiling was performed to enable better understanding of the heterogeneous stratification patterns (occurrence of clay lenses etc.) and to observe seasonal changes in groundwater physical chemistry in the Ljubljansko polje aquifer. The results clearly demonstrate the influence of infiltration of precipitation in first few metres of the logged wells. Below this uppermost zone, strong changes were suspected in the lower parts of observation wells, due to very heterogeneous sediments, but none were observed. Probably the method is not sensitive enough. The most interesting profile is from Navje observation well in December 2004, where T, EC and DO show elevated values between 27 m and 43 m in comparison with deeper part of the well (Figure 3.3.7). The water profile does not reflect the changes in sedimentary strata indicated in the borehole log, so the observed anomaly could be a result of a technical error - insufficient quantity of pumped water at the beginning of the measurements or, more likely, the elevated values could be the result of sewage influence. On account of frequent and intense rainfall at the end of November 2004 and in the beginning of December 2004, a sewage overflow could cause the anomaly observed in Navje well. EC vertical profiles of another well (Vodovodna observation well) show elevated values at 35 m, which was ascribed to an influx of salt water of unknown source. The source of stratification in the Stožice observation well, which is closest to the river Sava, could result from different seasonal recharge.

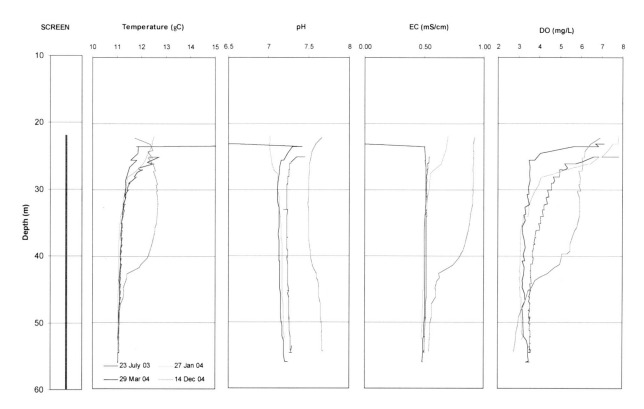

Figure 3.3.7. Vertical profiling - Navje, water table approx. 21 – 23m below surface.

An urban lysimeter was constructed in order to better understand the role and behaviour of the upper unsaturated zone in the alluvial gravel aquifer in the highly urbanized environment. It was equipped with a UMS environmental monitoring system. UMS supplied the lysimeter with sensors: 21 suction cups, 15 tensiometers and 9 TDR probes; a data recording system and a sampling system. On the right side of the lysimeter each layer was equipped with a single type of measuring probe: one tensiometer, one TDR probe and three suction cups (Figure 3.3.8). On the left side of the lysimeter each layer was equipped with a twin probe assembly (a tensiometer probe and a suction cup) and a single TDR probe. A scheme of probes disposition is presented in Table 3.3.4.

Table 3.3.4. Position of measurement probes on the right and left side of the Union urban lysimeter (SC – suction cup, TS – tensiometer and TDR – Time Domain Reflectometry probe).

Right side

Depth [m]	Layer label	Borehole name – probe type					
		1	2	3	4	5	6
0.3	R I	RI/1 – SC	RI/2 – SC	RI/3 – SC	RI/4 – TDR	RI/5 – TS	RI/6 – TS
0.6	R II	RII/1 – SC	RII/2 – SC	RII/3 – SC	RII/4 – TDR	RII/5 – TS	RII/6 – TS
1.2	R III	RIII/1 – SC	RIII/2 – SC	RIII/3 – SC	RIII/4 – TDR	RIII/5 – TS	RIII/6 – TS
1.8	R IV	RIV/1 – SC	RIV/2 – SC	RIV/3 – SC	RIV/4 – TDR	RIV/5 – TS	RIV/6 – TS
3.0	R V	RV/1 – SC	RV/2 – SC	RV/3 – SC	RV/4 – TDR	RV/5 – TS	RV/6 – TS
4.0	R VI	RVI/1 – SC	RVI/2 – SC	RVI/3 – SC	RVI/4 – TDR	RVI/5 – TS	RVI/6 – TS

Left side

Depth [m]	Layer label	Borehole name – probe type	
0.6	L I	LI/4 – TS+SC	LI/1 – TDR
1.2	L II	LII/5 – TS+SC	LII/2 – TDR
1.8	L III	LIII/6 – TS+SC	LIII/3 – TDR

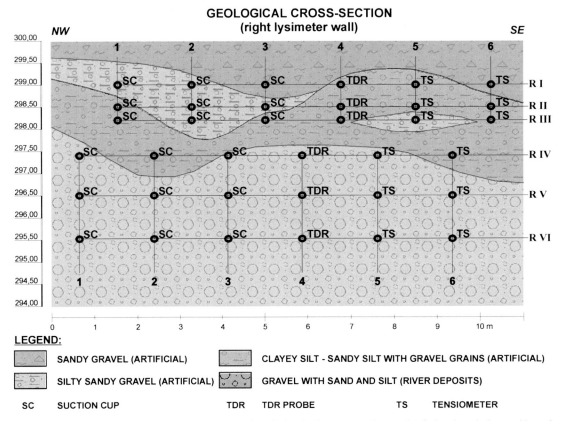

Figure 3.3.8. Geological cross-section on the right side of the lysimeter at the end of the boreholes, with scheme of measurement and sampling points (Veselič & Pregl, 2002).

On the basis of borehole core mapping and measurements of soil moisture and capillary pressure (Figure 3.3.8) we can infer that the unsaturated zone in the area of Union urban lysimeter is very heterogeneous. Probes with fairly fast reactions to precipitation events indicate fast preferential flow, whereas probes with small reactions indicate slow flow (Figure 3.3.9). Isotope analyses identified two important flow types - lateral and vertical flow. Lateral flow has an important role in the protection of groundwater in the Pleistocene alluvial gravel aquifer. However, the role of vertical flow is quite the opposite, because it is the main factor controlling contaminant transport towards the aquifer-saturated zone. Hence investigation of the occurrence and frequency of such rapid recharge events will represent one of the main topics of the next research phases.

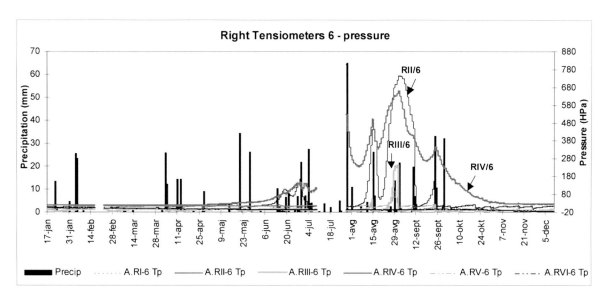

Figure 3.3.9. Capillary pressure values for 2003 in measuring points Ri/6 (i=I-VI); I – upper level, VI - lower level.

3.3.7 Models application to Ljubljana study area

3.3.7.1 Urban water volume and quality model UVQ

The data demands of the UVQ model are very high (over 115 separate spatially variable input parameters). Meeting these data requirements has been very problematic due to the lack of local primary sources for much of the data. So, required data has been gradually collected from a variety of sources, including the water utility VO-KA, the Environmental Agency of the Republic of Slovenia, the Surveying and Mapping Authority of the Republic of Slovenia, the Union brewery, and from other partner case-studies and literature.

3.3.7.2 Delimitating the study area and neighbourhoods

The study area for the model was originally proposed (in the Description of Work) as the whole city area. Later this was reduced to approx. 76 ha of central Ljubljana, in order to make data analysis and model application more feasible. GIS analysis was used to divide the modelling area into separate spatial neighbourhoods according to the sewer drainage pattern and land-uses. The distribution of neighbourhoods was primarily based on the flow of the sewer network, with neighbourhoods separated at significant changes in land-use and land-cover types. At the first stage, 17 neighbourhoods were defined in the study area. Collation of the water supply statistics indicated areas with very high consumption (due to industrial demand). So, neighbourhoods were slightly redefined to separate out high consumption areas and finally, two extra neighbourhoods were added to account for 10 septic tanks, which were discovered in the study area when trying to define number of customers connected to each pipe asset (data needed for connections file for NEIMO). Thus altogether 19 neighbourhoods were created (Figure 3.3.10).

3.3.7.3 Climate data

Daily precipitation, potential evaporation and average temperature were provided by the Slovenian environmental agency (ARSO) for the Ljubljana Bežigrad weather station, situated 1 km east of the project area centre, for the period from 1st January 2000 to 31st December 2004.

3.3.7.4 Water supply/consumption data

Water usage statistics were provided and computed from a combination of water supply data, average consumption rates and population data from municipal cadastres. This process proved quite problematic due to inconsistencies within the reported data and incompatibilities between water supply and population data (no common reference system, differing spatial points etc.). Considerable pre-processing and GIS analysis of data was required as a result. Figure 3.3.11 shows an example of the discrepancies in spatial point data between residential population and water supply cadastres. Many resident data points lack supply data, and vice-versa. Data was spatially linked, and residential units with no supply data assigned mean consumption rates of 190 L/c/d.

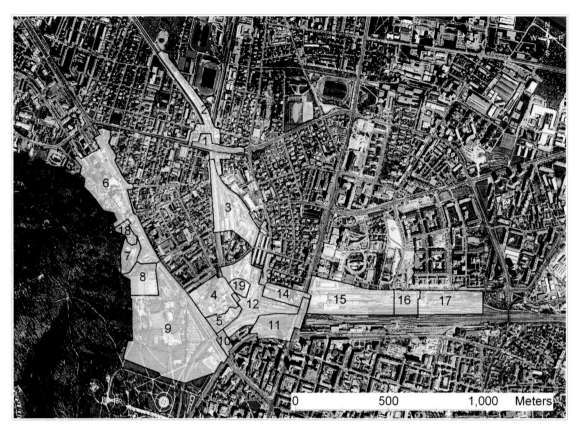

Figure 3.3.10. Neighbourhood distribution with 19 neighbourhoods.

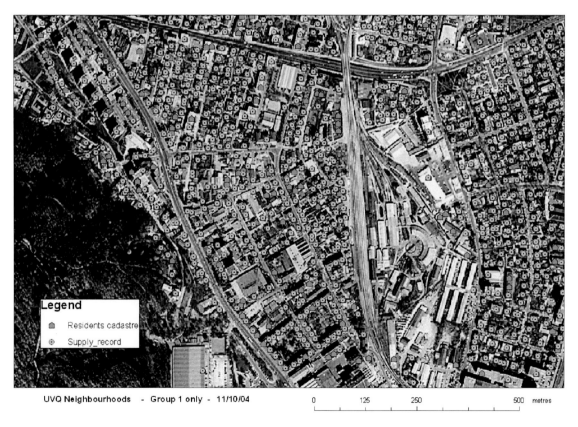

Figure 3.3.11. Thematic data overlays highlight discrepancies between population and water supply data.

Evaluation of these data after processing revealed a wide range of water consumption rates over a small spatial scale. This is due to the presence of very mixed land uses and many different water-using activities and industries within the study area (e.g. a large brewery, garages, petrol stations, railway works, print-works, dry cleaners, swimming pools, mixed residential and commercial units, etc.). Neighbourhood areas were redefined accordingly. Data for daily consumption per capita (L/c/d) were obtained by averaging supply data over the neighbourhood populations.

No local data were available for the breakdown of consumption between different indoor water uses (kitchen, bathroom, toilet and laundry). Ratios for indoor usage were taken from the UVQ tutorial example: 9.7% for kitchen, 41.8% for bathroom, 15.5% for toilet and 33.0% for laundry; and applied to the actual consumption data for each neighbourhood. Industrial water usage was input under 'bathroom'. Irrigation was set to 4% as it is assumed that just a small amount of water is used for garden irrigation.

Water leakage data were obtained from the water utility VO-KA. An approximation of 40% supply leakage was derived from the difference between abstraction for supply and metered consumption.

3.3.7.5 Contaminant data – indoor usage

No local site-specific data were available for contaminant profiles from indoor water use (kitchen, bathroom, toilet and laundry) or stormwater runoff (pavement, roof and road). So, the loads for indoor water use were recalculated from wastewater field measurements. For comparison, typical quality data for household discharge and stormwater categories have been sourced from CSIRO and UniKarl case studies and literature reviews. These have provided values for total nitrogen, total phosphorous, suspended solids, dissolved solids, sulphate, copper, lead, zinc, cadmium, COD, PAHs and oil and grease. Some of them, modelled in UVQ, are listed in Table 3.3.5. No data from the literature have been obtained for parameters of boron and chloride in household discharge or for boron in stormwater. Where the proportion for the parameter (e.g., boron, chloride) within the household was not known it was assigned equally to kitchen, bathroom, toilet and laundry.

For other non-residential units, typical residential contaminant loads have been applied, multiplied by the population equivalent water consumption.

Table 3.3.5: Loads for indoor water use.

Indoor usage (RESIDENTIAL) – mg/c/d								
	(1) Kitchen		(1) Bathroom		(1) Toilet		(1) Laundry	
	CSIRO data	Recalculation from ww*	CSIRO data	Recalculation from ww*	CSIRO data	Recalculation from ww*	CSIRO data	Recalculation from ww*
B		8.8		8.8		8.8		61.7
K^+		752.8		752.8		752.8		752.8
Na^+		3582.1		3582.1		3582.1		3582.1
Cl^-		6770.1		846.2		3385.0		5923.8
N_{total}	241.1	147.2	460.3	281.0	13698.6	8362.9	328.8	200.7
P_{total}	43.8	12.1	21.9	6.1	1567.1	434.1	153.4	42.5
SS	3990.0	4476.1	8303.9	9315.6	36240.0	40655.1	4858.0	5449.9
SO_4^-		2131.1		2131.1		2131.1		2131.1
TOC		9485.7		5399.3		2312.2		7479.5
Cu	1.096	0.12	7.671	0.87	1.096	0.12	2.192	0.26
Pb	0.219	0.006	0.526	0.012	0.022	0.001	3.452	0.08
Zn	19.726	0.3	96.438	1.4	10.959	0.2	4.384	0.1

ww - wastewater

For mixed and commercial units, water use was replicated from the typical residential profiles (kitchen, bathroom, toilets and laundry) – representing a general use by day-workers. For industrial areas, just bathroom contaminant values were used. This provides a predicted load, as if the same quantity of water was being consumed and discharged by a residential unit. It is assumed other specific industrial contaminants of concern are removed by on-site treatment before discharge to the public sewers. Obviously this applies a very different contaminant profile than actual present, and quite possibly overestimates the load. However, if no load profile was used, the contribution of the industrial areas would similarly distort results by contributing large volumes of nominally clean water that dilute the wastewater stream, hence underestimating contaminant concentrations in the predicted study area discharge.

3.3.7.6 Land-cover/Land-use data

No cadastral data for land-use and land-cover or surface sealing are available for the Ljubljana study area on a sufficiently detailed scale. As such, the only sources of data for calculating surface areas of road, pavement, roof, garden and open space within each neighbourhood are the general maps and aerial digital ortho-photos of the area. To calculate these areas, an approach of classifying land cover by spectral analysis of aerial photos using IDRISI and ArcView GIS software was attempted. This involved a process of supervised classification for assigning pixel colours to the different land-cover types (Figure 3.3.12), modified manually by trial and error reclassification and visual comparison of results with the original photos. Areas of each classification were then calculated from the output image (Figure 3.3.13), adjusting values manually where visual inspection showed mis-recognition of particular land covers. Results are not perfect, but provide some reasonable estimates of land-cover areas (especially distinguishing generally between hard/impermeable and soft/permeable surfaces which have strong water balance relevance).

3.3.7.7 Calibration and verification

Limiting data for observed wastewater flows are available for Ljubljana, making it impossible to credibly calibrate or validate the entire UVQ model. Some comparison was done with the results of 24-hour wastewater flow measurements (recalculated to the entire year) and output flows for a residential neighbourhood, the Union Brewery neighbourhood and for the entire area respectively. Surprisingly, the observed and calculated values matched very well. For the residential neighbourhood, the discrepancy between observed and calculated values was 2.5%, for Union brewery neighbourhood even lower and for the entire study area 8.2%.

Figure 3.3.12. Extract from IDRISI land cover analysis: 1-Open space; 2-Roads; 3-Shadows (re-classed visually); 4-Roofs.

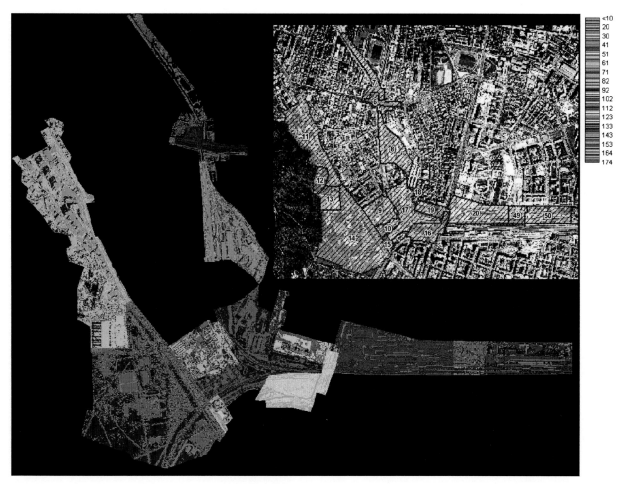

Figure 3.3.13. IDRISI classification results for the study area.

Model outputs for flows were compared to manually calculated data to verify that correct data input and realistic total flows were being modelled. Contaminant loads and concentrations were also separately calculated for one neighbourhood and compared to model outputs to partially verify the quality modelling. Several model edits and re-runs were required to remove errors and bugs in the model set-up and also within the model itself.

3.3.7.8 Network exfiltration and infiltration model: NEIMO

Data sources used to generate three input files (Pipes.csv, Connections.csv and Successor.csv) for the NEIMO are listed in Table 3.3.3; the fourth input file-**Uvq_nbhoods.csv** is generated by UVQ and contains wastewater volumes and contaminant loads for the 19 neighbourhoods in the study area. The fifth input file – **asset_nodes.csv** is required only if running MapObjects and can be generated from the GIS database (not used for the Ljubljana case-study).

Pipes.csv file gives information about pipe assets in the network. The required data for pipe type (PTYPE), diameter of the pipe (DN), length of the pipe (LENGTH), material of the pipe (MAT) and year of pipe installation (CDATE) were provided by water utility VO-KA. The slope of the pipe (GRADE) was calculated from the length of the pipe divided by the difference between initial and end elevation of the pipe asset. If the z-coordinates were not known, the default value of 100 m was used for the slope. The pipe material was known but joint type had to be deduced from installation year (VO-KA info). Soil type (GROUND) in which the pipe is installed is unknown. The neighborhood reference number (NBH_ID) was picked up from the UVQ output files. If an asset crossed two neighborhoods, it was assigned manually to one according to relative length. In the study area there are 263 assets of total length 14.115 km.

Some problems when preparing pipes.csv file were related to the pipe asset data, e.g. oval shape of pipes, pipe material that is not included in the material list but these have been successfully resolved with the help of the Australian project partner CSIRO.

Connections.csv file was prepared with the help of the water utility VO-KA, where their experts connected each land block (from a known address) to a corresponding pipe asset.

Successor.csv file gives the upstream/downstream relationship between pipe assets and requires all the pipes to be in a tree structure with one outlet pipe. This file was prepared manually using GIS.

In the Ljubljana case study the generic pipe defect curves were used as difficulties occurred trying to utilise local CCTV data in the NEIMO application. These included the lack of circumferential referencing of defects in the CCTV reports and no ID system to link reports to the asset database. Model parameters used in NEIMO were similar to those used in the Doncaster case study (DeltaL A (m) 0.015; DeltaL B (m) 0.015; DeltaL C (m) 0.011; DeltaL D (m) 0.006; Mannings N 0.013; K_cracks (m/s) 0.001, K_joints (m/s) 0.0001). With parameters used, the most leakage occurs through joints (97%) and only about 3% through cracks, which is not consistent with the Rastatt results. There it was considered that the two types of leakage should contribute similarly to the overall leakage. The total leakage volume is about 4% of the annual flow or related to the length of pipes: 1051 m^3/km/year or about 0.03 L/km/sec. As no volumetric field leakage measurements are available for Ljubljana case study, the leakage results should be treated with some degree of uncertainty, as they can only be very approximately cross-checked against the product of mass balance estimates.

3.3.7.9 Unsaturated transport models: POSI & SLeakI

The unsaturated flow and transport models POSI (Public Open Space Index) and SLeakI (Sewer Leak Index) were applied in the Ljubljana case study. They give contaminant transport rates and loads to the groundwater, taking into account infiltration rates from open space (gardens, parks) in the case of POSI, or sewer leakage rates in the case of SLeakI. Basic assumptions that should be followed are: a uniform two-layer soil and no preferential flow. The soil cover in the study area is relatively homogeneous. The unsaturated zone is more significant, comprising 10–30 m deep fluvial sandy gravels with uneven clay layers and perched aquifer conditions below the study area (Figure 3.3.14). So, the uniform soil in two layers assumption is hard to apply in the Ljubljana study area, as there are several poorly permeable clay layers within gravel and conglomerate, starting at the depth of 10 m. Therefore, for a sub layer, a first clay layer of 4 m thickness was employed as an approximation. As the study area is relatively small (76 ha), the soil properties are assumed to be uniform for all neighbourhoods (Table 3.3.6).

Table 3.3.6. List of soil and groundwater parameters used in POSI & SLeaki in the study area.

Neighbourhood Data (POSI & SLeakI)*			POSI infiltration data*	
	Soil layer	Sub layer		
Soil type	Sandy loam	Loam	Depth root zone (m)	1
Thickness (m)	1	4	Ground water temperature (°C)	12
Temperature (°C)	18	10	Ground water pH	6.5
pH	6	6		

*Uniform for all neighbourhoods

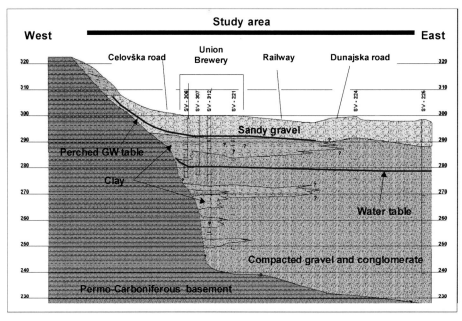

Figure 3.3.14. Geological profile through unsaturated and saturated zone under the study area. Scale: L 1:10000, H 1:1000.

3.3.7.10 Saturated zone flow and contaminant transport models

The saturated flow and transport models Visual ModFlow 4.0 and MT3D were applied in the Ljubljana case study. They provide groundwater flow and transport solutions, taking into account infiltration rates and loads from POSI and SLeakI. Basic assumptions that were followed were: two-layer model and general flow coming from northwest and west towards the east (Figure 3.3.4). The geology of the Ljubljansko polje area is relatively homogeneous in the centre but much more complex on the boundaries, especial in the transition from the plain to the hilly areas. The thickness of fluvial sandy gravels with uneven clay layers underlying Ljubljansko polje is in the range 60 m to 120 m. This made a uniform and homogeneous one layer approximation hard to apply for the polje area, so for groundwater modelling purposes an upper layer was employed as an approximation of the sandy gravels comprising the shallow zone of Ljubljansko polje aquifer and a lower partial layer (see Figure 3.3.14) was used to represent the conglomeratic horizons.

The modeling approach commenced with the conceptual hydrogeological model. This led to the decision to model a wider area than just the immediate environs of the study area, which is relatively small (76 ha) and lies in a zone under the influence of groundwater flows coming from both the Ljubljansko barje and the north. Employing a larger, basin-size modelling area enables hydrogeologically defensible boundary conditions to be applied. Another advantage of keeping the model boundaries with fixed conditions adequately separated from the scenario simulation area in the city centre is that it avoids these fixed head or fixed flow boundaries suppressing the induced changes from the application of different scenario conditions.

The selected area for modeling is shown in Figure 3.3.15 and is centred mainly on the geographical unit of Ljubljansko polje, limited by the River Sava in the north and northeast, by the low hills overlooking the districts of Stanežiče, Šentvid, Dravlje and Šiška in the west, by the area between Šišenski hrib, Rožnik and the Ljubljana Castle in the south and by eastern Zalog in the east. Only the northern part of the Ljubljansko barje was included in modelling area, near the watershed of the Mali Graben brook, so that the southern border of the 3D mathematical model area extends in a line from Šišenski hrib to the Ljublajna Castle. The largest and by far the most important surface stream in the research area (in the geomorphologic and hydrogeological sense) is the River Sava. In the southern part, the River Ljubljanica flows through the area between Rožnik in Šišenski hrib (Prule and Trnovo area), but does not affect the groundwater of the modelling area.

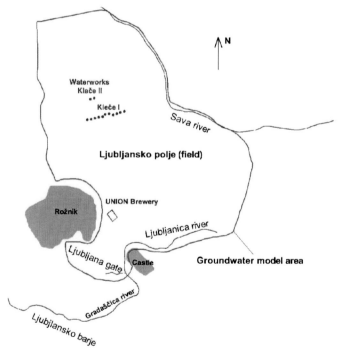

Figure 3.3.15. Groundwater modelling area with model boundaries.

According to the hydrogeological conceptual models the following hydrological boundary conditions were set:

Surface recharge:
- recharge   1700 mm/a;
- evapotranspiration 1100 mm/a.

Line recharge:
- for the River Sava a river boundary and
- general head boundaries to representation piezometric heads on the western and eastern part.

The river boundary condition is used to simulate the influence of a surface water body on the groundwater flow. Surface water bodies such as rivers, streams, lakes and swamps may either contribute water to the groundwater system, or act as groundwater discharge zones, depending on the hydraulic gradient between it and the groundwater system. The MODFLOW river package simulates the surface water/groundwater interaction via a seepage layer separating the surface water body from the groundwater system. The function of the general-head boundary package (GHB) is mathematically similar to that of the river, drain, and evapotranspiration packages. Flow into or out of a cell from an external source is provided in proportion to the difference between the head in the cell and the reference head assigned to the external source. The application of this boundary condition is intended to be general, as indicated by its name, but the typical application of this boundary conditions is to represent heads in a model that are influenced by a large surface water body outside the model domain but with a known water elevation. The purpose of using this boundary condition is to avoid unnecessarily extending the model domain outward to meet the element influencing the head in the model. As a result, the general head boundary condition is usually assigned along the outside edges of the model domain.

It was decided to use GHBs instead of constant head boundaries (CHBs) because GHBs can be better managed at the model calibration stage. The results of groundwater flow modeling are presented in Figure 3.3.16. The calibration results show that the maximum head error is smaller than 0.5 meters. The maximum calculated velocities in the model domain are high, between 40 and 60 m/day. These results match observed values from tracer tests.

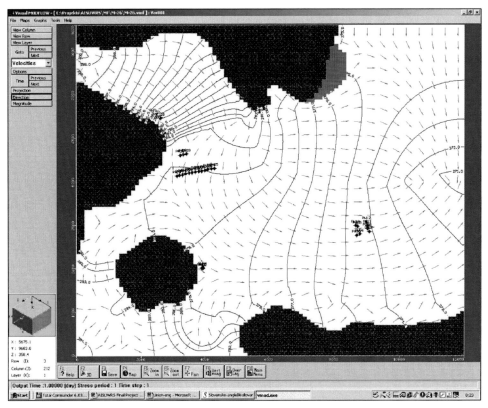

Figure 3.3.16. Flow directions and piezometric heads in the model domain.

On the basis of the flow model the transport model was constructed, using Cl⁻ as a conservative contaminant. The pollutant recharge was represented by concentration recharge boundaries in every neighbourhood. For dispersivity, values from a previously conducted tracer test were selected (30 L/m).

Figure 3.3.17 shows the results for a 730 day time period. The rapid movement down-gradient of pollution coming from the urban area of interest can be observed, and also the low dispersion of the plume as it spreads through Ljubljansko polje. The contaminant plume coming from the study area is mainly affecting the southeastern part of the polje and it would seem that the Hrastje water supply wellfield is not affected. The results demonstrate how an early warning system could be established using the results of the groundwater modeling exercise; for instance the effects of a spillage down-gradient of key industrial locations could easily be tracked, allowing first arrival and peak concentration times to be predicted.

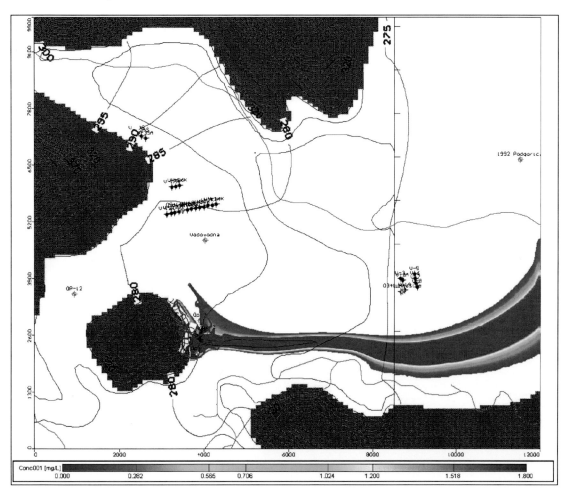

Figure 3.3.17. Direction and dispersion of a conservative pollution indicator (Cl⁻) spreading from study area through Ljubljansko polje after 730 days (base-case scenario).

3.3.8 Scenario modelling results and related topics

From the results of UVQ and NEIMO, a study area water balance for selected scenarios (Table 3.3.7) was calculated for the reference year 2003 (see also Figure 3.3.18-22). The strongest influence on quantity of wastewater in the combined system occurs for the infiltration scenarios. With infiltration basins installed, the volume of wastewater in the sewer system reduces by 11% in the study area. Other scenarios affect the total wastewater flows to a lesser extent. The groundwater recharge is most affected by water mains leakage. Leakage from water mains is potable water and represents no threat to groundwater quality. The wastewaters from leaky sewer pipes and spillage from septic tanks could pose a potential risk to the groundwater. They represent about 15% of total water volume, which infiltrates through the unsaturated zone in the study area (Figure 3.3.18).

Table 3.3.7. Modelled scenarios in the Ljubljana case study.

Action Scenario	Variant	Status quo
0- No additional action planned		01- Actual situation/Calibration
A- Sewer rehabilitation	Sewers constructed in 1965 or before rehabilitated	A1
B-Replacement of septic tanks by sewers connected to the Ljubljana sewage system		B1
C- Decentralized rainwater infiltration (infiltration scenario)	80% of roof and paved areas deliver stormwater to infiltration basin (residential areas); 50/25% of roof/paved areas deliver stormwater to infiltration basin (mixed areas)	C1i
	90% of roof and paved areas deliver stormwater to infiltration basin (residential areas); 75% of roof/paved areas deliver stormwater to infiltration basin (mixed areas)	C1ii

Infiltration scenarios give the worst results in respect of the groundwater quality, contributing more load than base line scenario, namely 12% more N, 17% more P and 2% more Cl⁻ (Table 3.3.8) to the unsaturated zone. In comparison with the base case scenario, the most favourable scenario, which has reduced the loads to the unsaturated zone (-6% N, -5% P and -2% Cl⁻) is replacement of septic tanks by sewers connected to the Ljubljana sewerage system (Table 3.3.8). The town of Ljubljana has c. 12000 active septic tanks; therefore their removal would probably improve the quality of groundwater. The sewer rehabilitation scenario assumes that all sewers constructed in the year 1965 or earlier are rehabilitated; the results showed negligible effect on total recharge and a minor impact on the unsaturated zone (Table 3.3.8).

Table 3.3.8. Yearly loads (in kilograms) and increase/decrease of loads in comparison with the Base case for nitrogen, phosphorus and chloride entering the unsaturated zone of Ljubljana case study urban area for different scenarios.

Study area	Nitrogen kg/year	Phosphorus kg/year	Chloride kg/year
Base case	4584.2	260.0	20261.8
Infiltration scenario (C1i, C1ii)	5185.3	312.9	20703.4
Replacement of septic tanks	4331.1	247.2	19771.2
Sewer rehabilitation	4512.7	256.3	20153.5
Increase/decrease of loads regarding to Base case			
Infiltration scenario	+12%	+17%	+2%
Replacement of septic tanks	-6%	-5%	-2%
Sewer rehabilitation	-2%	-1%	-0.5%

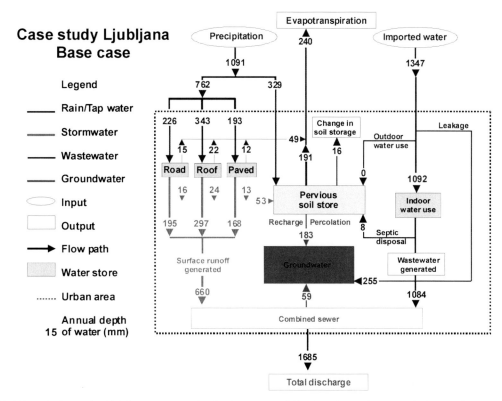

Figure 3.3.18. Water balance for Ljubljana case study: the base case (all values in mm/year equivalent recharge).

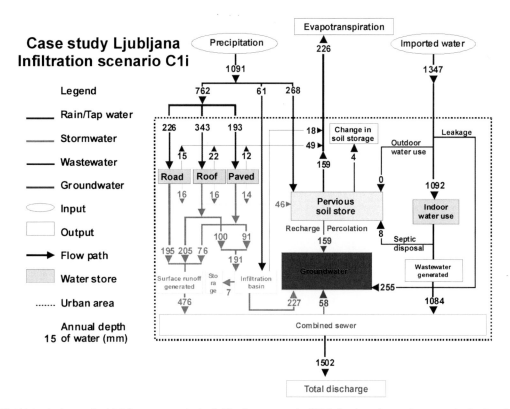

Figure 3.3.19. Water balance for Ljubljana case study: infiltration scenario C1i (all values in mm/year equivalent recharge).

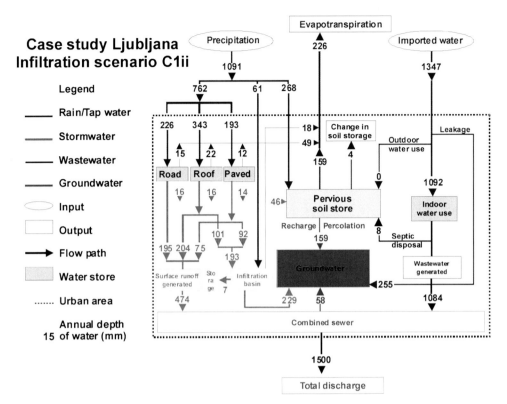

Figure 3.3.20. Water balance for Ljubljana case study: The infiltration scenario C1ii (all values in mm/year equivalent recharge).

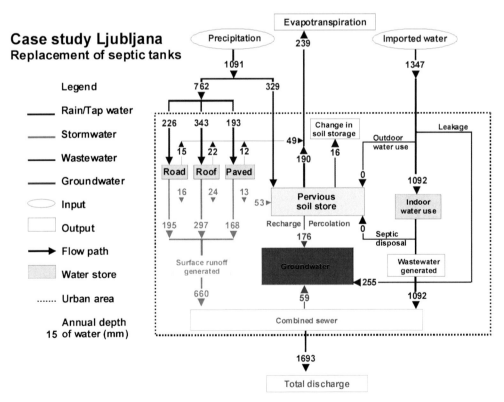

Figure 3.3.21. Water balance for Ljubljana case study: the removal of septic tanks and connection to piped sewerage system (all values in mm/year equivalent recharge).

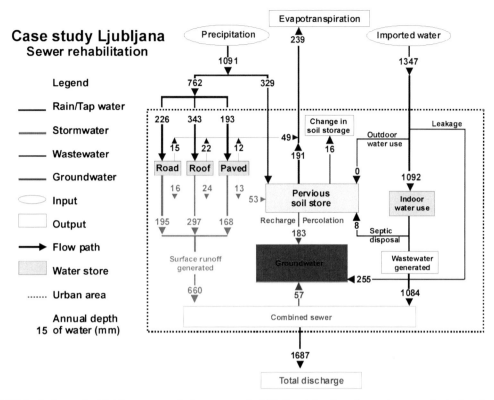

Figure 3.3.22. Water balance for Ljubljana case study: sewer rehabilitation (all values in mm/year equivalent recharge).

As illustrated when applying POSI and SLeakI negligible amounts of contaminants from the urban area reached the groundwater table because of the thick unsaturated zone in the study area (10-15 m), so there is no significant difference in contaminant loads to groundwater from the different scenarios. An additional approach was taken in regard to groundwater modelling, comprising comparison of the base case scenario with two extra scenarios. The first was to assume that all the study area is an industrial area and the second one is that none of the study area has industrial neighbourhoods.

The infiltration scenarios C1i and C1ii were modeled, but despite the resultant major increase in available recharge (Figure 3.3.19), no significant change in the piezometric levels was noted. This does not mean that changes were not present, but rather that the study area is too small to affect the piezometric levels. A much more significant impact on groundwater levels would be anticipated, for instance, if a major proportion of the city employed infiltration basins, and these effects could be rapidly predicted. A similar situation arises when trying to judge the effects of the infiltration scenarios on the conservative pollution spreading. An additional important factor is that the study area is neither wholly residential nor wholly industrial but instead comprises a mix of the two land uses. The neighbourhoods with dominant industry factors have concentration load values 30 times higher than the residential equivalent. So the main effect on urban recharge quality is coming from industry and not from the residential sector. For this reason it was decided include a scenario that models the study area as if it were entirely an industrial zone. The impact can be observed in the results shown in Figure 3.3.23 and compared with an all-residential equivalent in Figure 3.3.24. While concentrations in the all-industry scenario are higher than they would be if the area were entirely residential, they are still not so high, especially in comparison to the impacts coming from agriculture.

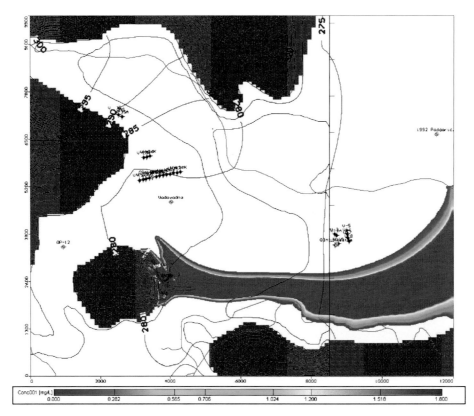

Figure 3.3.23. Effect of a conservative pollutant (Cl⁻) spreading from study area through Ljubljansko polje after 730 days if all the study area were an industrial zone.

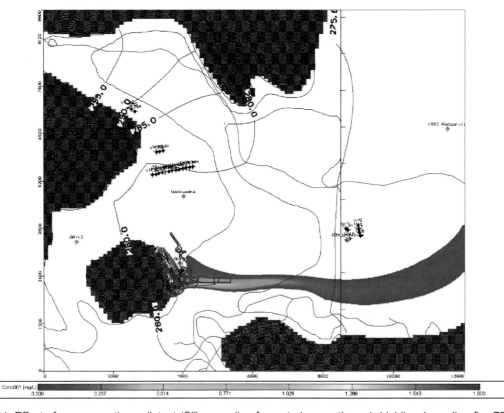

Figure 3.3.24. Effect of a conservative pollutant (Cl⁻) spreading from study area through Ljubljansko polje after 730 days if all the study area were residential; compare the very modest impact with that shown in Figure 3.3.23 for the all-industry equivalent.

3.3.9 Conclusions and observation outcomes

The scenarios selected for the Ljubljana case study were run through the whole DSS model chain but the obtained results should be treated with some degree of uncertainty. There are various reasons for this:
1. Ljubljana seems to be quite difficult to model with the UVQ because of the high percentage of industrial area within the study area. Water use and contaminant discharge figures are typically less available from industry.
2. The Ljubljana water authority has many problems with cadastral data that records the details of the assets comprising the sewerage and other pipeline systems. This meant that the first idea to capture the entire town of Ljubljana as a study area for the project was quickly shown to be too ambitious, as the area is too complex for testing the entire AISUWRS modelling chain. Consequently, the whole city area was reduced to a much smaller district with the expectation that it would correspondingly reduce data collation and processing tasks.
3. Nevertheless, even within the reduced area the same problems persisted, so the first task during the project was to build a database that could compensate for the missing data for sewage and pipeline network. This was very time consuming.
4. The next step was to improve data quality from the cadastre using raster GIS techniques and the last step was to improve the field data with detailed quality sampling of the unsaturated and saturated zones.

The improvement in the pipe asset data reduced the uncertainties of the input data sufficiently to allow the modelling to begin, but many uncertainties still remained. To point out just a few of them: no samples of water coming from roofs; lack of the data of sewage sediment and not enough results from the sewage wastewaters. All the missing parameters mentioned are vital for credible calibration of the UVQ model.

On the positive side, good quality groundwater data were available, which were used for the final evaluation of entire AISUWRS modelling chain. These allowed comparison of the results coming from pollution transport modelling with those from groundwater sampling. It was found that the calculated values were in some parts of the groundwater model a few times higher than expected from the groundwater quality monitoring. In contrast, using the model chain for modelling impacts on the groundwater levels (groundwater recharge) produced better results. So, the research has shown that greater confidence can be assigned to the water balance parts of the models than the data obtained for the pollution, mostly because the range of individual uncertainties is much wider in the latter case.

For the Ljubljana case-study the following observations can be made:
1. For the UVQ model there were very limited primary sources for many of the parameters (over 60% with no site-specific or local context values) and a lot of data needed to be sourced from other partner case studies and literature reviews. Reference data was required for typical water use amounts and ratios between different household uses (toilet, bathroom, kitchen etc.), contaminant profiles for different water discharges, and guideline values for calibration of stormwater, infiltration and wastewater components. With such a significant need to use non-local data the uncertainty in the modelling process is clearly increased.
2. The sewer drainage network itself provided several complications, due to its complexity (interweaving networks, separate networks close but not actually meeting within the study area, overflow connections and sewers actually splitting along the flow direction etc.) and missing data on flow directions and several z co-ordinates for asset nodes. The actual drainage network runs also over several km before entering the study area – making modelling of the full wastewater input difficult. Information on sewerage networks for a large central industrial area, originally intended to be part of the study was also unavailable.
3. No local data were available for the breakdown of consumption, nor for contaminant profiles between different indoor water uses (kitchen, bathroom, toilet and laundry). Reference values, which had to be taken from the literature, again increase the uncertainty of the model results.
4. More critically, considering the balance of consumption types and hence expected discharge, no contaminant data has been sourced for the specific industrial and commercial uses in the area. Many water users are present that produce very different contaminant loads than a typical residential area might. These include a major indoor swimming pool, an outdoor (summer) swimming pool, newspaper print works, textile works, a petrol station, a railway station, railway junctions and warehouses, garages (services and sales), carwashes, dry cleaners, restaurants and bars as well as various commercial units. No comprehensive cadastre of the activities present has been acquired. It has been possible to define some of these as separate UVQ neighbourhoods, whilst others are too mixed within other land-uses to separate out. Due their number and variety, time and resources were insufficient to source data on all these activities and some approximations unavoidably had to be made.
5. For the non-residential units, typical residential contaminant loads have been applied, multiplied by the population equivalent of the water consumption. For mixed and commercial units, water use was divided over the typical residential profiles (kitchen, bathroom, toilets and laundry) – representing a general use by day-workers. For

industrial areas, just bathroom contaminant data were used. This provides a predicted load, as if the same quantity of water was being consumed and discharged by a residential unit. It is assumed that other specific industrial contaminants of concern are removed by on-site treatment before discharge to the public sewers.

6. For groundwater impact a wider area than the study area was modelled. The main reason was that the model boundaries with fixed conditions need to be properly separated from the scenario simulation area in the city centre in order to avoid a suppression of the induced changes.
7. The maximum calculated velocities in the model domain are between 40 and 60 m/day. These results match the observations from tracer tests.
8. The pollutant loading was represented by concentration recharge boundaries in every neighbourhood. For dispersivity, values from a previously conducted tracer test were selected (30 L/m). The results for a 730 day time period demonstrated rapid movement down-gradient of pollution coming from the study area, mainly affecting the southeastern part of the polje rather than the Hrastje water supply wellfield.
9. The infiltration scenarios C1i and C1ii were modeled, but despite the resultant major increase in available recharge (Figure 3.3.19), no significant change in the piezometric levels was noted. This does not mean that changes were not present, but rather that the study area is too small to affect the piezometric levels. An additional important factor is that the study area is neither wholly residential nor wholly industrial but instead comprises a mix of the two land uses.
10. Ljubljana's town planning vision for the future is to move industry from the city centre to industrial zones on the borders of the city. Also a recognized water quality problem for Ljubljana city water supply is not the urban or industrial impact but the agriculture in surrounding periurban areas around the city. The initial results from this study suggest that residential land uses in an urban area may have significantly smaller impact on the groundwater than agriculture or industry. In this respect, use of sustainable urban development systems like on-site infiltration of roof runoff and improved sewer control and standards could produce in the next 50 years improvements to groundwater, a conclusion that was not expected at the beginning of this study.

3.4 A KARSTIC AQUIFER SYSTEM: MOUNT GAMBIER, AUSTRALIA

3.4.1 Background

3.4.1.1 Physical and geological setting

The regional city of Mount Gambier is located in the southeast corner of South Australia (Figure 3.4.1). It is situated approximately 30 km inland on the Gambier Plain on the flank of a volcanic complex formed around 28,000 years ago (Leaney et al. 1995). The rich volcanic soils and climatic conditions of the surrounding area are well suited to dairying, agriculture and forestry. As a result, the majority of native vegetation has been cleared or highly modified for primary production. Mount Gambier is often referred to as the "Blue Lake City" which is a reference to the lake of that name that each year during the warmer months turns a brilliant shade of blue. The Blue Lake is the best known of four lakes that occupy volcanic craters around the city.

Mount Gambier's drinking water supply is extracted from the Blue Lake, which is fed by the Gambier Limestone aquifer that underlies the city. It is a unique setting in that the aquifer is karstic and the bulk of stormwater is discharged directly into the aquifer via drainage wells, thus bypassing the unsaturated zone. Contaminant transport modelling in karstic aquifers is a significant challenge due to the large variation in possible flow rates and residence times within the aquifer, which in turn impacts on the potential for time-dependent attenuation. To deal with the uncertainty associated with the hydraulic properties of the aquifer, a methodology was developed to assess the potential risk of contaminants from the urban water system adversely impacting on groundwater quality and the supply of safe drinking water.

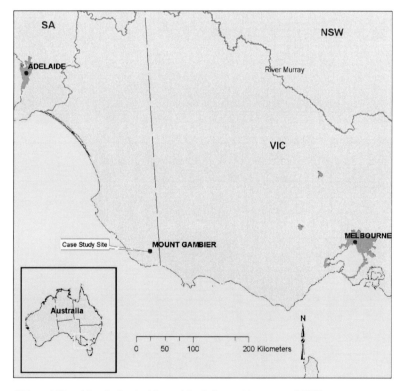

Figure 3.4.1. Location of Mount Gambier (adapted from Mustafa and Lawson, 2002).

The geological units of the Mount Gambier region were formed as recently as the Cainozoic era. The Dilwyn Formation, Gambier (Bryozoal) Limestone (comprising the Greenways, Camelback and Green Point Members), Bridgewater Formation and Holocene volcanic deposits that form the Mount Gambier region are displayed in Figure 3.4.2. The major structural features are related to rifting and post-depositional tectonics, which have formed a northwest-southeast structural trend through the region. Such faulting is thought to represent a zone of regional structural weakness through which the volcanic activity occurred (Lawson *et al.*, 1993). Karstic features within the Gambier Limestone, in the vicinity of Mount Gambier, are predominantly aligned in the northwest to southeast direction.

The geology within the heterogeneous Gambier Limestone consists of grey to cream bryozoal calcarenites with thin intervals of marl. Early work suggested three layers, an upper grey bryozoal calcarenite, a cream calcarenite with less fossil content and a lower calcisiltite (marl) (Mustafa and Lawson, 2002). More recently, Hill and Lawson (in press) have refined the stratigraphic record in the Mount Gambier region and describe alternating beds of bryozoal limestone and marl overlying a lower dolomitic Camelback member and again marl (Figure 3.4.3). Near the Blue Lake, the thickness of the Gambier Limestone is approximately 100 m (Waterhouse, 1977). The top of the dolomitic Camelback member is encountered at approximately -30 to -50 m Australian Height Datum (AHD) (Hill and Lawson in press) and this sub-unit is approximately 30 to 40 m thick. A confining layer of glauconitic and fossiliferous marls and clays separates the Gambier Limestone from the lower Tertiary Dilwyn Formation which consists of a series of unconsolidated sandstones and gravels with carbonaceous clay interbeds (Waterhouse, 1977).

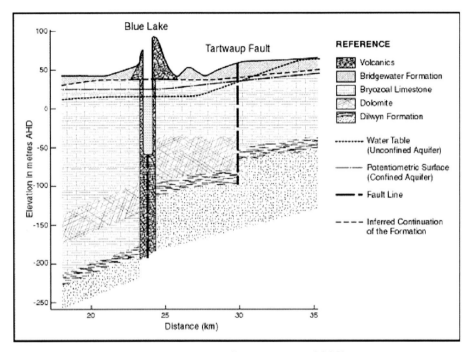

Figure 3.4.2. Geological units of the Mount Gambier region (Lamontagne, 2002).

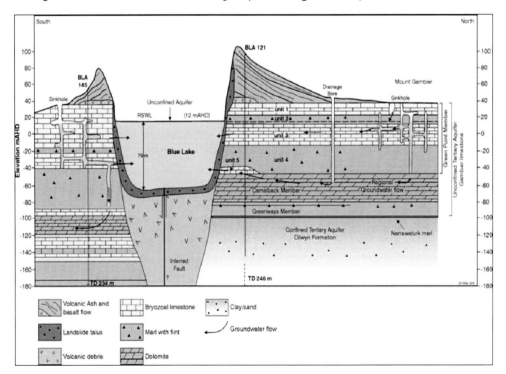

Figure 3.4.3. Recent stratigraphic assessment in the vicinity of the Blue Lake (Hill and Lawson, in press).

3.4.1.2 Hydrogeological setting

The important aquifer systems in the region are an unconfined aquifer contained within the Gambier Limestone and a confined sand aquifer in the Dilwyn Formation. The Gambier Limestone unconfined aquifer is the principal source of recharge to the Blue Lake (Waterhouse, 1977), which is the primary supply of water for Mount Gambier. The confined aquifer is used as an emergency potable supply for Mount Gambier.

The unconfined aquifer has characteristics of a dual porosity medium with secondarily-developed karstic flow supplementing primary intergranular porosity. Due to the karstic nature of the aquifer, pumping-test derived transmissivities vary over two orders of magnitude from 200 m^2/day to greater than 10,000 m^2/day, with porosity estimated between 30% and 50% (Love *et al.*, 1993). Thus there is considerable potential for variation in groundwater flow rates to Blue Lake via pathways through the porous matrix of the aquifer and preferential flow through karst features.

The regional groundwater flow direction in both the unconfined and confined aquifers is toward the southwest (Figure 3.4.4), with groundwater discharges via coastal and offshore springs. However, in the lake vicinity, the hydraulic gradient is low (10^{-4}) and local flow systems and the karstic domain can reverse the flow direction (Richardson, 1990; Schmidt *et al.*, 1998). The dolomitic Camelback Member is believed to be the dominant source for groundwater flow into the lake. However, there is a lack of hydraulic and water quality data for the subunits of the Gambier Limestone unconfined aquifer that could be used to ascertain flow directions and rates.

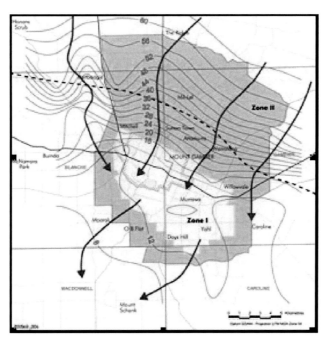

Figure 3.4.4. Water level contours and groundwater flow direction for the Gambier Limestone unconfined aquifer (Hill and Lawson, in press).

3.4.1.3 Urban Development Setting

The population of Mount Gambier was 22,751 at the time of the 2001 census. The population has grown steadily over the last decade with increases of 4% over the 5-year interval from 1991 to 1996 and 3.1% between 1996 and 2001, (ABS, 2001). Population projections predict this rate of population growth will not be sustained, with the Mount Gambier population projected to reach 24,421 by 2016 (Planning SA, 2005). The demographic structure of the Mount Gambier population is marginally younger than that for South Australia as a whole; with 22% of the population aged less than 15 years and 12% aged 65 or more, compared to South Australia as a whole where 19% of the population is less than 15 years and 14% are 65 or older (ABS, 2001). The age profile of Mount Gambier (Figure 3.4.5) is generally indicative of families with children. The dip in the age cohort 20-24 may be due to out-migration of young adults for employment or education opportunities.

It is an important regional centre for the south east of South Australia and Victoria's western districts and supports a significant rural economy with timber, dairy, fishing, horticulture and agriculture being some of the major industries in the region (ABS, 2001). Figure 3.4.6 shows that most industry sectors grew in the period 1991 to 2001, with a notable exception in the electricity, gas and water supply sector where the workforce declined by almost half, influenced by major restructuring and privatisation within these sectors. There is significant retail, business, health and education infrastructure that services the surrounding population. The unique character, associated with the volcanic landscape, including the spectacular Blue Lake, and the many heritage-listed buildings constructed from locally sourced dolomite and limestone are important for the local tourist industry.

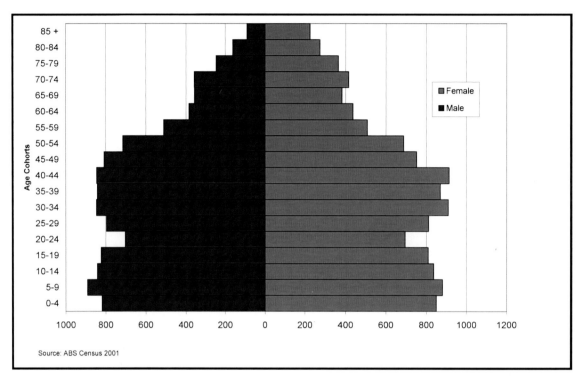

Figure 3.4.5. Population of Mount Gambier (2001) by Sex and Age Cohort (ABS, 2001).

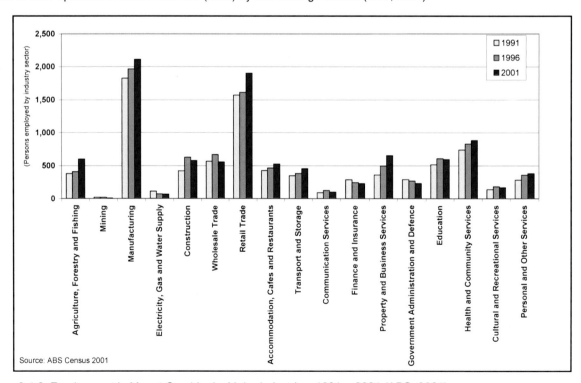

Figure 3.4.6. Employment in Mount Gambier by Major Industries, 1991 – 2001 (ABS, 2001).

Figure 3.4.7 provides an aerial overview of the Mount Gambier case study area. The urban area occupies 27 km², comprising approximately 40% (11 km²) of impervious surfaces. The major land use is residential (13 km²), characterised by separate houses situated on land blocks that are approximately 980 m². Public open space occupies 5.2 km², with a significant portion of this situated around the volcanic crater lakes. There are two main industrial areas (total area 2.7 km²), situated on the western and eastern fringes of the city. Timber processing plants are the major heavy industry in these areas.

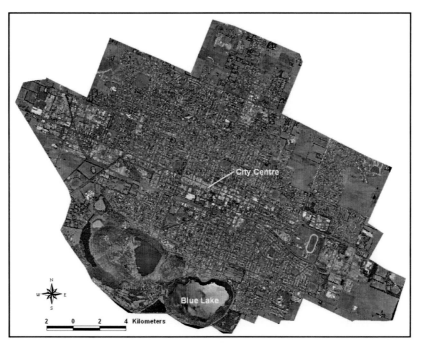

Figure 3.4.7. Aerial overview of Mount Gambier study area (Data source: Mount Gambier City Council, 2004).

3.4.2 Urban Water Infrastructure

3.4.2.1 Sewer System

The majority of Mount Gambier's urban area has been served by a reticulated sewer system since the 1970s. The sewer system comprises approximately 177 km of gravity sewers. The type of pipe material is often related to the period in which the pipe was laid. Figure 3.4.8 shows that the more recent urban developments on the outskirts of Mount Gambier, which were developed in the 1990s or later, are predominately serviced by PVC, while older areas are mostly served by reinforced concrete and vitreous clay pipes. Properties outside the urban area have still retained the use of septic tanks.

The sewer pipe materials and the approximate years they were laid are listed below (SA Water, 2004):

AC:	Asbestos Cement	(1964 – 1978)
CI:	Cast Iron	(1964 – 1967)
Conc:	Concrete	(1964 – 1966)
MSCL:	Mild Steel Cement Lined	(1977)
PE:	Polyethylene	(2003)
PVC:	Polyvinyl Chloride	(1965 – 2004)
RCRJ:	Reinforced Concrete Rubber Jointed	(1964 - 1971)
UNKN:	Unknown	
VC:	Vitrified Clay	(1964 – 2000)

Mount Gambier wastewater is treated at the Finger Point Wastewater Treatment Plant (WWTP), which is located on the coast to the south of Mount Gambier and treated effluent is discharged directly to the sea. The WWTP was commissioned in 1989 and has been designed for a plant flow of approximately 6000 m³ per day (SA Water, 2004). In the 2001/02 financial year 1.8 Mm³ of treated effluent, which included an estimated load of 1.6×10^7 g of phosphorous and 1.1×10^7 g of nitrogen (SA DEH, 2005), was discharged from Finger Point WWTP. Table 3.4.1 provides an overview of volumes of water supply and treatment for two reference years (1995 and 2005).

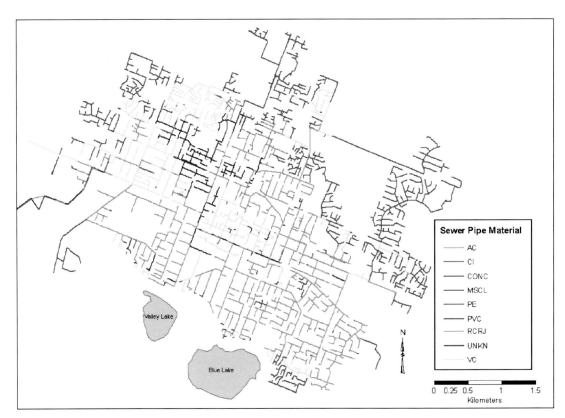

Figure 3.4.8. Mount Gambier sewer network (Data Source: SA Water, 2004).

Table 3.4.1. Summary of water supply and treatment (1995 and 2005).

Volume (Mm^3/year)	1995	2005
Waste water entering Finger Point WWTP	1.8	1.7
Potable supply extracted from Blue Lake	3.3	3.6

3.4.2.2 Stormwater

Owing to the lack of natural surface drainage, Mount Gambier's stormwater has been discharged to drainage wells and natural karstic features since settlement by Europeans in the 1850s. The high permeability of the Gambier Limestone allows direct drainage of stormwater to the unconfined aquifer. This is predominantly achieved via drainage boreholes but in some cases, natural solution features in the karst landscape, such as sinkholes, have been utilised for the discharge of stormwater.

There has been a marked increase in impervious cover in Mount Gambier, from 4.2 km^2 in 1985 (Emmet, 1985) to the current estimate of 11 km^2 (section 3.4.1.3), which is testament to the ongoing population growth and associated urban development in the area. The volume of stormwater discharged to drainage wells annually is estimated at 2.9-4.2 Mm^3, based on rainfall of 625-870 mm/year, evaporation of 1270-1320 mm/year and 184-207 rainfall days/year. Stormwater is discharged to the unconfined aquifer by approximately 500 stormwater drainage wells in Mount Gambier (Figure 3.4.9). Well log records indicate that around 70% of the drainage wells are between 50-80 m deep, which means that most wells penetrate 25-55 m into the unconfined aquifer (Emmett, 1985). A large proportion of these drainage wells target the dolomitic Camelback Member, which also provides recharge to the Blue Lake.

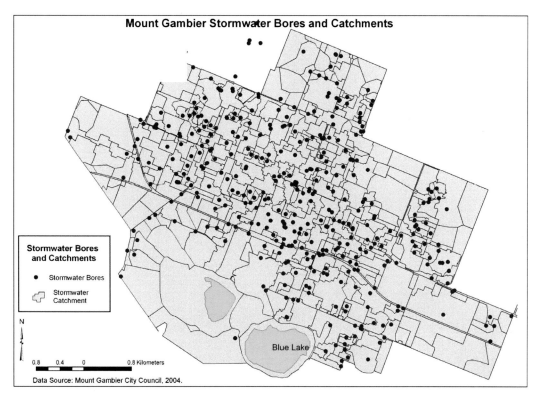

Figure 3.4.9. Stormwater drainage bores and catchments (Data Source: Mount Gambier City Council, 2004).

3.4.2.3 Water Supply

The drinking water supply for Mount Gambier is sourced entirely from the Blue Lake. Reticulated supply commenced in the late 1800s and there are approximately 305 km of water assets to service the population (Figure 3.4.10). The majority of the older pipes are asbestos cement, but since the middle of the 1990s installed water supply pipes are predominately made of PVC.

The water pipe materials and the approximate years they were laid are listed below (SA Water, 2004):

AC:	Asbestos Cement	(1883 – 1989)
CI:	Cast Iron	(1911 - 2003)
DICL:	Ductile Iron Cement Lined	(1967 – 1998)
HDPE:	High Density Polyethylene	(2000 - present)
MS:	Mild Steel	(1954)
PVC:	Polyvinyl Chloride	(1997 – present)
UNKN:	Unknown	

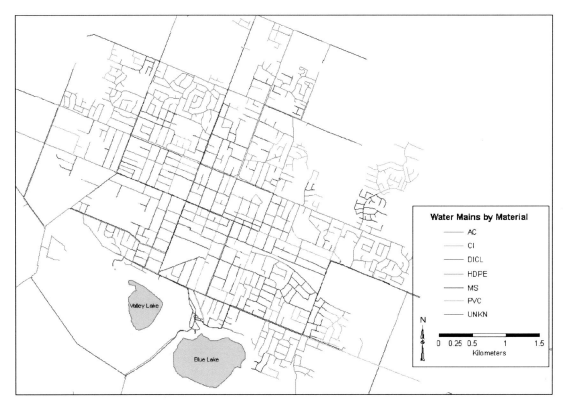

Figure 3.4.10. Mount Gambier water mains network (Data Source: Mount Gambier City Council, 2004).

3.4.3 Resource development studies

3.4.3.1 Water Balance – Blue Lake

The steep walls of the Blue Lake result in a surface catchment area (0.86 km^2) only slightly greater than the surface area of the lake itself (0.61 km^2). The depth of water and volume of the lake are approximately 70 m and 36 Mm^3 respectively. The hydrology of the Blue Lake has been modified substantially since European settlement in the 1840s. Prior to this date, the residence time of water in the lake was thought to be in excess of 20 years, and the main water losses were from evaporation and groundwater outflow. At the peak of extraction, during the 1970s, about 15% of the lake volume was removed on an annual basis. Pumping has resulted in a large increase in groundwater inflow to the lake and a reduction of water residence time to 8±2 years (Herczeg et al., 2003). Withdrawal rates have stabilised at approximately 3.6 Mm^3 annually, which is comparable to the volume of stormwater runoff recharging the Gambier Limestone unconfined aquifer. While the lake may be close to a hydrological steady-state, the water level in the Blue Lake still declined by approximately 6 m between 1900 and 1997 (Lamontagne, 2002). Long-term climatic data spanning from the 1860s to the present indicates that Mount Gambier has been in a period of below-average rainfall for the past decade and that lake levels have declined by approximately 2 m during this period (BLMC 2001).

Groundwater is the predominant inflow to the Blue Lake, with minimal input from precipitation to the surface. Estimates for groundwater recharge to the Blue Lake of 3.5-4.5 Mm^3 per year (Turner et al., 1984) are affected by the uncertainty associated with quantification of the lake outflow (is it a groundwater sink or a flow-through system?) (Lamontagne and Herczeg, 2002). Isotopic tracers have indicated a 9:1 ratio in the contribution to Blue Lake from the unconfined and confined aquifers respectively (Turner et al., 1984; Ramamurthy et al., 1985). The annual stormwater contribution to groundwater recharging the lake has been estimated as 35-55%. However this was based on the behaviour of chloride and there is considerable variability in the groundwater end-member chloride concentration which could alter this calculation.

Lamontagne and Herczeg (2002) report that the water balance of the Blue Lake (Equation 3.4.1) is not completely understood, but estimates are as follows:

$$\frac{d[V_L]}{dt} = I_P + I_G - E - O_W - O_G \qquad \text{Equation 3.4.1}$$

where $\dfrac{d[V_L]}{dt}$ = the change in the Blue Lake volume with time

I_P = precipitation input 0.5 Mm³/year
I_G = groundwater input 3.5-4.5 Mm³/year
E = evaporation 0.7 Mm³/year
O_W = pumped withdrawals 3.6 Mm³/year
O_G = groundwater outflow 0-1 Mm³/year

3.4.3.2 Data Sources

Data for running scenarios in the DSS model chain has predominantly been obtained from unpublished sources, as contributions from the various government agencies in Mount Gambier (South Australia Environment Protection Authority [SA EPA], South Australia Water Corporation [SA Water], Department for Water, Land and Biodiversity Conservation [DWLBC] and City Council of Mount Gambier [CoMG]). Where Mount Gambier-specific data were not available for certain model parameters the value has been based on the reviewed literature. Existing data sources have been complemented by supplementary data, collected as part of the AISUWRS project and are described in the following section.

Table 3.4.2: Data sources for application of models.

Model	Data	Source	Comments
UVQ	Spatial dimensions (land use, block size, pervious/impervious surfaces etc)	CoMG (unpublished): GIS layers (cadastre, aerial imagery, roads, land use etc)	Neighbourhoods and their characteristics were defined using this data, particularly the aerial images
	Household occupancy	ABS (2001) population census	
	Indoor water consumption	SA Water (unpublished) WSAA (2003)	Due to lack of data indoor water consumption was considered uniform between all neighbourhoods
	Climate Data	BOM (2005): Mount Gambier airport station	Baseline climate was based on 35 year climate averages
	Climate Projections	CSIRO Atmospheric and Marine Research OzClim software (Jones et al., 2001)	Climate change factors were generated using OzClim and then applied to baseline climate data
	Contaminant loads – indoor wastewater	Gray and Becker (2002) Muttamara (1996) Christova-Boal et al. (1996) Magara (1998)	Data specific to Mount Gambier were not available
	Contaminant loads – rainfall, run-off, water supply, wastewater etc	DWLBC (unpublished) SA Water (unpublished) CSIRO (See section 3.4.2.5)	First flush values were not available for run-off, therefore it was assumed that first flush concentrations were double that of subsequent run-off (Mitchell and Diaper, 2005)
	Volume of flows – wastewater, water supply and stormwater	SA Water (unpublished) BLMC (2001) Emmett (1985)	
NEIMO	Wastewater asset information	SA Water (unpublished): GIS layer with associated database	CCTV data was not available for Mount Gambier
	Water supply leakage	SA Water (unpublished)	Unaccounted water was calculated as the difference between supplied (extraction from Blue Lake) and metered volumes
SLeakI and POSI	Soil data	DWLBC (unpublished) Hamblin and Greenland (1972)	
	Depth of unsaturated zone	Rosemann (2004)	

Risk Assessment	Contaminant loads (stormwater and sewer leaks)	Emmett (1985) SA Water (unpublished) SA EPA (unpublished) Makepeace et al. (1995) ARQ (2003) Rosemann (2004) Komarova et al. (2005)	Integrates loads, pathways, attenuation and target concentrations showing relative risk and effects of interventions on risk
	Target concentrations	NHMRC and NRMMC (2004) SA EPA (2003)	
	Attenuation coefficients	Vogel (2005)	
	Concentrations in groundwater and Blue Lake	SA Water (unpublished) SA EPA (unpublished)	

3.4.3.3 Field and Laboratory Programme

The objectives of the field and laboratory programmes for Mount Gambier were:
- to obtain additional sewage quality data for the suite of indicator parameters used in the AISUWRS programme (Cl, Na, K, SO_4, B, P, N and various organics) to allow calibration of the UVQ model;
- to assess site-specific stormwater-based contaminant concentrations for assessment of the risk associated with stormwater disposal;
- to obtain concentrations in the Blue Lake and the unconfined aquifer for comparison with simulated/calculated concentrations;
- to quantify the potential for contaminant attenuation in a carbonate aquifer; and
- to assess residence time in the saturated zone, using both applied tracers and anthropogenic organic species markers, from which to base the calculation of contaminant attenuation.

Water quality data

SA Water provided local sewage quality data collected between 1995 and 2004. The monitoring location chosen to gauge the quality of sewage entering the Finger Point WWTP was the number 1 pumping station. The list of parameters monitored in this period included some of the nominated AISUWRS tracers (N, P, organic indicators) but not all. Therefore effluent entering the Finger Point WWTP was sampled and analysed for Cl, Na, K, SO_4 and B on 8 occasions between December 2004 and May 2005.

Stormwater quality monitoring was undertaken between 1978 and 1982 (Emmett, 1985) and more recently between 1999 and 2002 (URS, 2000; 2003). Additional stormwater sampling was undertaken to determine the temporal and spatial variability of contaminants and to quantify a wider suite of analytes than previously considered, in particular a range of organic contaminants. Two stormwater drainage bores (SWB141 and SWB125) were utilised in a detailed stormwater quality study. Stormwater sampling combined rainfall event sampling (Rosemann, 2004) with integrated or 'passive' sampling for organic species (Komarova et al., 2005).

SA EPA and SA Water monitor the quality of Blue Lake, and an extensive data set beginning in the late 1960s was available. SA Water currently utilise the pumping station to collect a lake surface sample. The frequency of sampling varies from fortnightly to annually depending on the parameter under consideration. In addition, the EPA monitor water quality in the Blue Lake on a 3 monthly basis at four depths (surface, 20 m, 40 m and 60 m).

A sample taken from the Blue Lake at surface level on 15 April 2004 was concentrated using a reverse osmosis (RO) unit for analysis of trace organic species. Effectively this process concentrated the sample by a factor of greater than 30, to reduce the detection limits (i.e. 3 ng/L becomes about 100 ng/L). Also submitted for analysis was an unconcentrated Blue Lake surface water sample and a blank, prepared by passing demineralised water through the RO unit. Two samples from the Blue Lake, collected on 16 May and 12 July 2005, were also submitted to laboratories (Technical University Berlin and Southern Nevada Water Authority) with the capacity to measure organics, including sewage indicators, to ng/L detection limits.

A groundwater quality data set beginning in 1995 was obtained from the SA EPA. Groundwater samples were collected from 9 locations in January 2005, for total petroleum hydrocarbons (TPH) (n=4) and boron (n=7) analysis. In addition, the samples collected in January 2005 as part of the SA EPA's regular groundwater monitoring programme were submitted for boron analysis (n=22).

Contaminant Attenuation

Contaminant attenuation constants were predominantly taken from the existing literature. A laboratory batch study was undertaken to assess the removal of boron in carbonate aquifer material (Vogel, 2005). In the batch study, 10 g portions of aquifer material and 50 mL boron solutions were shaken for 24 hours, decanted, centrifuged, and filtered with subsequent boron analysis by ICP. This analysis was performed for three intervals from the Gambier Limestone aquifer varying in calcite and dolomite content. Boron solutions were prepared in water from the Blue Lake to maintain a suitable sample matrix, with B concentrations from 0 to 3 mg/L (typical sewage concentrations) and 0 to 100 mg/L (for comparison with literature). Each batch study sample was conducted in duplicate. The amount sorbed was calculated from the difference between the initial and final boron concentration in solution.

Estimating residence time within the aquifer and the Blue Lake

To assess attenuation with time, it was necessary to estimate the likely residence time within the aquifer and Blue Lake. The annual extraction from Blue Lake is approximately one tenth of the lake volume, which suggests a residence time of around 10 years, which has been confirmed by isotopic analysis of lake waters at 8 ±2 years (Herczeg et al., 2003).

The input variables for calculating aquifer residence time (see section 2.5.4.3.5) in the karstic aquifer are uncertain (Table 3.4.3). Stormwater recharge is directed to the dolomitic Camelback member of the Gambier Limestone aquifer, thus for a worst-case scenario, the thickness of this aquifer horizon was used (approximately 30 m) rather than the entire aquifer (100 m). Geophysical data indicates an effective porosity of approximately 0.1 within the matrix, increasing to 0.25 in areas of strong fracturing (J. Lawson 2004, pers. comm.). Minimum travel times vary significantly with different aquifer thicknesses (Table 3.4.4) indicating the importance of a better understanding of the hydrostratigraphy of the unconfined aquifer. Based on an aquifer thickness of 30 m, an estimated range for minimum residence time (t_{min}) of 1 to 20 years was selected for use in the risk assessment (see equation 6 in section 2.5.4.3.5).

Table 3.4.3. Aquifer parameters to be used within the risk assessment.

Parameter	Value	Reference
Aquifer thickness [D (m)]	30-100	J. Lawson 2004, pers. comm.
Effective porosity [n_e]	0.1-0.25	J. Lawson 2004, pers. comm.
Pumping rate [Q (m^3/d)]	3.6×10^3 m^3/year = 10^4 m^3/d	

Table 3.4.4. Calculated t_{min} values based on aquifer parameters.

Distance between injection and recovery [L (m)]	t_{min} (years) with D=30m	t_{min} (years) with D=100m
1000	0.9-2.2	2.9-7.7
2000	3.5-8.6	12-29
5000	22-54	72-180

Given the uncertainty regarding travel time in the karstic aquifer and the importance of this parameter for quantifying time dependent attenuation, estimates from hydraulic data can be combined with estimates from hydrogeochemical tracers. An example of a suitable hydrogeochemical tracer is the identification of a solute, such as an anthropogenic species, that can be linked to specific events or timeframes. Organic iodine compounds, including triiodinated benzene derivatives, have been employed as X-ray contrast agents in Australia (including Mount Gambier) since around the 1950s and were chosen as a suitable tracer for this case study. Thus a positive identification of X-ray contrast media within the Blue Lake would indicate recharge post 1950s, or a combined unsaturated and saturated zone residence time of less than 50 years. Their concentrations can be measured collectively as total organic iodine (Adsorbable Organic Iodine (AOI) or as the specific active ingredient. The Blue Lake samples collected from the centre of the lake in May 2005 had 0.7 µg/L AOI at 40 m and 60 m, but the surface and 20 m samples were <0.5 µg/L. In addition, all samples contained iopromide, an active ingredient in X-ray contrast media, at concentrations of 12 ng/L at the surface, 34 ng/L at 20 m, 19 ng/L at 40 m and 17 ng/L at 60m. However, a sample collected in July 2005 did not identify any X-ray contrast media. Detection of X-ray contrast media, shown by AOI and iopromide concentrations, confirmed that recent water (<50 years) has recharged Blue Lake. Given that the use of X-ray contrast media is increasing, it would be recommended to include an annual analysis for these types of markers, with sampling recommended when the lake is fully mixed (August).

To further define groundwater travel times, an applied tracer test, utilising sulphur hexafluoride (SF_6) was undertaken in August 2005. SF_6, a harmless, non-toxic gas, can be used to 'tag' large volumes of water and is detected at extremely low concentrations (~10^{-14} M), which makes is suitable for the extensive dilution within this system. Approximately 400L of water tagged with SF_6 was injected into each of 24 boreholes located in Mount Gambier. Boreholes were between 1 and 3 km distance from Blue Lake and penetrated to the depth of the dolomitic layer. Each injection was followed by at least two well volumes of water to displace the tracer into the formation. The subsequent SF_6 monitoring program within the Blue Lake includes weekly sampling from a pontoon at the pumping station (taken approximately 5 m below the surface) and monthly at four depths (surface, 20 m, 40 m, 60 m) in the centre of the lake. Results to date (May, 2006) indicate SF_6 had not reached Blue Lake, which confirmed the minimum residence time within the aquifer for stormwater to reach Blue Lake is greater than 9 months. The initial monitoring results are critical as attenuation half-lives for a number of species are of the order of months.

3.4.4 AISUWRS model set-up for Mount Gambier

This section describes the set-up of the AISUWRS model chain. The DSS provides a graphical user interface that allows the user to easily set-up and run the model chain. The DSS also facilitates information transfer between the models. At this stage, the saturated zone risk assessment is run manually with data provided from the DSS. The set-up and definition of scenarios occurs within the UVQ model, which is the initial model in the chain. UVQ defines the characteristics of the water use system and provides the inputs to drive the other AISUWRS models. The results of different scenarios can be compared from within the DSS.

3.4.4.1 Scenario Definition

Scenarios were defined for Mount Gambier in consultation with local stakeholders, and were based on common scenarios developed for the AISUWRS project. The four scenarios analysed within the DSS for Mount Gambier were simulated over a 30-year period (2005-2035) to allow for the potentially long time between a pollution event and contamination of the groundwater body occurring, and are described briefly below:

1. Status quo: continuation of the current water demand and climate pattern for Mount Gambier. The climate is based on 30-year climate averages collected from the Mount Gambier airport weather station.
2. Climate change: a decrease in rainfall and increases in average temperature and evaporation based on predictions using the global emission scenarios from the Intergovernmental Panel on Climate Change (Jones *et al.*, 2001).
3. Population increase: an increase in the Mount Gambier population of 15%. This assumes that all existing vacant land (1.4 km^2) and 70% of agricultural land (1.05 km^2) within the study area will be converted into residential development.
4. Greywater recycling: 20% of all residential households install a greywater recycling system for garden irrigation and toilet flushing.

The focus of the scenarios is on the quantity of urban water flows rather than quality. There is no difference in the concentration of contaminants between scenarios 1-3, but there are significant differences in the flow volumes, which then impacts on the total contaminant load reaching the groundwater resource. In the greywater recycling scenario the contaminant concentrations in the wastewater output are higher than the baseline scenario, due to the removal of the less contaminated wastewater streams (e.g. laundry) for greywater recycling. Possible management interventions are considered for the scenarios, including wellhead stormwater treatment, proactive inspection and repair of sewer leaks and additional treatment of potable water extracted from Blue Lake.

3.4.4.2 UVQ

UVQ models urban water flows and contaminant balances. The following section details the set-up of key parameters for the scenarios, and also the assumptions that were made in defining Mount Gambier's urban water system. Detail of the UVQ modelling approach itself is covered in Chapter 2.

Spatial Dimensions

In Mount Gambier the study area was defined as the extent of the stormwater catchments and comprises an area of approximately 27 km^2, representing the total Mount Gambier urban area. The neighbourhoods were defined by aggregating similar land uses that were relatively homogeneous in terms of water use, impervious surface coverage, block size and occupancy rates. Eleven neighbourhoods have been defined that represent the dominant land uses. The neighbourhoods are not necessarily spatially contiguous (Figure 3.4.11).

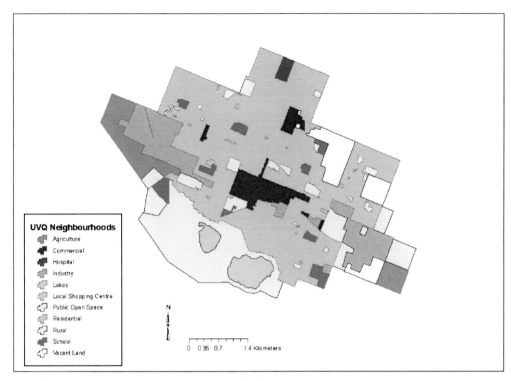

Figure 3.4.11. UVQ Neighbourhoods defined for Mount Gambier.

UVQ requires a range of physical data that describe the characteristics of the neighbourhoods and their land blocks. The majority of this data was acquired through the GIS analysis of aerial imagery supplied by the Mount Gambier City Council. Defining the impervious and pervious surface dimensions is important in order to realistically simulate the quality and quantity of stormwater. In Mount Gambier representative land blocks were selected for each of the neighbourhoods in order to calculate the average garden, roof and paved areas. Table 3.4.5 summarises the key characteristics defined for Mount Gambier's neighbourhoods.

Table 3.4.5. Neighbourhood and land block summary.

Neighbourhood	Area (km^2)	Ave. occupancy per block	Block area (m^2)	No. of blocks	Total impervious area (km^2)*	Total pervious area (km^2)
Agriculture	1.5	2.6	6977	209	0.1	1.4
Commercial	1.2	9	1474	657	1.2	0
Hospital	0.2	200	153684	1	0.1	0.1
Industry	2.7	12	6275	360	2.4	0.3
Lakes	0.8	0	0	0	0	0.8
Local Shopping Centre	0.1	3	869	74	0.1	0
Public Open Space	5.2	0	62637	83	0.2	5.0
Residential	13.1	2.6	988	11266	6.8	6.3
Rural Living	0.2	2.6	6057	36	0	0.2
School	0.7	400	40445	15	0.3	0.4
Vacant Land	1.4	0	10233	133	0	1.4

*Total impervious and pervious areas represent the entire neighbourhood area, including block area and also ancillary land uses, such as roads and public open space

Climate Inputs

UVQ requires the following climate data on a daily time step for the period that the model is to be run:
- Precipitation
- Pan evaporation
- Average temperature

In Mount Gambier the DSS, and therefore UVQ, was run over the period 2005 to 2035. Climate projections for the thirty-year period were made based on two different future climate patterns (baseline and climate change).

The baseline scenario, as the name suggests, looks at simulating a climate pattern that is similar to what has been experienced in Mount Gambier in the recent past. The scenario was developed using historical climate records from the

Bureau of Meteorology (BOM), collected from the Mount Gambier airport weather station (station no. 26021) from the period 1970 to 2004 (SILO, 2004). The daily average temperature was calculated using the mean of the daily maximum and minimum temperatures. To project values for average daily temperature and potential evaporation, the 35-year mean was calculated for each day of the year and then projected as a repeating annual data series to 2035. This approach allowed the simulation of daily climate variability, which was useful in ensuring realistic simulation of weather patterns.

To project rainfall, a mean of historical daily climate was not used, as this would have the effect of spreading rainfall events evenly over the projected period. The simulation of rainfall needs to account for the episodic nature of rainfall events and also the probability of high rainfall days and days without rain (Chapman, 1998; Gyasi-Agyei, 2005). This enables a realistic simulation of stormwater run-off and groundwater infiltration, as well as likely irrigation demands. To address this issue for the baseline climate the rainfall pattern from the last 30-years was replicated for the next 30-years.

The second climate scenario looks at a potential shift in the climate due to global warming. This scenario uses climate inputs based on modelling undertaken by CSIRO Atmospheric Research (Jones *et al.*, 2001). The climate change scenario for Mount Gambier simulates a decrease in rainfall, and an increase in average temperature and evaporation based on global emission scenarios from the Intergovernmental Panel on Climate Change (IPCC). This scenario allows the exploration of the possible consequences of future climate change on issues such as likely changes in groundwater recharge.

Climate change factors were produced for five-year intervals up to 2035 using the OzClim software developed by CSIRO Atmospheric Research (Jones *et al.*, 2001). The driving forces behind the different climate change scenarios are greenhouse gas emissions and the sensitivity of the climate to these emission levels. The scenario used for this project simulates medium climate sensitivity to increased greenhouse gas emissions and is based on the assumptions that globalisation is dominant, economic growth is rapid, population growth is low, uptake of new technology is rapid and there is maximum use of available fossil fuel resources (http://www.dar.csiro.au/impacts/ozclimemissions.html).

The change in climate is displayed at a spatial resolution of 1:500,000 using 5 km grid cells. The value for each location is extracted by performing a linear regression of the global climate change model versus the observed base climatology, which is spatially interpolated from individual weather stations.

Three climate variables were generated in OzClim:
- Monthly precipitation
- Monthly average temperature
- Monthly point potential evaporation

All of these climate variables were generated as change from a base climate at five yearly intervals from 2005. There was a linear interpolation of values between the five-year periods in order to simulate the pattern of potential climate change. Figure 3.4.12 depicts the projected change in climate from the baseline climate over the 30-year period.

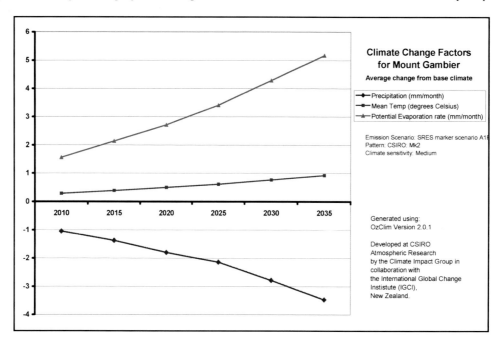

Figure 3.4.12. Climate change factors for Mount Gambier – Change from base climate.

Water Consumption

Indoor water consumption figures specific to Mount Gambier were not available. Therefore, figures were taken from the Water Services Association of Australia (WSAA, 2003), which collects data from water authorities throughout Australia to provide an overview of trends in water supply and consumption. In South Australia, average water consumption is 0.33 m^3/capita/day, with approximately 70% of this consumption being for indoor purposes (WSAA, 2003). A lack of empirical data made it difficult to accurately estimate indoor water consumption for non-residential neighbourhoods. This resulted in water consumption for non-residential neighbourhoods being over-estimated.

UVQ requires indoor water consumption for neighbourhoods to be assigned, based on usage in the kitchen, bathroom, toilet and laundry. The selected breakdown was based on figures from SA Water for the average suburban home (http://www.sawater.com.au/sawater), and these are consistent with WSAA figures.

Contaminant Data

Contaminants modelled in UVQ for Mount Gambier were:
- Boron
- Potassium
- Sodium
- Chloride
- Nitrogen
- Phosphorus
- Suspended Solids

Where possible the calibration of UVQ contaminants was based on the field assessment undertaken as part of this project. Contaminant loads for specific flow paths (road, roof and pavement runoff) were obtained from a review of the literature (Christova-Boal et al., 1996; Gray and Becker 2002).

Rainfall quality was based on unpublished data collected from a residential house lot in Mount Gambier (DWLBC, unpublished). Contaminant concentrations through different flow paths in residential water systems were based on Australian data from Gray and Becker (2002) and unpublished data from Gray. This, in particular, was used to determine the proportions of nitrogen, phosphorus and suspended solids for different indoor water streams (Gray and Becker, 2002). Where explicit information was not available regarding the proportion of contaminant flows through indoor streams, information this was inferred from other sources based on international data (Muttamara, 1996; Christova-Boal et al., 1996; Magara et al., 1998). The total contaminant load for residential wastewater was verified by SA Water to ensure that it was representative of the contaminant loads reaching their wastewater treatment facility.

Contaminant concentrations for imported drinking water were based on water quality information from the Blue Lake. Boron and sodium were considered conservative so there was no difference between treated and untreated water. The majority of suspended solids were removed through filtration as part of the treatment process, while values for nitrogen and phosphorus were based on figures for Mount Gambier from SA Water. The first flush runoff values were based on UVQ defaults, where first flush is double the concentration of subsequent roof runoff (Mitchell and Diaper, 2005).

3.4.4.3 Network Exfiltration and Infiltration Model (NEIMO)

The NEIMO model calculates exfiltration and infiltration from stormwater or wastewater pipe assets. In Mount Gambier NEIMO is only applied to wastewater assets as stormwater is directly discharged to the subsurface via stormwater drainage boreholes. NEIMO simulates defects, such as cracks or displaced joints, on each pipe asset in the network, the wastewater flow through the system, the position of groundwater relative to the pipe and determines the leakage from (exfiltration) or into (infiltration) each asset. The size and distribution of defects can be related directly or indirectly to CCTV damage inspection reports. In the case of Mount Gambier, CCTV data was not available for the wastewater network, therefore generic defect data were used. These defect data are not specific to a particular pipe network and provide an estimate of defect size and distribution based on pipe characteristics (age, material, connections). The wastewater flow and contaminant load is passed to NEIMO from a UVQ output file.

The following section briefly describes the preparation of NEIMO input files for the Mount Gambier case study.

Pipes file

This file provides the basic information about each of the pipe assets in the network, with data derived from the database of sewer assets maintained by SA Water. The neighbourhood ID, which identifies the UVQ neighbourhood each asset is located in, was generated by performing an INTERSECT operation within ArcView GIS, which returned all the pipe assets within each of the neighbourhoods. Where a pipe asset was present across two neighbourhoods one record was deleted (as each asset had to be exclusive to one neighbourhood) with selection based on which neighbourhood contained the largest proportion of the asset.

Connections file
This file lists the number of property connections for each asset. The information for Mount Gambier was sourced from SA Water.

Successor file
This file specifies the upstream and downstream relationship between pipe assets. In Mount Gambier such information was not readily available from the water authority, and so was generated from a manual analysis of the GIS layer. This was a time consuming process, but was feasible for Mount Gambier as the entire network has less than 3,500 assets.

Depth below Groundwater (DBG) file
This file gives the position of the pipe relative to the water table. If a pipe is below the water table pipe infiltration is calculated, while if it is above the groundwater table exfiltration is calculated. In Mount Gambier all assets were above the water table, so pipe infiltration was not considered.

3.4.4.4 Unsaturated Zone Transport Models (SLeakI and POSI)

The set up of SLeakI and POSI within the DSS requires a number of parameters relating to soil and groundwater conditions that are important in determining the movement and attenuation of contaminants through the unsaturated zone. Details of the requirements are given in Chapter 2.

SLeakI
The two soil layers represented in SLeakI must account for the whole of the unsaturated zone, which for Mount Gambier consists of volcanic ash deposits with a significant proportion of material from the underlying Gambier Limestone Formation (Smith, 1980). At present, the two layers that can be defined for the unsaturated zone must be described as one of eight soil types for the purposes of representation in the unsaturated models. Therefore, assumptions had to be made in order to assign the unsaturated zone to two soil types. The depth of the unsaturated zone was based on the analysis of borehole drilling logs undertaken by Rosemann (2004). It varied from around 22 m to more than 80 m due to topographical relief. For the purposes of the unsaturated zone models, the depth of the unsaturated model was assumed to be 30 m, which in most cases was an underestimate. Table 3.4.6 lists the values that have been used for SLeakI soil parameters.

Table 3.4.6. Soil characteristics used in SLeakI model for Mount Gambier.

Layer	Characteristics
Surface	Soil type: Sandy Loam (Hamblin and Greenland, 1972)
	Thickness: 12 metres (Rosemann, 2004)
	Temperature: 13°C (BOM, 2005)
	pH: 6.1 (Hamblin and Greenland, 1972)
Sub-Surface	Soil type: Sandy Clay Loam (Hamblin and Greenland, 1972)
	Thickness: 18 metres (Rosemann, 2004)
	Temperature: 12°C (BOM, 2005)
	pH: 6.5 (Hamblin and Greenland, 1972)

POSI
The parameters required by POSI include rooting depth of vegetation. The rooting depth is needed as POSI considers infiltration from below the root zone. The behaviour of contaminants in the root zone, particularly with respect to nitrogen, is a gap in the model chain that is not covered by AISUWRS. The groundwater temperature and pH are important in determining the behaviour of contaminants within the unsaturated zone. The following values were used for POSI parameters.

Depth of root zone: 1 metre   Groundwater temperature: 12.5   Groundwater pH: 6.3

3.4.4.5 Risk Assessment (RA)

In Mount Gambier, due to the complex nature of the karstic aquifer, modelling of the saturated zone using an intergranular flow model and associated solute transport module was not appropriate, and instead a different approach was needed to consider transport and attenuation of contaminants within the aquifer. The methodology applied was a risk assessment, which acts as one of the contributing elements in a Hazard Analysis and Critical Control Points (HACCP) framework for managing drinking water in Australia (NHMRC and NRMMC, 2004). In this system, the hazards associated with the urban water systems are identified initially, the likely concentrations of potential contaminants are estimated and attenuation mechanisms are identified. This allows the concentration of each contaminant to be calculated, taking into account the attenuation possible within a given travel or residence time. Comparison of the calculated concentration (allowing for attenuation) with the guideline or trigger value indicates the risk of pollution.

Potential contaminants and concentration targets

The first stage of this risk assessment was to identify potential contaminants in source waters and then determine the appropriate targets (guidelines or trigger values) with which to compare them for Blue Lake. For Mount Gambier, the potential contaminants include those modelled in the DSS (Cl, Na, K, SO_4, N, P, B) together with site-specific parameters, which include additional metals and trace organics. Site-specific parameters for the stormwater and sewage of Mount Gambier were determined following a review of existing data, in combination with knowledge of point sources in the region and species typically found in stormwater runoff (Table 3.4.7-Table 3.4.9). As the significance of Blue Lake in Mount Gambier is not only its function as a potable water supply, but also its aesthetic value, it was considered appropriate to assess the risk against both potable and the aquatic ecosystem water quality targets. Thus the suite of likely contaminants is compared with the Australian drinking water quality guidelines (NHMRC and NRMMC, 2004) and the SA EPA (2003) aquatic ecosystem water quality criteria to define target values to be adopted within the risk assessment (Table 3.4.8 and Table 3.4.9).

Table 3.4.7. Potential contaminants in Mount Gambier's sewage and stormwater (after BLMC, 2001).

Contaminant	Point sources in Mount Gambier	Origin in stormwater runoff
Nutrients	Dairy industry, sewage, gasworks	Roadside fertiliser use, animals (cats, dogs), leaf litter
Bacteria	Dairy industry, sewage	Animals (cats, dogs)
Metals	Service stations, fuel storage sites, gasworks, railway sites, timber industry	Tyre wear, vehicle exhaust, roofs and gutters, road wear
Herbicides	Railway sites	Roadside use
Total petroleum hydrocarbons (TPH)	Service stations, fuel storage sites, gasworks, railway sites, timber industry	Vehicle exhaust
Volatile organic compounds (VOC)	Service stations, fuel storage sites, gasworks, railway sites	Vehicle exhaust
Phenolic compounds	Gasworks, railway sites, timber industry	
Polynuclear aromatic hydrocarbons (PAH)	Service stations, fuel storage sites, gasworks, railway sites, timber industry	Vehicle exhaust

Table 3.4.8. Target value definition for inorganic species based on Australian Drinking Water Guidelines (NHMRC and NRMMC, 2004) and the SA EPA (2003) aquatic ecosystem values.

Parameter	Guideline value (mg/L)			Target value (mg/L)
	Health	Aesthetic	Aquatic	
Total dissolved solids		500		500
Chloride		250		250
Sodium		180		180
Sulphate		250		250
Potassium				no value available
Phosphorus			0.5 (0.1 soluble)	0.5
Nitrate as N	11 (infants)		5 (total N)	11
Aluminium		0.1 (soluble)	0.1 (soluble)	0.1 (soluble)
Arsenic	0.007		0.05	0.007
Cadmium	0.002		0.002	0.002
Chromium	0.05 as Cr(VI)		0.001 as Cr(VI)	0.005
Copper	2		0.01	0.01
Iron		0.3	1	0.3
Lead	0.01		0.005	0.005
Nickel	0.02		0.15	0.02
Zinc		3	0.05	0.05
Boron	4			4

Table 3.4.9. Target value definition for organic species based on the Australian Drinking Water Guidelines (NHMRC and NRMMC, 2004) and the SA EPA (2003) aquatic ecosystem values.

Parameter	Guideline value (µg/L)			Target value (µg/L)
	Health	Aesthetic	Aquatic	
Benzene	1			1
Toluene	800	25		25
Ethylbenzene	300	3		3
Xylene	600	20		20
Naphthalene			3	3
Acenaphthylene			3	3
Acenaphthalene			3	3
Fluorene			3	3
Anthracene			3	3
Phenanthrene			3	3
Fluoranthene			3	3
Pyrene			3	3
Chrysene			3	3
Benzo(a)pyrene	0.01		3	0.01
Benzo(b&k)fluoranthene			3	3
Benzo(g,j,i)perylene			3	3
Indeno(1,2,3-cd)pyrene			3	3
2-Chlorophenol	300	0.1	7	0.1
2,4-Dichlorophenol	200	0.3	0.2	0.2
2,4,6-Trichlorophenol	20	2	18	2
Pentachlorophenol			0.5	0.5
Atrazine	0.1 or detection limit	0.3	0*	0.5
Simazine			0*	0.5

* target for pesticides

Where more than one possible water quality criterion existed, the target values set for this assessment were generally the more stringent value. However, there are regulatory anomalies. For instance, the target value adopted for chromium of 0.05 mg/L was greater than the SA EPA (2003) aquatic ecosystem target value for chromium (VI) of 0.001 mg/L. This generic value is lower than the current analytical detection limit for total chromium. Furthermore, this stringent target value would not be achievable within the risk assessment as the median concentration within Blue Lake and the groundwater within the Gambier Limestone were both greater than 0.001 mg/L. In this case the historical chromium concentrations were adopted as an indicator of aquatic ecosystem health.

The target value for aluminium refers specifically to the soluble portion, whereas other species are total. Therefore, in the risk assessment only soluble aluminium is considered, whereas other species are examined in both fractions.

While there is potential for nitrogen to be added to the Gambier Limestone as nitrate and as ammonium, it is assumed that under aerobic conditions, nitrate will be dominant. Therefore total nitrogen inputs are compared to the nitrate guideline value. At this stage, the target value adopted for mineral N is the drinking water guideline for nitrate as N. Concurrent ongoing research is aiming to develop site-specific aquatic ecosystem targets for nitrogen and phosphorus in Blue Lake, which may be adopted at a later date.

Most of the targets set for the organic species are in relation to the more stringent aquatic ecosystem water quality criteria. The target value adopted for atrazine and simazine, of 0.5 µg/L, relates to the prevalent detection limit reported in the current data set as provided by the Australian Water Quality Centre (AWQC).

Testing water quality data against concentration targets
Three possible source concentration and attenuation scenarios are possible when testing a potential contaminant against a concentration target. Case A is where the source concentration is at an acceptable level when compared with the concentration target and further attenuation is not required. Case B is where the source concentration breaches the concentration target, but with the natural sustainable rate of dilution and attenuation in the aquifer, Blue Lake water quality would be expected to comply with the guideline value. Case C is where the contaminant concentration breaches the target value for Blue Lake and the sustainable rate of dilution and attenuation in the aquifer would be unlikely to enable Blue Lake water quality to remain below the guideline value.

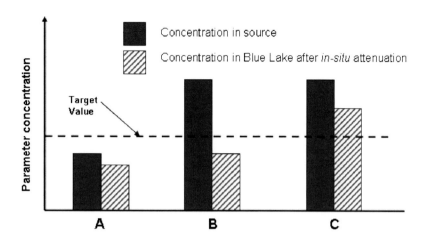

Figure 3.4.13. Possible scenarios for long-term water quality impacts of urban water systems on Blue Lake.

The output from the unsaturated zone transport models (SLeakI and POSI) is an important input to the risk assessment where the concentration of a contaminant in groundwater recharge, which includes contaminant loadings from sewer leaks, exceeds typical concentrations in the aquifer. Thus the quality of groundwater recharge provided by the DSS was compared to the existing data set for groundwater quality. Where predicted contaminant concentrations exceed the range reported by existing data, these predicted concentrations were used to set the upper bound for groundwater quality in the risk assessment.

Summaries of water quality data for Mount Gambier's stormwater, wastewater, the unconfined Gambier Limestone aquifer and the Blue Lake are presented in Table 3.4.10 to Table 3.4.13 (note that the availability of organic water quality data was limited.)

Table 3.4.10. Source concentrations for inorganics in stormwater.

Parameter	Target	Literature range (mg/L)		Detection limit (mg/L)	No. samples	Stormwater		
		Makepeace et al., 1995	ARQ, 2003			No. detections	Median (mg/L)	Peak (mg/L)
Cl	250			1	35	35	5	24
Na	180			1	10	10	8	12
K				0.1	7	7	1.4	2.3
SO$_4$	250			1	10	10	8	14
TP	0.5	0.01-7.3	0.1-3.0	0.01	122	120	0.28	9.2
FRP	0.5				39	37	0.09	0.74
TN	11	0.07-16	0.5-13	0.01	122	122	1.6	9.0
Al-sol	0.1	0.1-16		0.005	7	7	0.12	0.42
As-total	0.007	0.001-0.21		0.001 and 0.005	34	8	<0.005	0.1
As-sol	0.007				30	2	<0.005	0.005
Cd-total	0.002		0.002-0.05	0.001 and 0.0005	10	0	<0.0005	<0.0005
Cd-sol	0.002				2	0	<0.0005	<0.0005
Cr-total	0.005	0.001-2.3	0.02	0.001 and 0.005	90	67	0.005	0.06
Cr-sol	0.005				30	3	<0.001	0.008
Cu-total	0.01	0.00006-1.41	0.4	0.001	99	96	0.01	0.23
Cu-sol	0.01				32	30	0.003	0.014
Fe-total	0.3	0.08-440		0.1 and 0.005	10	10	0.68	2.4
Fe-sol	0.3				2	0	<0.03	<0.03
Pb-total	0.005	0.00057-26	0.01-2	0.001 and 0.0005	99	98	0.01	0.88
Pb-sol	0.005				32	8	<0.001	0.003
Ni-total	0.15	0.001-49		0.001	85	63	0.003	0.059
Ni-sol	0.15				32	1	<0.001	0.003
Zn-total	0.05	0.0007-22	0.01-5	0.001	99	97	0.1	2.4
Zn-sol	0.05				32	32	0.03	0.29
B	4			0.005 and 0.04	10	5	0.08	0.13

Table 3.4.11. Source concentrations for inorganics in sewage.

Parameter	Target	Detection limit			Sewage	
		(mg/L)	No. samples	No. detections	Median (mg/L)	Peak (mg/L)
Cl	250	1	8	8	280	704
Na	180	1	8	8	200	270
K			8	8	21	25
SO_4	250	1	8	8	120	250
TP	0.5	0.01	108	108	13	20
TN	11	0.01	110	110	58	80
Al-total	0.1	0.005	109	109	0.5	3.6
Cd-total	0.002	0.001 and 0.0005	109	52	<0.0005	0.002
Cr-total	0.005	0.001 and 0.005	108	57	0.03	1.6
Cu-total	0.01	0.001	111	109	0.10	0.34
Fe-total	0.3	0.1 and 0.005	108	108	0.38	2.3
Pb-total	0.005	0.001 and 0.0005	114	108	0.006	0.039
Ni-total	0.15	0.001	109	109	0.005	1.6
Zn-total	0.05	0.001	108	108	0.2	1.1
B	4	0.005 and 0.04	8	8	0.12	0.16

Table 3.4.12. Source concentrations for organics in stormwater.

Parameter	Target (µg/L)	Literature range (µg/L)	Detection limit (µg/L)	No. samples	No. detections	Stormwater Median (µg/L)	Average* (µg/L)	Peak (µg/L)
Benzene	1	35-130	1	75	0	<1		nil
Toluene	25	9-12	2	75	0	<2		nil
Ethylbenzene	3	1-2	2	75	0	<2		nil
Xylene	20		4	75	0	<4		nil
Naphthalene	3	0.0036-2.3	0.1 and 2	79	2	<2		0.2
Acenaphthylene	3		0.1 and 2	79	0	<2		nil
Acenaphthalene	3		0.1 and 2	79	0	<2	0.0009	nil
Fluorene	3	0.006-1	0.1 and 2	79	0	<2	0.002	nil
Anthracene	3	0.009-10	0.1 and 2	79	0	<2	0.0005	nil
Phenanthrene	3	0.045-10	0.1 and 2	79	1	<2	0.008	0.2
Fluoranthene	3	0.03-56	0.1 and 2	79	1	<2	0.005	0.7
Pyrene	3	0.045-10	0.1 and 2	79	1	<2	0.004	0.6
Chrysene	3	0.0038-10	0.1 and 2	79	0	<2	0.002	nil
Benzo(a)pyrene	0.01	0.0025-10	0.1 and 2	79	0	<2	0.0002	nil
Benzo(b&k)fluoranthene	3	0.0012-10	0.1 and 4	79	0	<4	0.0006	nil
Benzo(g,h,i)perylene	3	0.0024-1.5	0.1 and 2	79	0	<2	0.0009	nil
Indeno(1,2,3-cd)pyrene	3	0.31-0.5		79	0	<2	0.0001	nil
2-chlorophenol	0.1		0.1	8	0	<0.1		nil
2,4-dichlorophenol	0.2		0.1	8	0	<0.1		nil
2,4,6-trichlorophenol	2		0.1	8	0	<0.1		nil
Pentachlorophenol	0.5	1-115	0.1	8	0	<0.1		nil
Atrazine	0.5		0.2 and 0.5	75	10	<0.5		12.4
Simazine	0.5		0.2 and 0.5	75	9	<0.5		5.3

* calculated from passive sampling

Table 3.4.13. Concentrations of inorganic and organic species within the Gambier Limestone unconfined aquifer and the Blue Lake.

Parameter	Target	Gambier Limestone aquifer				Blue Lake			
		No. samples	No. detections	Median (mg/L)	Peak (mg/L)	No. samples	No. detections	Median (mg/L)	Peak (mg/L)
Cl	250	206	206	75	233	770	770	86	105
Na	180	73	73	58	99	760	760	59	74
K		82	82	2	23	770	770	3	5
SO_4	250	67	87	12	33	750	750	18	30
TP	0.5	212	210	0.1	3.5	929	772	0.01	2.7
FRP	0.5	209	185	0.01	2.5	645	244	<0.005	0.17
NO_3-N	11	149	149	9.2	88	1545	1539	3.4	5.5
Al-sol	0.1	176	129	0.02	2.2	136	104	<0.02	0.17
As-total	0.007	175	121	0.002	0.02	158	109	<0.001	0.005
Cd-total	0.002					62	23	<0.001	0.0005
Cr-total	0.005	176	103	0.005	0.82	169	108	<0.005	0.03
Cu-total	0.01	175	111	0.005	2.9	172	111	<0.005	0.05
Fe-total	0.3	191	187	1.7	40	729	411	0.01	4.3
Pb-total	0.005	177	166	0.004	0.06	42	31	0.001	0.008
Ni-total	0.15					35	33	0.001	0.07
Zn-total	0.05	177	169	0.03	1.4	138	138	0.02	0.10
B	4	7	7	<0.04	1.4	107	82	<0.04	0.22
Atrazine	0.5	130	60	<0.5	1	74	66	<0.5	0.5
Simazine	0.5	111	38	<0.5	0.5	74	66	<0.5	0.5

Summary of attenuation coefficients and assumptions
Attenuation was considered to occur by the following processes:
- Biodegradation of organics in the aerobic aquifer with an exponential rate of decay defined by a half-life (Aq t_{50}) determined by experiments using Gambier Limestone aquifer material and groundwater, or in their absence, by literature values under similar conditions.
- Exponential decay of organics in Blue Lake with half lives (BL t_{50}) taken from the literature in similar environments that combine the effects of biodegradation and volatilisation. Alternatively biodegradation and volatilisation can be assessed individually using the equations described in section 2.5.4.3.
- Nitrogen loss within Blue Lake as a result of algal production and calcite precipitation depositing organic matter on the floor of the lake that allows subsequent denitrification.
- Physical removal via adsorption of metals to the solid phase (wellhead pre-treatment) or precipitation as insoluble species (in aquifer).
- Inactivation in the aquifer and lake of micro-organisms that is pathogenic to humans.

Recharged stormwater or sewage effluent that has leaked is diluted in groundwater that has recharged from non-urban origins. The relative contribution of urban stormwater to the groundwater that flows into Blue Lake is considered to be between 35 and 55%. Knowledge of adsorption of metals and organics could be used to account for retardation of their migration through the aquifer thereby prolonging the time available for biodegradation. Similarly adsorption to organic carbon stripped from the water column in Blue Lake at times of calcite precipitation is another mechanism that is likely to reduce the concentrations of metals and trace organics in Blue Lake. Although sorption coefficients for assumed linear adsorption isotherms are presented in Table 3.4.14, the quantitative risk assessment, conservatively, did not take these processes into account, as they do not necessarily imply permanent removal from the system. As the only rates for attenuation in lakes reported in literature were for the combined effects of biodegradation and volatilisation, to be conservative in the case of Blue Lake, the separate effects of volatilisation were not calculated. SF_6 results show that the residence time of pathogenic microorganisms in this aerobic aquifer should be sufficient for inactivation of all pathogens sourced from the stormwater or sewer leaks prior to reaching the lake. Contamination from surface runoff and other sources is relatively a much higher risk and these are addressed by chlorination of drinking water supplies drawn from the lake. Hence no further analysis is given to microbial pathogens here.

Given the paucity of monitoring data for trace organics in stormwater, the quantitative risk assessment (QRA) was based on concentrations fitting a uniform distribution and ranging from zero to the maximum solubility for each species. The solubility limit for the organic species was used as the maximum concentration to simulate a worst-case scenario.

Table 3.4.14 summarises the attenuation coefficients used for the suite of contaminants in the risk assessment.

Table 3.4.14. Summary of attenuation coefficients applied within the risk assessment.

Parameter	Sol (mg/L)	Aq t_{50} (d)	BL t_{50} (d)	I (mg/L)	K_d (L/kg)
TP					130-180[D]
N				2[G]	
Al-sol					
Cr					1[H]
Cu					7[H]
Zn					0.1[H]
Benzene	1790[A]	10-16[D]	5-16[D]		0.1[H]
Phenanthrene	1.2[A]	30-200[A,B]	16-200[D]		20[B]
Fluoranthene	0.3[A]	280-400[A,B]	140-400[D]		40-70[B]
Pyrene	0.1[A]	420-1900[A,B]	210-1900[D]		40-80[B]
Pentachlorophenol	14[A]	50-180[A]	20-180[D]		13[H]
Atrazine	30[B]	100-180[B]	120-365[E]		0.04-0.1[I]
Simazine	6[C]	90[B]	65-150[F]		0.6[H]

Sol=solubility, Aq t_{50}=degradation half-life in aquifer, BL t_{50}=half-life in Blue Lake (combination of degradation and volatilisation), I=additional in-lake removal processes, K_d=sorption coefficient.
[A] SRC, 2005, [B] Vogel, 2005, [C] Knuteson et al., 2002, [D] Howard et al., 1991, [E] Chung and Gu, 2003, [F] Vink and Van der Zee, 1997, [G] Lamontagne and Herczeg, 2002, [H] Oliver et al., 1996a, [I] Oliver et al., 1996b.

3.4.4.6 Scenario Results

The scenarios implemented within the DSS for Mount Gambier do not encompass the full range of issues that need to be considered in developing strategies for the sustainable management of urban groundwater resources. Evaluation of the DSS scenario outputs, however, does indicate the relative risks or impacts of different hydrologic scenarios (such as climate change) and also the potential benefit from intervention strategies on the quality and quantity of groundwater recharging the Blue Lake (Figure 3.4.14). The assessment of risk considers attenuation via natural processes available within the aquifer and Blue Lake and the likely benefits of adding intervention strategies to the system (Figure 3.4.15).

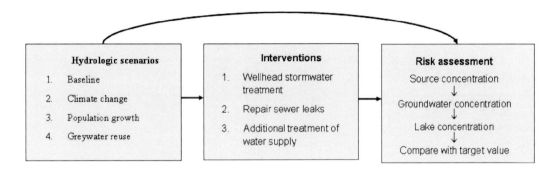

Figure 3.4.14. Assessing the impact of hydrologic scenarios and possible interventions on the Blue Lake.

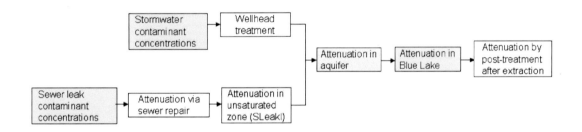

Figure 3.4.15. Schematic representation of potential for attenuation of stormwater or sewage based contaminants, either via natural processes in the aquifer or Blue Lake or through possible interventions.

Analysis of NEIMO results shows that leakage of wastewater pipes is on average 4.2% of the total wastewater volume, representing an exfiltration volume of approximately 207 m³/day. Given a lack of monitoring data on pipe condition, the leakage is assumed to occur predominantly through joints (98%), with the remainder occurring through cracks (2%). Figure 3.4.16 shows the distribution of leaks across the Mount Gambier sewer network. The assets with higher leakage volumes are associated with concrete pipes that were laid more than 40 years ago (Figure 3.4.16). The area to the northeast, which has very low annual leakage, is served by PVC pipes of 150 mm diameter laid less than 20 years ago. There was no infiltration into wastewater pipes in Mount Gambier, as they are situated above the water table.

While the volume of wastewater exfiltration is a very small component of total groundwater recharge (annual average of <2%, but ranging daily from 0.46-45% depending on precipitation and irrigation inputs), the modelling assumptions in the baseline scenario result in this contributing 57% of the nitrogen load due to the high contaminant load contained within wastewater (Figure 3.4.17). When comparing the spatial extent of the sewer leaks to land use via the planning zones, two areas stand out as potential priorities for infrastructure rehabilitation. The first is the main commercial centre to the north of Blue Lake where high leak rates are concentrated and the second is the industrial zone to the northwest of the lake.

Figure 3.4.16. Annual pipe leakage volume – Mount Gambier sewer network.

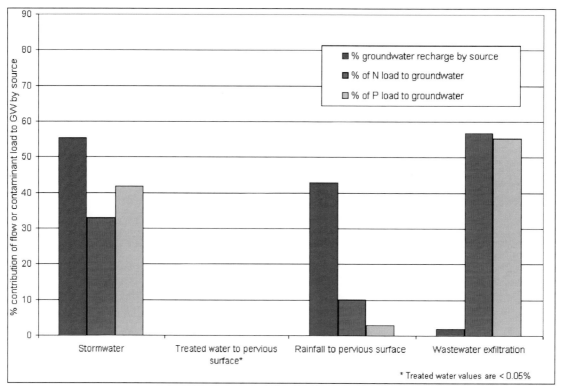

Figure 3.4.17. Relative contribution of different sources to recharge, N and P loads to groundwater (Baseline Scenario).

The contaminants modelled within the DSS are mostly conservative and therefore there is little or no attenuation of contaminant load through the unsaturated zone. The exception is phosphate, where there is significant removal in the unsaturated zone under all scenarios. These results are consistent with the lysimeter data obtained from an area close to Mount Gambier where a residence time of a minimum of 100 days was estimated in a soil of pH 8.5 and no phosphate was detected in the drainage water (Miller *et al.*, 2005).

Worst and best case scenarios were simulated when considering leakage from wastewater pipes. Best case is when exfiltration is occurring at multiple small leaks, while worst case simulates exfiltration from one large defect in the asset. Leaks from sewers typically form a colmation layer that is rich in organic matter. This colmation layer can act to increase attenuation of some contaminants in the exfiltrating wastewater. Results from SLeakI showed the concentration of phosphate to increase significantly between best and worst case (0 mg/L to ~ 0.85 mg/L), while for conservative contaminants there was no significant difference between best and worst case scenarios.

Nitrogen loads may currently be overestimated, as SLeakI neglects any N removal in the colmation layer, converting all the nitrogen from sewer leaks to nitrate of approximately 65 mg/L N. In addition POSI considers infiltration from below the root zone. The behaviour of solutes in the root zone, particularly nutrient species such as nitrogen, is a gap in the model chain that is not covered by the DSS.

Figure 3.4.18 to Figure 3.4.21 depict the water balance (in mm/yr) for Mount Gambier's urban water system for a representative year under the assumptions of the four scenarios. Table 3.4.15 compares the key climate inputs and DSS outputs for the year 2015. The DSS outputs are over-estimating wastewater volume for Mount Gambier, when compared to actual wastewater discharge from Finger Point WWTP due to a lack of information regarding water consumption behaviour in non-residential neighbourhoods.

The climate change scenario shows a significant decrease in precipitation and increase in evapotranspiration when compared to the baseline scenario (Table 3.4.15). This reduces the amount of groundwater recharge. The population increase scenario sees a significant rise in the consumption of imported water and consequently the generation of wastewater. The greywater recycling scenario has the obvious impact in that implementation of greywater recycling reduces both the demand for imported water and the volume of wastewater generated.

The climate change scenario demonstrates a marked reduction in groundwater recharge, which will impact on the Blue Lake water level. Recharge via drainage wells in impervious areas is reduced less than recharge through pervious soils. The level of Blue Lake has declined significantly since the early 1900s, caused in part to lower than average rainfall but exacerbated by variable extraction rates for potable supply (BLMC, 2001). Analysis of historical data reveals there is a strong correlation between periods of below average rainfall and a decline in lake level. There is however a perception in the community that the falling lake level can be attributed to increasing groundwater extraction within the Blue Lake capture zone. A shift to a drier, hotter climate for Mount Gambier, which is projected under most climate change scenarios, would result in a further decline in the level of the Blue Lake. This may require costly modifications to the Blue Lake pumping station in order to adjust for a declining water level (BLMC, 2001).

The population increase scenario shows a significant rise in contaminant loads to groundwater. The conversion of predominately pervious land use, such as rural, to more intensive urban land use increases runoff from impervious surfaces and the volume of stormwater being recharged directly to the aquifer via drainage wells. Infiltration of urban runoff through the shallow subsurface has the advantage that it permits natural contaminant attenuation processes to occur within the soil (Pitt *et al.*, 1999). Future urban development within the Blue Lake capture zone will need to consider the likely impact on groundwater quality, and may require alternative stormwater management practices which provide pre-treatment, such as incorporation of permeable pavement or roadside swales.

The scenario simulating a strategy to increase recycling of household and industrial greywater shows that it would have a significant impact on water demand and also wastewater output. Greywater recycling has the potential to be an important part of a management strategy to reduce or stabilize the amount of water pumped from the Blue Lake, but would require confirmation that groundwater quality would not be impaired through greywater irrigation. The implementation of a greywater recycling strategy would require consideration of the appropriate level of treatment required before the greywater could be used for toilet flushing or garden irrigation. Greywater treatment systems range from simple diversion system that redirect untreated greywater from wastewater pipes to the garden, to filtration and disinfection treatment systems and more sophisticated technologies such as reverse osmosis. The removal efficiency of contaminants will depend on the technologies employed. Testing of different technologies has shown that boron, potassium, sodium and chloride will not be removed unless reverse osmosis technology is used, while the removal of phosphorous is in the range of 10–20% depending on the system. Removal of nitrogen is relatively high, in the range of 50–80%, while most technologies will remove upwards of 80% of suspended solids (C. Diaper 2006, pers. comm). The contaminants removed are mostly discharged as sludge to the sewer system. The use of greywater recycling for garden irrigation has the benefit that the volume supplied is consistent throughout the year, while rainwater capture for irrigation is often highest when irrigation demand is lowest. The possibility for centralised advanced wastewater treatment for subsequent recharge to the aquifer could also be considered as a future means of buffering against climate change.

Table 3.4.15. Scenario comparison: Climate inputs and key DSS outputs.

Scenario	Precipitation (mm/yr)	Potential Evaporation (mm/yr)	Stormwater Output (Mm^3/yr)	Wastewater Output (Mm^3/yr)	Groundwater Recharge* (Mm^3/yr)
Baseline	703	1277	4.22	3.30	9.49
Climate Change	682	2300	4.06	3.25	4.05
Population Increase	703	1277	4.63	3.62	9.13
Greywater Recycling	703	1277	4.50	2.82	9.31

* Groundwater recharge refers to infiltration through the unsaturated zone

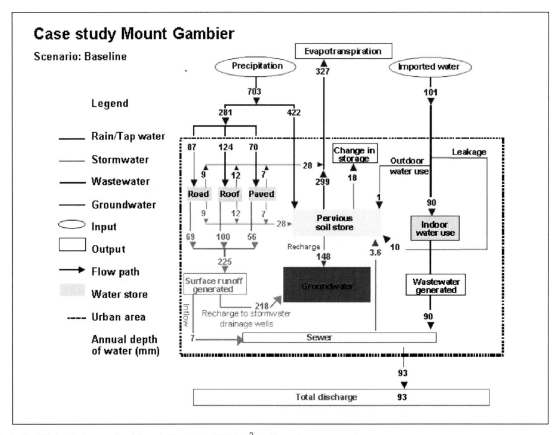

Figure 3.4.18. Water balance for Mount Gambier (27 km^2) – Baseline Scenario.

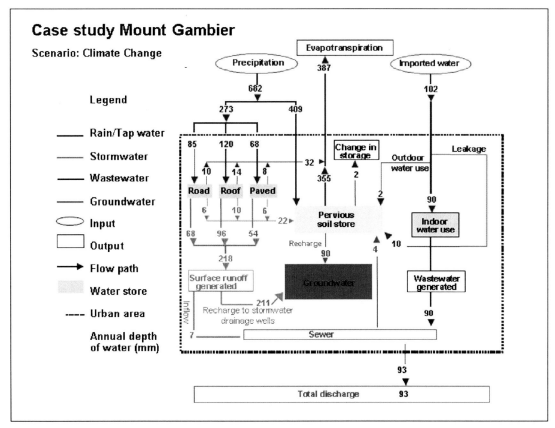

Figure 3.4.19. Water balance for Mount Gambier (27 km^2) – Climate Change Scenario.

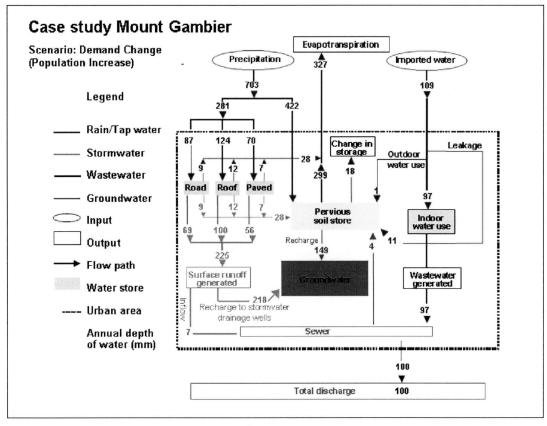

Figure 3.4.20. Water balance for Mount Gambier (27 km^2) – Population Increase Scenario.

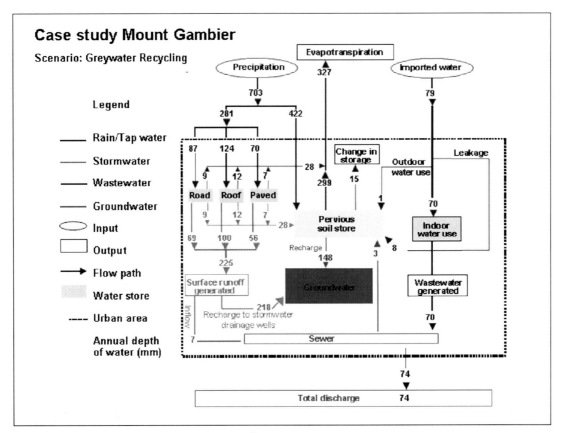

Figure 3.4.21. Water balance for Mount Gambier (27 km^2) – Greywater Recycling Scenario.

The different components of the urban water system can be assessed in relation to the quality of groundwater recharge they provide (Table 3.4.16). Only the first hydrologic scenario, the baseline condition, was analysed in this risk assessment as the effects of other scenarios were less than the uncertainty in some attenuation and source concentrations. Based on the recharge concentration calculated by the DSS, there was no indication of any potential for water quality in the Blue Lake to breach drinking water guideline values for chloride, sodium, sulphate or the aggregate measure of total dissolved solids (there is no specific target value for potassium). Freshening of the groundwater by stormwater recharge more than compensates for any impact on these species by leaky sewers.

Of the species considered by the DSS, nitrogen and phosphorus from sewer leaks pose the greatest threat to target values (Figure 3.4.22). The indicator for risk is expressed by showing the measured or simulated median concentration as a ratio with respect to the target value. Values greater than 1 exceed the target value and can be considered as high risk, while those below 1 are less than the target value and are thus low risk. However when sewer leaks are combined with infiltration, the quality of groundwater recharge via infiltration meets all target values and is lower than (Cl, Na, B, N) or similar to (K, SO$_4$, P) urban groundwater quality. Additional dilution occurs for most parameters, as recharge to Blue Lake is a mixture of groundwater and stormwater that is discharged directly into the aquifer. This is most evident for N, where stormwater discharge adds lower concentrations than present within the aquifer. In addition to dilution, efficient P removal is expected in the carbonate aquifer (Oliver *et al.*, 1996b; Miller *et al.*, 2005), while Lamontagne and Herczeg (2002) report 2 mg/L removal of NO$_3$-N in Blue Lake itself (possibly via denitrification). As sewer leaks have a small impact on the quality of annual recharge to the groundwater and Blue Lake, there is little justification for repair. However opportunities for upgrades in the two priority areas discussed previously could be considered in more detail.

Table 3.4.16. Simulated concentrations in recharge to groundwater via stormwater discharge, sewer leaks (worst case) and infiltration in comparison with groundwater quality and target values.

	Simulated median concentration in groundwater recharge (mg/L) from:				Median measured groundwater quality (mg/L)		Target (mg/L)
	Stormwater discharge	Sewer leaks (worst case)	Infiltration	Sewer leaks + infiltration	Urban	Non-urban	
Cl	5	328	12.6	35	75	85	250*
Na	8	206	8.9	23	58	57	180*
K	1.4	18	1.2	2.4	2	2	-†
SO_4	8	119	5.2	13	12	11	250*
B	0.12	0.14	0.16	0.13	0.4	na	4
TN	1.6	65	2.4	6.2	9.2	8.6	11
TP	0.28	3.1	0.001	0.16	0.1	0.02	0.5

* water quality targets are based on aesthetics only (see Table 3.4.8)
† no water quality target value

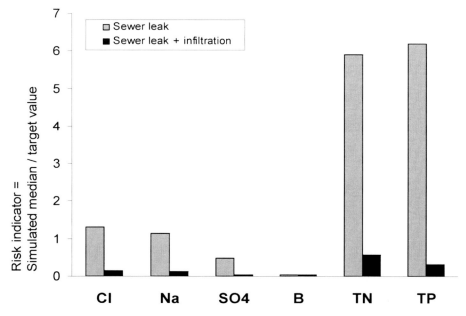

Figure 3.4.22. Risk to quality of groundwater recharge from sewer leaks alone and sewer leaks as a component of total infiltration. The risk indicator = measured median concentration/target value in Blue Lake.

The site specific assessment of water quality from the urban water systems was largely based on the quality of stormwater discharged directly to the unconfined aquifer as this component is a significant part of the recharge to Blue Lake. Water quality data for the full suite of parameters was examined in comparison with water quality targets and with potential attenuation mechanisms in mind.

The highest priority inorganic species were the metals with a soluble component, aluminium, chromium, copper and zinc. The median soluble aluminium (0.12 mg/L) was slightly above the target value (0.1 mg/L), which was set based on aquatic ecosystem protection (

Figure 3.4.23). Dilution is sufficient to reduce Al concentrations calculated for recharge to Blue Lake to below the target value. In addition, under the neutral pH and oxygenated conditions in the carbonate aquifer and Blue Lake, Al is predominantly present as the highly insoluble $Al(OH)_3$. Precipitation of aluminium oxide is a mechanism for attenuation within the saturated zone and Blue Lake and explains the low concentrations measured.

Stormwater medians for total Cr and Cu were equal to their respective target values, while the median for total Zn was twice the target value. However, assessing the risk based on the soluble or mobile portion of the metal reduced the median concentrations for Cr, Cu and Zn to below the target value. Total groundwater and Blue Lake concentrations for Cr, Cu and Zn were assumed to be in the soluble phase, again taking a conservative approach. Over-estimation of groundwater concentrations may occur if solids were disturbed during sampling. Based on the soluble metal fractions in stormwater, the concentrations of Cr, Cu and Zn calculated for recharge to Blue Lake are all below the target value. This indicates the importance of the aquifer in removing the particulate fraction. Although the sustainable capacity for particulate removal within the aquifer is thought to be large, it is unquantified and the extra barrier of particulate removal at the wellhead should be considered as best practice for stormwater management.

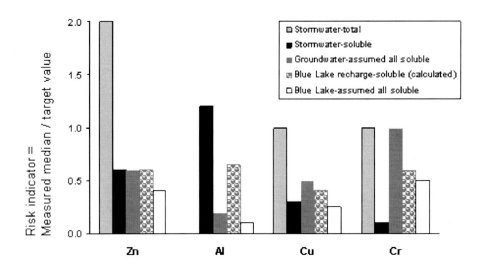

Figure 3.4.23. Change in risk as stormwater based inorganics are attenuated in groundwater and Blue Lake. The risk indicator = measured median concentration/target value in Blue Lake.

As mentioned previously, there was limited data to assess the risk posed by trace organics in the urban water systems. Despite having source concentrations below the current aquatic ecosystem target value, phenanthrene, fluoranthene and pyrene were chosen as representative PAHs to be considered in the quantitative risk assessment. Atrazine and simazine were also examined, as the only organic species to exhibit peaks in the stormwater quality data above the target value. Benzene and pentachlorophenol have not been detected locally but were also considered, as the literature suggests these species could exceed target values.

To represent a worst-case scenario, the solubility limit was used as an upper concentration representing an extremely conservative approach. In all cases the stormwater quality used in the risk assessment was far in excess of either measured concentrations or literature values for stormwater (

Table 3.4.12 and Table 3.4.17). The groundwater concentration used for dilution was assumed to be zero, based on the lack of detection of any trace organics in the ambient groundwater.

Benzene begins as the stormwater species with the highest risk due to the potential for high concentrations in solution (Figure 3.4.24). However, the degradation of benzene is so rapid that predicted concentrations in the groundwater are approaching zero, which in turn results in zero concentrations in Blue Lake. Benzene thus becomes the lowest risk contaminant considered.

Despite starting with median concentrations orders of magnitude greater than those recorded for Mount Gambier, calculated medians for the three PAHs (phenanthrene, fluoranthene and pyrene) in both the groundwater and Blue Lake are below the aquatic ecosystem target of 3 µg/L. Pyrene, with the highest resistance to degradation (Table 3.4.14), produced concentrations that are close to the current analytical capabilities for trace concentrations (0.1 µg/L). As a result of these simulations, pyrene is ranked as the highest risk organic for the groundwater and the Blue Lake, but still presumably low risk in practical terms.

Pentachlorophenol is also efficiently degraded in both the aquifer and the lake, resulting in predicted concentrations well below detection limits and at least 7 orders of magnitude below the target value.

Atrazine and simazine are soluble pesticides and begin with high simulated stormwater concentrations. However both are subject to rapid degradation, which produces groundwater concentrations several orders (>5) of magnitude below the analytical detection limit and target value (0.5 µg/L).

While not incorporated into the quantitative risk assessment, sorption and subsequent removal of solids at the wellhead could remove a significant proportion of PAHs and pentachlorophenol (PCP) before entering the aquifer. Considering the upper solubility limit for pyrene (0.1 mg/L) sorption would reduce the aqueous concentration within the simulated stormwater to 2 µg/L (using K_d=40 L/kg), which is below the target value. For phenanthrene (solubility 1.2

mg/L and K_d=20 L/kg), fluoranthene (solubility 0.3 mg/L and K_d=40 L/kg), and PCP (solubility 14 mg/L and K_d=13 L/kg), sorption gives aqueous concentrations of 60, 7, 1000 µg/L each.

Table 3.4.17: Simulated concentrations for assessing the impact of stormwater based organics on the groundwater recharging Blue Lake and the quality of Blue Lake.

Parameter	Target (µg/L)	Simulated concentrations* (µg/L) for:		
		Stormwater	Median Blue Lake recharge	Median Blue Lake
Benzene	1	1800000	1×10^{-20}	$<1 \times 10^{-20}$
Phenanthrene	3	1200	2×10^{-8}	3×10^{-19}
Fluoranthene	3	300	2×10^{-2}	7×10^{-6}
Pyrene	3	100	2.0	0.2
Pentachlorophenol	0.5	14000	3×10^{-7}	2×10^{-18}
Atrazine	0.5	30000	3×10^{-5}	2×10^{-9}
Simazine	0.5	6000	2×10^{-10}	4×10^{-19}

*based on dilution of 45-65% with the ambient groundwater, aquifer residence time of 1-20 years and Blue Lake residence time of 6-10 years

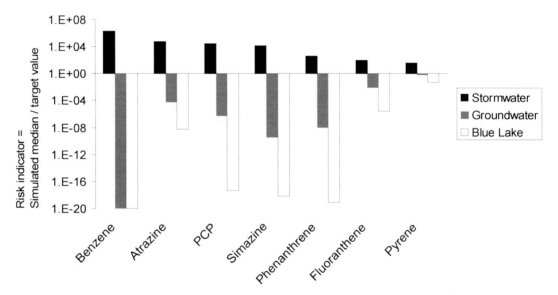

Figure 3.4.24. Change in risk as stormwater based organics are attenuated in groundwater and Blue Lake. The risk indicator = simulated median concentration/target value in Blue Lake.

3.4.5 Conclusions and recommendations

The reliance of Mount Gambier on water extracted from the Blue Lake means that identifying and managing potential threats to groundwater quality in the unconfined aquifer is of critical importance. This project has attempted to quantify the impact of recharge from stormwater and the piped wastewater system. Evaluation of the DSS scenario outputs indicates the relative risks or impacts of different strategies or potential uncertainties, such as climate change, on the quality and quantity of groundwater recharging the Blue Lake. Impacts on groundwater quality under each of these scenarios were negligible.

The results show that under the assumptions of the scenarios modelled in the DSS Mount Gambier's groundwater quality is not threatened. However, the dependence of Mount Gambier on groundwater for potable water supply means that a precautionary approach should be adopted in managing risk of groundwater contamination. Management interventions outlined below should be considered, in particular wellhead treatment of stormwater in industrial areas and the proactive inspection of sewer assets with a high probability and consequence of failure. Based on existing knowledge, the current water attenuation processes, provided by the saturated zone and in-lake processes such as the annual carbonate precipitation cycle, is considered adequate for protection against the current risk of potential contamination from the urban area. However, reinforcement against specific identified risks would be a small cost in comparison with the value of ensuring the perpetuation of the high quality water in Blue Lake. This project is intended to assist Mount Gambier authorities in identifying their preferred management strategy to protect groundwater quality and also develop best practice response to potential issues arising from a range of likely city development scenarios.

3.4.5.1 *Future Monitoring and Assessment*

The key inputs to the risk assessment model are source concentrations and attenuation mechanisms. The quantitative risk assessment could be improved with site-specific attenuation coefficients for precipitation or sorption. Future assessment should be undertaken of the mechanisms for removal of inorganic and organic species within the Blue Lake and their permanency. This may include analysis of lake sediment, collected during the annual precipitation and sedimentation cycle and from cored samples from the lake bottom. Stormwater quality monitoring should continue to include total and soluble Al, Cr, Cu and Zn. Given the large variability in groundwater quality, greater understanding of the hydrostratigraphic units and their contribution to Blue Lake is desirable.

Future monitoring of organic species within the stormwater and groundwater needs to be improved to provide input data. Suggested species to focus on include selected PAHs (phenanthrene, fluoranthene and pyrene) in order to quantify concentrations in stormwater and groundwater; relevant pesticides in stormwater and groundwater including analysis of atrazine to current Australian Drinking Water Guideline value of 0.1 µg/L; and analysis of atrazine's main metabolites of degradation (desethylatrazine [DEA] and desiopropylatrazine [DIA]), to gain site specific information regarding degradation rates.

3.4.5.2 *Intervention options*

The three intervention options to be considered for the urban water systems of Mount Gambier are pre-treatment of urban stormwater at the wellhead, repair of leaky sewers and post-treatment of water extracted from Blue Lake (Figure 3.4.14-Figure 3.4.15).

Wellhead stormwater treatment is considered desirable in reducing the total metal concentrations reaching the aquifer and thereby reducing dependence on the unknown capacity of the aquifer to sustainably remove metals sorbed on particulates. Existing wellhead treatments and stormwater pollution control methods in current use appear to be effective in managing hydrocarbons and other organics. The introduction of grass swales and further pollution abatement as proposed by SA EPA (2006) in new subdivisions will provide further barriers to protect the quality of groundwater and the Blue Lake.

Wellhead treatment techniques need to be cost-effective and suited to both the catchment size (<1 to 60 ha) and the contaminants requiring removal. Any retrofitting programme should begin with the highest priority catchments, based on size, traffic density and the presence of major industry.

Given the sewer infrastructure in Mount Gambier is recent (post 1960s), it is understandable that sewer leaks are a small contribution to the annual recharge to Blue Lake. Despite the potential for addition of high nutrient loads, there is sufficient dilution to reduce concentrations within recharge to the lake to below target values. Thus on a city wide basis proactive sewer replacement in Mount Gambier is not justified. However, given the large potential contribution of sewer leaks to the nitrogen load in recharge, and that nitrogen concentrations in Blue Lake have been increasing, a strategy to reduce sewer leakages in specific areas may be warranted. It may be feasible to analyse the sewer network to identify assets that have a high probability of failure based on physical pipe characteristics and also identify assets where the consequence of failure is high, such as those proximal to the Blue Lake. Assets with a high probability and consequence of failure could be proactively inspected by CCTV to determine pipe condition and the need for pipe maintenance or replacement.

The water quality data available for Blue Lake does not indicate the need for post-treatment of water following extraction from Blue Lake beyond the current practice of chlorination. Any additional treatment would not provide any protection to the lake as an aquatic ecosystem. Nitrate has been identified as the greatest risk to drinking water guidelines for Blue Lake (Lamontagne and Herczeg, 2002), due to the concentration range exhibited within the unconfined aquifer. While groundwater N concentrations are influenced by stormwater discharge (adding low N) and leaky sewers (adding high N), the impact of these may be similar to the effects of primary production in the catchment, however, this would need to be confirmed by additional studies.

4
Socio-Economics and Sustainability

GKW CONSULT & FUTUREtec

4.1 OBJECTIVES, SCOPE AND APPROACH

4.1.1 Environmental Sustainability Analysis

Sustainability has been defined by the Brundtland Commission as meeting the needs of the present generation without compromising the ability of future generations to meet their own needs (World Commission on Environment and Development 1987). Sustainable development leads to intergenerational and intragenerational justice. A sustainable urban water system should over a long term perspective provide the services a society requires whilst minimizing the use of scarce resources and protecting human health and the environment (Lundin 1999, p25). Taking the urban water system, especially groundwater, as the centre of our concern, the AISUWRS analysis is about how to use water and the urban environment in ways that lead towards such environmental, societal and economic sustainability. The objectives of the AISUWRS socio-economic-environmental analysis are therefore to analyse problems encountered in urban water systems and to propose further actions in the context of a sustainable development. The analysis herein focuses on the actual situation of the urban water system with the intention to take measures for the future into consideration.

Interactions of the urban water system with the environment, society and the economy are considered within the AISUWRS approach. For future application of the socio-economic environmental sustainability analysis approach, especially in Eastern/South-Eastern Europe or in the Mediterranean region, the objective was to develop a pragmatic and effective methodology for data collection, analysis and testing of procedures, taking into account the AISUWRS urban water computer model application. Thus, the socio-economic-environmental analysis focuses on key aspects in the AISUWRS approach, for example leakage from urban waste water systems, and the AISUWRS computer models developed within the project.

The methodology aims to link technical and environmental information gained on the urban water system with its socio-economic context in order to approach a qualitative Triple Bottom Line assessment. "The idea behind the 3BL paradigm is that a corporation's ultimate success or health can and should be measured not just by the traditional financial bottom line, but also by its social/ethical and environmental performance" (Norman / McDonald 2003). Within the AISUWRS project however, it is not the corporations i.e. the water utilities, which are the centre of our concern, but rather the urban water resources and systems as a whole. This adds complexity which prevents the application of the 3BL in a standard way, and so the socio-economic-environmental assessment of different urban water management scenarios analyses for better environmental outcomes, response to water users' preferences, more sensitive cost recovery through user charges, a better appreciation of the water users' ability and willingness to pay (both current charges and increased charges for improved services), and the prediction of the likely effect on water consumption patterns after price increase.

The AISUWRS sustainability assessment approach is decision-oriented. It aims to help municipalities and municipal councils as well as other stakeholders understand the complex situation within an urban water system by providing simplified assessment tools. The approach furthermore aims to support geologists and engineers in defining future action scenarios, as well as evaluating these for further decision making.

4.1.2 Approach and methodology

4.1.2.1 Framework: stakeholders and perspectives

The project objectives need to consider the interests of the main stakeholders with their different perspectives on the urban water system. The typical array of stakeholders have been identified and grouped as shown in Figure 4.1.1 below. Some have a direct and primary involvement in decision-making on water production and use, as for example public water producers and owners of public installations (utilities, municipal councils, municipality), private water producers and consumers and regulatory agencies. Other stakeholders are indirectly involved, such as the municipal administration, business associations, labour unions, research and development agencies, engineering companies, surveying institutions, political parties, and landowners' associations (secondary stakeholders).

Relation to water \ Involvement	Water producers and owners of public installations	Water producers and consumers with / without private disposal and irrigation	Water consumers
Direct involvement	Water Utilities Municipal Council Municipality Regional Environment Agency	Businesses Industrial Manufacturers Food and beverage industries Agriculture etc. Land owners	Private households Businesses
Indirect involvement	Municipal administration (part)	Bus. Associations Labour Unions Research & development	Surveying institutions Political parties Land owners' associations

Stakeholder systematic and involvement

Business here defined as all kinds of activities and employments, regardless of private or public ownership

Green: potential contaminators

Figure 4.1.1. Stakeholder systematic and involvement matrix.

Stakeholders have different perspectives on the water system, as shown in Figure 4.1.2. When considering the impacts on the environment, the social situation and the economy, a practical assessment needs a framework of values and indicators. This framework, still being rather generic, has to be translated into usable indicators (which will be shown later in this chapter).

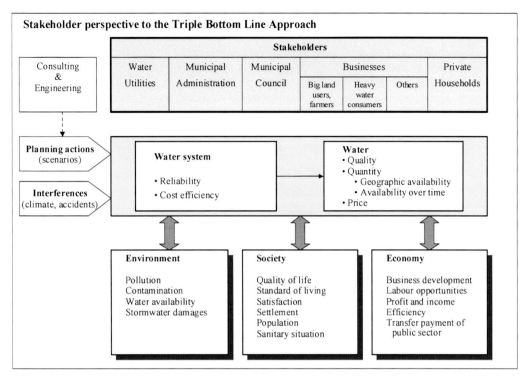

Figure 4.1.2. Stakeholder perspectives on the Triple Bottom Line Approach.

4.1.2.2 Analysis perspectives

The AISUWRS sustainability analysis uses two perspectives: a scientific-technological perspective with focus on environmental and technical aspects of the urban water system (see Figure 4.1.3), and, secondly, a socio-economic perspective considering the interrelations between the urban water usage and society and economy (see Figure 4.1.4). In each perspective on the water system it is important to consider the influences by and on the Environment, Society, and the Economy.

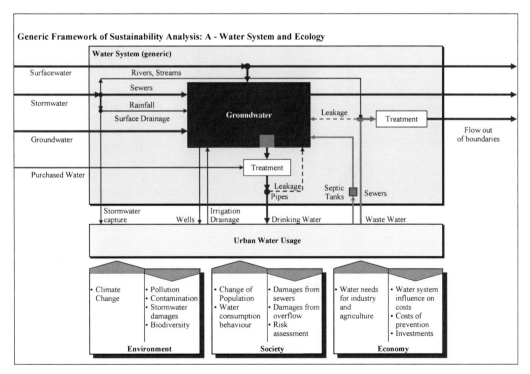

Figure 4.1.3. Generic framework of sustainability analysis: Perspective on water system and ecology.

In the technological environment perspective (see Figure 4.1.3) the analysis mainly centres on groundwater quality and groundwater quantity, the effectiveness of the urban water service systems, the efficiency of some aspects of the urban water service systems and the society's sufficiency with regard to drinking water consumption. For an analysis of these aspects, data are collected from the water utilities as the central institution affected by the quality and quantity of their primary raw material water, and from the scientific AISUWRS partners which deliver their results from the AISUWRS urban water models on the effects of proposed technical measures on the environment. The investment costs of such technical measures have been calculated case specifically.

The second, socio-economic perspective considers the urban water service system as a black box that interacts with the "urban system" of water users; the urban settlement, people and their way of life, businesses that give employment and wealth to the population etc. (see Figure 4.1.4). The socio-economic analysis aims at finding how the quality and availability of urban water interacts with society and the economy in each city. Such effects are analysed by questionnaire surveys and stakeholders' interviews.

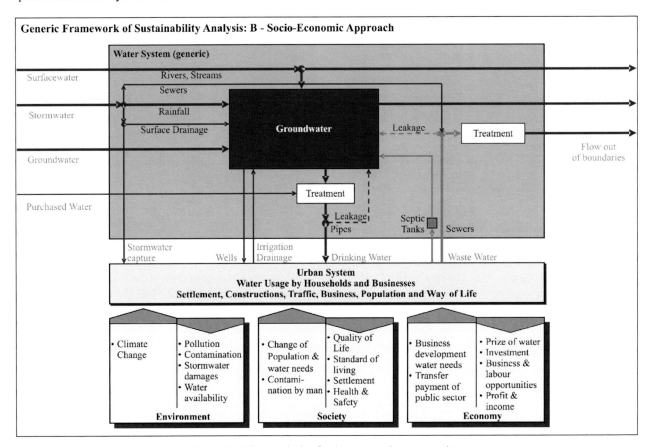

Figure 4.1.4. Generic framework of sustainability analysis: Socio-economic perspective.

4.1.2.3 *"Decision on Scenario" Approach*

The AISUWRS approach can help guide stakeholders on what to discuss and decide upon when analysing urban water sustainability and proposed improvements (see Figure 4.1.5). The plausibility and applicability of the AISUWRS models can best be assessed when applied to real problem solving tasks, especially as communication with stakeholders needs practical scenarios for illustration purposes when discussing the consequences of different action plans. Otherwise the assessment remains rather theoretical, with the danger that people become unwilling to discuss and decide because the issues seem remote. From a practical point of view, it is thus an integral part of the approach to define improvements or action scenarios that deal with problems to be solved in each city, and assess these according to the particular environmental, social and economic situation.

The action scenarios may derive from the urban water model analysis or may be proposed by the municipality or the water utilities as being relevant for their decision-making. The selection and definition of action scenarios is a difficult and iterative process. The AISUWRS methodology integrates this process and Figure 4.1.5 illustrates the necessary steps for sustainability analysis and decision-making.

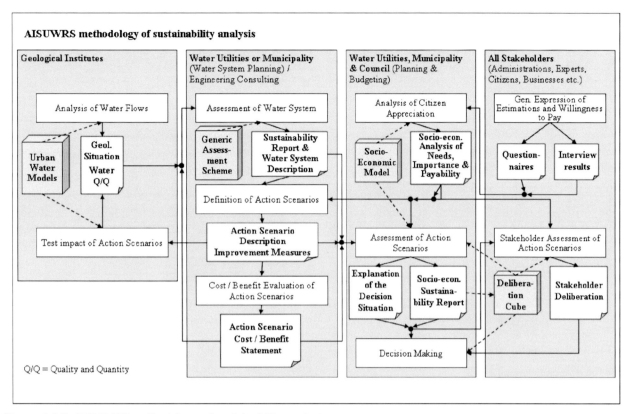

Figure 4.1.5. AISUWRS methodology of sustainability analysis.

The diagram shows the actions of the stakeholders involved at a given stage (large green boxes). The stakeholders and their roles may vary from country to country: the Geological Institutes and/or Academic Research Partners analyse the urban water flows and test the impact of action scenarios under discussion using the urban water models; the water utilities & municipality jointly with engineering consultants (and possibly a geological institute) are involved in the water system's assessment and define alternative action scenarios. These scenarios are then being analysed in different ways: firstly they are tested with the AISUWRS urban water models for their physical effects. Secondly the costs involved will be determined. Finally the whole situation will be assessed by a broad group of stakeholders considering the socio-economic impact, the social needs and importance as well as the payability. This final assessment and decision making is generally done by the water utility or municipality. The AISUWRS methodology can provide additional support to this process by a more systematic analysis of the citizen appreciation via a questionnaire survey, structured interviews and workshops.

4.1.3 Methodology

The perspective analyses follow a step by step procedure and are supported by two guidance tools: the Socio-Economic-Environmental Sustainability Analysis of Urban Water Systems Model (SEESAW Model) structures the data gathering and data processing and calculates the output; the AISUWRS Deliberator assembles the stakeholder assessment showing the degree of similarity of the individual deliberations during the process of decision making (which involves different people at different stages).

4.1.3.1 Perspective I - AISUWRS Environmental Sustainability Analysis

The AISUWRS environmental sustainability analysis supports the assessment by geologists and engineers using a computer-based tool (the SEESAW Model), together with guidelines and reports and examples from the pilot cities. The methodology proposes several steps that were applied within the AISUWRS project:

(1) Step 1 - Data collection: The analysis of water flows by the geological/research institutes provides data on the aquifer resource and the groundwater quality and quantity. In addition, data are required from the water utilities on the drinking water system, public water consumption and the waste water system. The data are collected in the SEESAW questionnaires developed for the AISUWRS project.
(2) Step 2 - Assessment of the water system and the environment: the data can be entered directly into the SEESAW Model, which assists in computing indicators for a numerical analysis. Interpretation is necessary to clearly draw conclusions from the resulting indicators.

(3) Step 3 - Elaboration and evaluation of action scenarios: primary stakeholders agree together on action scenarios needed to improve or ensure long-term sustainability of the water system. The description of the initial action scenarios should clearly state the improvement measures, to allow the estimation and calculation of costs and benefits.

4.1.3.2 Perspective II - AISUWRS Socio-Economic analysis

The AISUWRS socio-economic analysis supports the final assessment and decision making using two computer-based tools (the SEESAW Model and the AISUWRS Deliberator Cube), together with practical guidelines, reports and examples from the pilot cities. The methodology of the socio-economic analysis proposes several steps, which were applied within the AISUWRS project:

(1) Step 1 - Questionnaire survey among private households: this provides general background on citizens and their willingness to pay for changes in the water system quality and quantity. It might also be appropriate to involve a certain number of individual interviews with technical experts. A minimum of 50 questionnaires should be the objective. Before starting the survey it is recommended to raise interest in the project and in making a personal contribution. This could be done by a publication in the local newspaper, website, or radio/TV station or by means of a leaflet distributed to the households about a day in advance. This information should be written in an easy to understand language style, explaining briefly the project, its relevance to the local situation and the action scenarios. Instead of the paper-based form of the questionnaire an Internet version can be employed (see the German version at www.futuretec-gmbh.de/aisuwrs/) or the PC-Questionnaire developed by CSIRO and used in the town of Mount Gambier.

(2) Step 2 - Interviews with stakeholders: interviews to disseminate information about the decision situation and the socio-economic indicators and to gather the stakeholders' assessment of the action scenarios well before the final decision meeting. Right from the beginning the local AISUWRS project partner selects the most important stakeholders of the urban water system using the systematic stakeholder scheme. Then he will ask each stakeholder for an interview of 1 to 2 hours. During this interview the socioeconomist will inform the stakeholder about the project and the action scenarios under discussion. He will furthermore ask the stakeholder for his views of the situation and the action scenarios. All meetings with the stakeholders would ideally take place over a short period, typically 1 week maximum. With preference the stakeholders of the water utilities, water system planning department of the municipality, and the municipal council should be interviewed first.

(3) Step 3 - Analysis of citizen appreciation: analysis of the questionnaires of the stakeholder interview results to produce statistical data on needs, importance, and payability. A selection of the stakeholders with technical proficiency in, and knowledge of, the water systems would be asked to fill in an expert questionnaire. The SEESAW Model is an analytical tool also designed for the purpose of socio-economic analysis of the urban water system. Once the questionnaire data are entered, the SEESAW Model will produce results for further use in the project.

(4) Step 4 - Socio-economic-environmental assessment of the actual situation and the proposed action scenarios: these assessments aim to produce decision support information, especially a comprehensive explanation of the decision situation for the target group stakeholders and a set of socio-economic-environmental indicators. The socio-economic-environmental analysis using the SEESAW Model combines all data collected and evaluated within the previous steps.

(5) Step 5 - Stakeholder workshop for assessment and discussion: support of the decision making process by means of one or more workshops for assessment and discussion, supported by the AISUWRS Deliberator, a special tool to explain the action scenarios' assessments of different stakeholders which considers relevant decision criteria in a three dimensional cube. This may serve as an important basis for the urban decision-making. From the point of view of the AISUWRS project this workshop serves the joint purpose of assessment of usability of the methodology and of the models' output.

4.2 GUIDELINES FOR DATA COLLECTION AND ANALYSIS – THE SEESAW MODEL

4.2.1 Introduction

The SEESAW Model is an application in Excel that has been chosen for its flexibility and adaptability to the ever-changing situations under study. Although there is an obvious danger that the user might be unwilling to change parts of the model, it is should be pointed out that it is important to adapt the model to needs as required. The SEESAW Model is available in English and German, and other languages may easily be added by filling in the language table. In the current version, four cases can be dealt with simultaneously. This could be four cities (like in the AISUWRS project) or four parts of an urban area. The number of action scenarios is limited to 30, but it could also be increased. The user is free to enter the currency and unit of surface measurement used in his country. This will then affect the input forms and output reports at all relevant places.

The Model consists of the following parts:
(1) Input forms for input on the situation from the different points of view of the water utility and the groundwater resource modellers, a scenario description, questionnaires to private households, technical/system experts and to businesses
(2) Service tables on linguistic equivalents, scenario administration, input and statistical analysis of questionnaires
(3) Output and reporting tables: assessment of the actual situation from a technical point of view, reporting on environmental sustainability indicators including changes induced by the proposed action scenario, overview, assessment and reasoning behind the scenario, evaluation of the questionnaire sets from private households and technical/system experts, with comparison of their answers, comparison between objectives of water consumers and of action scenarios, calculation of costs and the new water price for the action scenario, change in costs of living as a consequence of the scenario.

4.2.2 Assessment of the environmental sustainability

The actual situation of the urban water system is analysed according to a scheme of indicators elaborated for the project. The data can be gathered by using the SEESAW Model questionnaires developed within the AISUWRS project. The input files In_Util (Input from water utility) and In_Geo (Input from geological/research institutes) collect the key data relevant for an application. The output table EvaSusInd (Evaluation of Sustainability Indicators) within the SEESAW model deals with the aggregation of technical and environmental information from which experts can draw conclusions.

4.2.2.1 Life cycle concept

The indicator scheme is based on a life cycle assessment approach according to existing iterative concepts for the analysis of urban water systems (Lundin 1999; Lundin et al. 2002). As this analysis should on one side focus on the sustainable use of groundwater resources, but on the other side value processes of water services, key indicators have been chosen to account for both. Sustainable development is assessed by using time series data in order to analyse development trends. The application should compare information at different points in time. Lundin recommends for her analysis timespans of approximately 20 to 30 years for retrospective analysis.

Some of the AISUWRS indicators account for information aggregation with regard to the PRESSURE on groundwater caused by anthropogenic activity, with regard to the actual STATE of the natural resource and with regard to the RESPONSE chosen by civil society and legislation to react on such pressure and state. The classification of pressure, state and response indicators as developed in the OECD PSR framework (OECD 1998) takes into account that human activities exert pressures on the environment, which can induce changes in the state of the environment, and society then responds to such changes with environmental and economic policies and programs intended to prevent, reduce or mitigate pressures and/or environmental damage.

Other indicators assess sustainablitiy by using ratios indicating aspects of effectiveness, efficiency or sufficiency within the present system.

The indicators aim to represent key aspects in the life cycle. The scheme contains 17 indicators in five groups of relevance for the urban water system life cycle:

(1) group I: "groundwater"
(2) group II: "drinking water production"
(3) group III: "consumption"
(4) group IV: "waste water treatment"
(5) group V: "off-products and byproducts".

With an additional indicator group S "changes induced by the case study cities' scenarios", a forecasting capability is added to the analysis of the present situation, taking into account the effects of improvement measures for the case study cities via specific scenarios. Figure 4.2.1 is a schematic illustrating for the case study town of Mount Gambier which part of the urban water system can be covered by the five indicator groups listed above scheme and where the scenarios S1 and S2 tend to induce improvements.

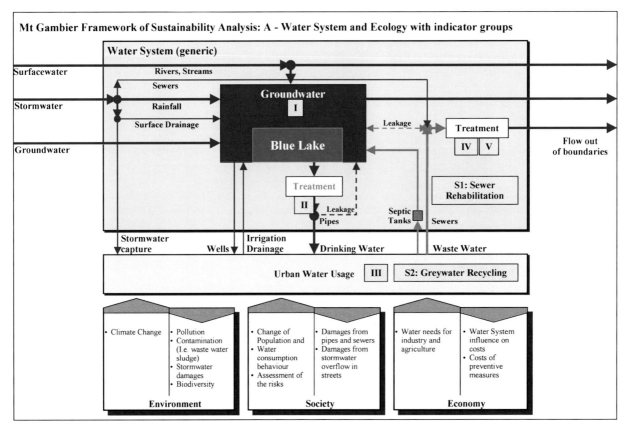

Figure 4.2.1. Mt Gambier framework of sustainability analysis: A – water system and ecology with indicator groups.

4.2.2.2 Drawing conclusions-a cautionary note

When drawing conclusions on the sustainability of the urban water system, special attention has to be given to the problem of system boundaries: the sustainability analysis is a generic approach and cannot distinguish between effects to the urban water system within its system boundaries or effects that partly or fully act on the system outside the system boundaries. Therefore it is recommended that each data input is assessed thoroughly during application.

The problem is inherent in the way that available data are provided for the population of input tables. For example, a water utility will provide data related to operational units such as metered water supply districts or wastewater treatment works sewer and stormwater disposal catchments; in a city they will overlap of course, but there is no engineering or utility administration reason why these districts should coincide (they use different engineering skills and staff cadres) and in practice they do not. Then again, a municipality will provide data appropriate to its functions such as town planning or service provision, while political/administrative boundary datasets will be the most readily available. They need bear no relation to the coverage of the water utility datasets, especially if the utility is a large private company serving a larger area that just the city limits (an increasingly common situation). Meanwhile the resource assessments will use hydrologic or geological boundaries (and corresponding datasets) to provide catchments that are amenable to modeling. The resultant datasets underlap, overlap but rarely coincide, and measured judgment skills are needed to reconcile this inevitable occurrence. This was the case during the AISUWRS project, when system boundaries used by the research institutes for the AISUWRS models mostly did not correspond to the catchment or to the utilities' administrative areas. Approximations can be made but it is a process that needs to be consciously undertaken as part of the assessment process.

A related problem is that indicator calculation may need to cast its net wide to compile from various sources and it has to be verified that the data rely on the same background figures (e.g. population numbers). It is important to carefully treat the data from different sources - and in case of deviations to be aware of such deviations when drawing conclusions.

4.2.2.3 Indicator scheme

In order to aggregate information on the environmental and technological aspects within the life cycle of urban water, the following key indicators were developed (see Table 4.2.1):

Table 4.2.1. Indicators for the environmental and technological assessment.

Indicator Group	Life cycle component	No.	Indicator	Sustainability Indication
I	Groundwater	1	Groundwater footprint	Pressure
		2	Trend in availability (quantity) - groundwater level	State
		3	Trend in availability (quality) - groundwater quality	State
		4	Trend in protection	Response
II	Drinking water production	5 A B	Drinking water availability from utility	Effectiveness
		6	Effectiveness of water services	Effectiveness
		7 A B	Efficiency of production process in the utility	Efficiency
		8	Utility water losses	Efficiency
III	Consumption	9	Sufficiency and scale effects from water consumption	Sufficiency
IV	Wastewater treatment	10	Exfiltrated loads from wastewater leakage	Pressure
		11 A B	Effectiveness of wastewater treatment	Effectiveness
		12	Remaining pollution loads after wastewater treatment	Pressure
		13 A B	Efficiency of wastewater treatment process	Efficiency
V	Off & By-products	14	Recovery and recycling	Response
S	Changes induced by scenario	S1	Scenario 1 specific indicator	
		S2	Scenario 2 specific indicator	
		S3	Scenario 3 specific indicator	

Indicator Group I "groundwater" assesses the urban groundwater resource with regard to water quality and quantity. The indicators aim to show whether the usage of the resource is balanced over time. A further point considered in this block is the protection of groundwater resources, taken as an indicator for awareness and effort.

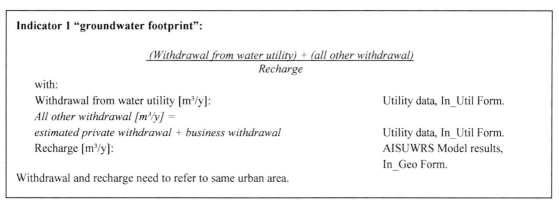

Indicator 1 "groundwater footprint":

$$\frac{(Withdrawal\ from\ water\ utility) + (all\ other\ withdrawal)}{Recharge}$$

with:
Withdrawal from water utility [m³/y]: Utility data, In_Util Form.
All other withdrawal [m³/y] =
estimated private withdrawal + business withdrawal Utility data, In_Util Form.
Recharge [m³/y]: AISUWRS Model results, In_Geo Form.

Withdrawal and recharge need to refer to same urban area.

Figure 4.2.2. Indicator Group I: Calculation of indicators.

(1) Indicator 1 "groundwater footprint" (see Figure 4.2.2): the main focus of this ratio is the equilibrium between the total water abstraction and consumption in the urban area and the water input on the same urban area, i.e. the urban recharge. Thus, the city itself is the centre of concern and framework conditions such as detailed hydrogeological settings resulting in heterogeneous pollution patterns and groundwater levels are not considered. Balance or disequilibrium might occur on a daily as well as on a seasonal or long-term time scale. However, as sustainability analysis refers to long-term developments, the annual scale has been chosen as reference to draw conclusions from. In order to evaluate the groundwater footprint at an annual scale, total groundwater withdrawal in the urban area is compared to the recharge to the same area. The indicator is defined as the ratio between the withdrawal and the recharge, given in %. In case this indicator is below 100%, the groundwater abstraction is considered as being sustainable and decreasing trends indicate a development towards sustainability. Total annual withdrawal has been split up into a) withdrawal by the

water utility and b) all other withdrawal via private wells, business wells etc. gathered from the water utility in the "In_Util Form". Total annual recharge is delivered via the "In_Geo Form" resulting from the groundwater models. It is evident, that within the data collection process, both parties need to refer the numbers to the same system boundary. For interpretation, it has to be taken into account, that the footprint indicator simplifies the assessment and therefore does not cover the detailed hydrogeological setting, which may have causing heterogeneous flow and thus complex pollution patterns.

(2) Indicator 2 "trend in availability (quantity) - groundwater level": another measure in order to consider groundwater balance is the trend in groundwater table development. Decreasing groundwater levels indicate that withdrawal (or flow out of boundaries) exceeds the groundwater production through recharge in the catchment area. Increasing groundwater levels show excess of recharge to the aquifer in comparison to withdrawal or flow out of the boundaries. As groundwater levels vary annually and on the local scale the "In_Geo Form" that is collecting the data does ask for groundwater levels at observation wells that are significant, meaning that they show the long-term temporal trend in the area. In case such obvious trends cannot be registered, water level changes at several indicator observation wells can be used. Average groundwater levels and average yearly changes in groundwater level should also be estimated. Emphasis is given to general trends, where more approximate figures can be accepted.

(3) Indicator 3 "trend in availability (quality) – groundwater quality": availability of groundwater depends also on quality aspects. In general, groundwater quality has to be valued with regard to existing national or international quality standards, to likely end-use requirements or less satisfactorily to stakeholders' perception. Possible standards are: a) national standards on the quality of drinking water or the worldwide applicable standard as given by the World Health Organisation (WHO 2004); b) the standards on bathing waters e.g. those of the European Union that impose quality standards with regard to leisure and health. Concentrations of nitrate, boron and *E. coli* bacteria etc. are collected by using the "In_Geo Form". The parameters are interpreted as markers of groundwater contamination. However, the processes of attenuation etc. cannot be taken into consideration with these concentrations. The values from significant observation wells are primarily compared to the WHO drinking water standards. Stakeholder perception of their water with regard to their own personal needs can be assessed during a survey, but care should be exercised in the interpretation, as testified by the inexorable rise in sales of bottled water in Western Europe despite generally excellent tap water quality. Decreasing groundwater quality, as indicated by any concentration rising, indicates an excess of groundwater usage compared to natural decomposition potentials or anthropogenic efforts to remediation. This indicator needs to be compared to the loads to receiving water bodies from leaky sewers, overflow and remaining pollution load after treatment, in order to gain information on the causes of deteriorating groundwater. As groundwater quality can vary on the local scale, the "In_Geo Form" does not ask for complete data, but uses that from indicative observation wells.

(4) Indicator 4 "trend in protection": this strives to represent the societal awareness of technological environmental problems related to groundwater and the effort to solve them. National or local groundwater protection policies or simple statements by the municipality or the water service utilities can be seen as a basic political approach to increase the sensibility in society. A calendar of when first milestones in protection were obtained is helpful. Another indication of the trend in protection is the protection of wells. Indicating protection zones around wells, simple well head protection policies or even statements by the utility can give information about sensibility of the stakeholder filling in the form.

Indicator Group II "Drinking water production" focuses on the utility responsible for public water supply (see Figure 4.2.3). Being the principal stakeholder of water abstraction and consumption via purchase, the utility itself should have a main interest in sustainable and environmentally sound processes in order not to increase pressure on their resource.

(5) Indicator 5 "drinking water availability from utility": this indicator strives to represent the present availability of drinking water from the utility to society. Data on the annual total consumption from the public system and total drinking water production are collected from the utility. The ratio of consumption to production indicates, whether more drinking water is produced than actually consumed from the utility supply. This indicates a kind of "water efficiency" of water output for consumption to water resources input in production. The reasons for higher production than consumption are manifold: unaccounted for water, water sales etc.

(6) Indicator 6 "effectiveness of water utility services": effectiveness is the degree to which a system's features and capabilities meet the user's needs. Thus, the total societal demand needs to be taken into account. It is assumed: in case of an effectiveness < 1, societal demand is higher than production by utility. The trend of this "effectiveness of water utility services" has been chosen to indicate, in the European context, a development towards an effective water supply, as such centralised services can more easily be monitored either with regard to compliance with legal regulations or with regard to an implementation of economic

instruments that aim for an internalisation of external pollution effects via the market. Decentralised private abstraction is assumed to result in much higher monitoring costs.

Indicator 5A "drinking water availability from utility":

$$\frac{\textit{Annual drinking water consumption from public system}}{\textit{Annual drinking water production by utility}}$$

with:
Annual drinking water consumption from public system [m³/y],
Annual drinking water production by utility [m³/y]:   Utility data, In_Util Form.

Indicator 5B "excess water production per capita":

$$\frac{\textit{(Annual drinking water consumption from public system)} - \textit{(Annual drinking water production by utility)}}{\textit{Population}}$$

with:
Annual drinking water consumption from public system [m³/y],
Annual drinking water production by utility [m³/y],
Connected population: Utility data, In_Util Form.

Population figures for the service area needs to correspond to those of the system boundaries.

Indicator 6 "effectiveness of water utility services":

$$\frac{\textit{Annual drinking water production by utility}}{\textit{Total annual demand from society}}$$

with:
Annual drinking water production by utility [m³/y]:   Utility data, In_Util Form.

Total annual demand from society =
Annual drinking water production by utility + water purchase from outside + all other withdrawal – (leakage and water losses (unaccounted for water) + water sales out of region)
     Utility data, In_Util Form.

Indicator 7A "efficiency of the production process in the utility":

$$\frac{\textit{Annual drinking water production by utility}}{\textit{Annual consumption of energy}}$$

$$\frac{\textit{Annual drinking water production by utility}}{\textit{Annual consumption of chemicals}}$$

with:
Annual drinking water production by utility [m³/y],
Annual consumption of energy [kWh/y],
Annual consumption of chemicals [kg/y],
Annual drinking water consumption from supply [m³/y]:   Utility data, In_Util Form.

Indicator 7B "per capita chemical and energy use from the utility":

$$\frac{\textit{Annual consumption of chemicals}}{\textit{Population}}$$

$$\frac{\textit{Annual consumption of energy}}{\textit{Population}}$$

with:
Annual consumption of energy [kWH/y],
Annual consumption of chemicals [kg/y],
Connected population   Utility data, In_Util Form

Indicator 8 "utility water losses":

$$\frac{\textit{Water losses (unaccounted for water)}}{\textit{Volume of produced water in network system}}$$

with:
Water losses (unaccounted for water) [m³/y],
Volume of produced water in the network system [m³/y] = Annual drinking water production by utility [m³/y] + Water purchase from outside [m³/y] – Water sales out of region [m³/y].
     Utility data, In_Util Form

Figure 4.2.3. Indicator Group II: Calculation of indicators.

(7) Indicator 7 "efficiency of production process in the utility": efficiency herein is used as comparison of water production outcome to the demanded input of capital resources, which in this case is energy and chemical input. Annual drinking water production by the water utility is related to the annual total input of chemicals and of energy. The trend of this indicator over time might show a reduction of chemical and energy input, e.g. through technological improvement measures, indicating a step forward to sustainability.

(8) Indicator 8 "utility water losses": Technical water losses in the utilities' network result in different unsustainable effects as wastage of scarce drinking water or water prices increase in order to account for losses or for repair investment. Data on water losses collected from the utility can include all water losses resulting from leakage, unmetered water supply or errors in the administrative processes of a water utility.

Indicator Group III "consumption and scale effects" compares data on water consumption to increasing demand in order to gain information on the trend in sufficiency of the society and scale effects (see Figure 4.2.4).

(9) Indicator 9 "sufficiency and scale effects from water consumption": the water consumption pattern for domestic, commercial, institutional and industrial stakeholders can indicate trends towards a reduction in per capita water consumption over time, leading to sustainability through sufficiency. However, due to scale effects also the total domestic and non-domestic demand needs to be taken into consideration. Total domestic annual demand has been calculated based on the estimated daily per capita consumption for domestic households and on available population figures. The total non-domestic annual demand can be deduced from the total annual societal demand minus the total annual domestic demand. These figures can indicate scale effects meaning that although per capita consumption decreases, the groundwater resources can be over-exploited e.g. by population growth due to increased birth rates, migration etc. or by increased rhythm of industrial and commercial activities due to economic development.

Indicator Group IV "waste water treatment" analyses the efficiency of treatment processes in the wastewater treatment plant and the sources of contaminant loads related to sewer exfiltration and treated wastewater into groundwater (see Figure 4.2.4). As the calculation of pollution loads is also part of the modelling, these coarse estimations resulting from raw and treated wastewater concentrations can be compared to the model results.

(10) Indicator 10 "exfiltrated loads from waste water leakage": the utility was asked to deliver exfiltration rates from the sewers. Based on raw wastewater concentrations, also provided by the utility, the pollution loads infiltrating into groundwater from the leaking sewers can be estimated. This is also calculated by the AISUWRS model chain. When available, the values can be compared to each other and can thus be used as a general plausibility control to the model output.

(11) Indicator 11 "effectiveness of wastewater treatment": the volume of water treated does not correspond to the volume of produced or consumed water as the volume of infiltrating water as well as collected stormwater has to be considered. Effectiveness of treatment processes can either be valued with regard to quality or with regard to quantity. We distinguish thus: a) quantitative effectiveness: the total annual volume of water passed through the whole urban water system is evaluated. However, it is evident that not all water in the urban system is contaminated or contaminated in the same way, so it does not have the same need for treatment. This shows the complexity encountered herein. The annual volume of water needing treatment is determined as annual total societal water demand, total annual effective stormwater (from UVQ) and infiltrating water into the sewer network. The effectiveness-ratio is then calculated as the annual volume treated divided by the volume where treatment is deemed necessary. The ratio indicates the percentage of water not considered at all in the processes of central treatment plants; b) qualitative effectiveness: this is indicated showing concentrations of wastewater parameters, e.g. BOD, N and P, before and after treatment. This comparison shows the performance, given in % of reduction of the individual parameters of the wastewater treatment plants. As the effectiveness of treatment processes further depends on the raw wastewater concentrations entering the treatment plant, the ratio of concentration difference before and after treatment per concentration before treatment is used. Biological treatment plants usually have ratios of 95-98 % for BOD removal, sometimes referred to in sanitary engineering as "treatment efficiency". However within the AISUWRS sustainability assessment the term efficiency is only used when output is compared to a certain resource input needed for the process. We therefore use the term effectiveness indicating the improvements from treatment with regard to a general objective to remove all pollution. When drawing conclusions, it has to be taken into consideration, that raw wastewater characteristics might differ from one case study to another and that the treated wastewater standards for the discharge into the water receiving body might differ from one country to another.

Socio-Economics and Sustainability

Indicator 9 "scale effects":

$$Annual\ domestic\ demand\ [m^3/y] =$$
$$Estimated\ daily\ per\ capita\ consumption\ in\ domestic\ households \times 365\ days \times population.$$

$$Annual\ demand\ other\ than\ domestic\ [m^3/y] =$$
$$Annual\ societal\ demand - Annual\ domestic\ demand.$$

with:
Estimated daily per capita consumption for domestic households [l/cap x day],
Population figures for the town: Utility data, In_Util Form

Indicator 10 "exfiltrated loads":

$$Exfiltrated\ loads\ from\ waste\ water\ leakage\ [mg/y] =$$
$$Estimated\ sewer\ exfiltration \times$$
$$Concentration\ of\ chemical\ parameter\ in\ raw\ wastewater \times 1000$$

with:
Estimated sewer exfiltration [m³/y],
Concentration of parameter in raw wastewater [mg/l] for N & P   Utility data, In_Util Form

Indicator 11 "quantitative effectiveness of wastewater treatment":

$$\frac{Volume\ of\ annual\ treated\ wastewater}{(Total\ annual\ demand\ from\ society + Annual\ effective\ stormwater + Annual\ infiltrating\ waters)}$$

with:
Total annual demand from society [m³/y] =
Annual drinking water production by utility [m³/y] + Water purchase from outside [m³/y]
+ All other withdrawal [m³/y] − Leakage and water losses (unaccounted for water) [m³/y]
− Water sales out of region [m³/y].

Treated wastewater [m³/y],
Annual infiltrating water [m³/y]:   Utility data, In_Util Form.
Annual effective stormwater data [m³/y]:   UVQ models.

Indicator 11B "qualitative effectiveness of wastewater treatment":

$$\frac{Load\ of\ parameter\ before\ treatment}{Load\ of\ parameter\ after\ treatment}$$

$$\frac{(Load\ of\ parameter\ before\ treatment) - (Load\ of\ parameter\ after\ treatment)}{Load\ of\ parameter\ before\ treatment}$$

with:
Load of parameter before treatment [kg/y] =
concentration of parameter in raw wastewater [mg/l] x treated wastewater [m³/y] x 0.001.
Load of parameter after treatment [kg/y],
Concentrations of parameter in raw wastewater [mg/l],
Volume of treated wastewater [m³/y]:   Utility data, In_Util Form.

Indicator 13 "efficiency of the wastewater treatment":

$$\frac{Annual\ reduction\ of\ load\ of\ parameter}{Annual\ consumption\ of\ energy}$$

$$\frac{Annual\ reduction\ of\ load\ of\ parameter}{Annual\ consumption\ of\ chemicals}$$

with:
Annual reduction of load of parameter [kg/y] = load before treatment [kg/y] − load after treatment [kg/y];
Annual consumption of energy [kWh/y] & annual consumption of chemicals [kg/y]
                                                                                   Utility data, In_Util Form.

Figure 4.2.4. Indicator Group III & IV: Calculation of indicators.

(12) Indicator 12 "loads after treatment process to receiving water bodies": loads after treatment that are directed to receiving water bodies can be calculated using the concentration values and the volume of treated wastewater ("In_Util Form"). The sustainability analysis is based on a generic approach and therefore cannot consider whether the loads after the treatment process that enter into receiving water bodies actually enter the urban water system within its system boundaries or affect the environment outside the system boundaries.

(13) Indicator 13 "efficiency of treatment process" is calculated as the ratio of annual reduction of load for certain parameters (here BOD, nitrogen, phosphorus) to the annual input of chemicals or energy for the treatment process, assuming the same chemical has been used at the same chemical concentration to aim for a water quality defined by legal regulations. It thus shows either improved or reduced remediation efficiency according to the input needs for treatment. An increasing rate over time shows that more loads have been reduced per unit of chemicals or energy applied during treatment. As removal of BOD and N in communal treatment plants occurs through aeration inducing biological processes, only energy input is taken into consideration for these parameters. Removal of P occurs partly through aeration and partly through chemical precipitation and thus both ratios are used. When drawing conclusions, it has to be taken into consideration, that the pollution parameters treated and the degree of contamination in the raw wastewater as well as the required treated wastewater discharge standards have an effect on the ratio. A further indicator is the per capita chemical and energy use from the utility side.

Indicator Group V "off-products and by-products" includes indicators for recovery and recycling.

(14) Indicator 14A "energy recovery from wastewater treatment" is used to show which potential of energy recovery is used or not used by the utility. The annual energy recovery is calculated as value "per capita". However, in our analysis, we do not have exact figures of citizens employing the wastewater treatment plants but only population figures for the city and there is a potential risk that the use of these numbers would distort the ratio, as people that are in reality not connected to the treatment plant e.g. households on septic tank drainage would decrease the rate for energy recovery. However, in Western Europe, coverage rates of public wastewater treatment plants are generally > 95%. It has to be considered case study specific whether general population figures can be used as an estimate or not. The values come from the data given directly by the utility.

(15) Indicator 14B "recovery of nutrients N and P": This indicator is used to show which potential of nutrient recovery is used or not used by the utility. The annual nutrient removal is calculated as value "per capita". Within the AISUWRS case study cities, however, nutrient removal for recycling so far does not occur. Nevertheless, the indicator is left in the AISUWRS approach as in future it will be of more relevance than at present. Ecosan concepts, which can cover such aspects have partly been analysed in the scenario analysis (e.g. for Rastatt).

Indicator Group S "scenario assessment" analyses the potential changes after action. For each scenario one indicator is recommended. As such indicators are scenario specific, examples from the AISUWRS case study cities are listed in Figure 4.2.5. With these indicators improvements induced by the scenario or increased pressure on the environment from the action scenario can be shown.

(1) Indicator S1 "resulting estimated loads from sewer exfiltration after rehabilitation of sewers" can be compared to indicator 10 of the analysis in order to estimate potential improvements. It aims to measure the reduced loads of BOD, N and P that exfiltrate from the sewer pipes into the groundwater after rehabilitation of the sewer system. In order to support decision makers, investment demand for the scenario needs to be taken into consideration thereafter.

(2) Indicator S2 "increase in inefficient stormwater generation and wastewater treatment disposition effectiveness" calculates the ratio of annual volume of wastewater treated by the utility per total annual objective. The indicator can be compared to indicator 11 for eventual improvements in the effectiveness of treatment. However, it has to be taken into consideration that with rainwater infiltration a certain amount of pollution might enter the ground, soil and groundwater. The indicator shows improved effectiveness of the wastewater treatment. This indicator would therefore be limited to a scenario employing only stormwater collected from residential areas with a low contamination risk.

(3) Indicator S3 "efficiency increase of water production and waste water treatment by increased recovery of nutrients N and P": estimates of N and P recovery were not applicable in the AISUWRS case study cities. Indicator S3 can be compared to indicator 16.

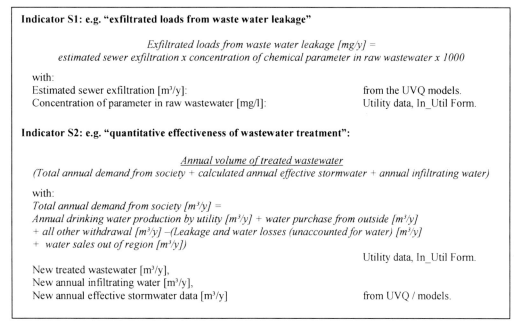

Figure 4.2.5. Indicator Group S: Calculation of Indicators.

4.2.3 Assessment of socio-economic sustainability

4.2.3.1 The Assessment steps

The socio-economic assessment following the AISUWRS methodology has two stages. Firstly the actual situation and the proposed actions are analyzed. Then in-depth discussion occurs of the decision situation and the assessment needed for decision making (see Figure 4.2.6).

The first analysis is based upon several inputs. The Environmental Sustainability Report with a thorough description of the water system should be available. The Action Scenario Description should contain a description of the improvement measures and a cost-benefit statement. Finally the perspectives of the stakeholders should be known from stakeholder interviews and from a (preferably larger) questionnaire survey among the private households. All this information is collected and formally entered in the SEESAW Model for further analysis as described below. Figure 4.2.6 summarizes the flow of information during the assessment steps.

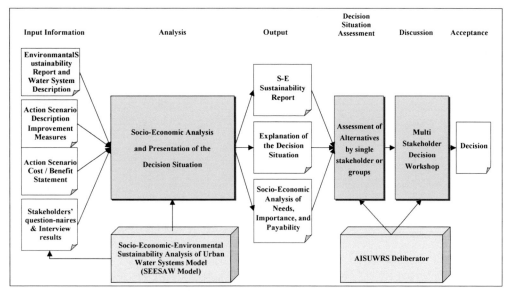

Figure 4.2.6. Socio-economic assessment steps.

The output of the first step provides material for the comparative assessment of the alternative actions in the second step. The information needs to be prepared in a jargon-free and simple style as it needs to be understood by a broad spectrum of stakeholders and decision makers, and should include an explanation of the decision situation, a socio-economic analysis of the sustainability, the needs, importance, and payability of the proposed action scenarios.

Each decision maker or group of stakeholders will then assess the alternatives using the AISUWRS Deliberator as a tool for expressing the weight and estimation of each of the decision criteria for each alternative. Once the individual assessments are completed and entered, the AISUWRS Deliberator shows the comparison of judgments. The stakeholders may now come together in a workshop meeting to discuss the controversial assessments with the aim of reaching a consensus conclusion and decision.

4.2.3.2 The Quality-Importance Concept

The socio-economic analysis comprises many aspects and collects a great variety of information in order to scan each situation (case), trying not to neglect any facts that might be of specific importance. Especially the interviews with stakeholders are an opportunity to raise open questions and to hear unexpected arguments.

The core of the analysis centres on a set of systematic questions leading to a structured analytical model that has been proven successful in the AISUWRS project. The aim is to know what people (stakeholders) think about the system under consideration – here the urban water system - in terms of quality, importance and the willingness to pay for improvements. The water system and its interrelation with the environment, society (and also individuals), and to the economy is described by a set of indicators (shown in Figure 4.2.7), in 9 groups (see the following Table 4.2.2):

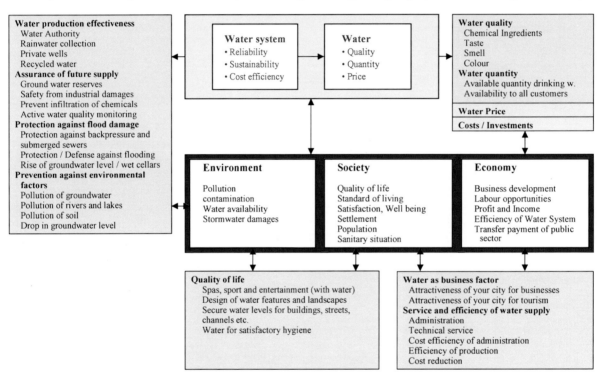

Figure 4.2.7. Indicators used in the SEESAW analysis.

Table 4.2.2. Grouping of indicators.

No.	Indicator class
1	Water quality
2	Water quantity
3	Water Production Effectiveness
4	Assurance of future supply
5	Quality of life
6	Protection against flood damage
7	Prevention against environmental factors
8	Water as business factor
9	Service and efficiency of water supply

The classes of indicators have been chosen in order to cover all aspects relevant to the AISUWRS field of investigations. They cover objectives and interests in the present (such as water quality and quantity) and in the future (such as protection against future damage). They can be either measurable or intangible, but stakeholders may nonetheless judge them. Some may need specific knowledge to be understood and carefully assessed (such as prevention against environmental factors), but even if someone is not knowledgeable in a professional sense, he or she will probably have an opinion and may even participate in a democratic vote for future policies budget allocation, so it is therefore important that every opinion can be expressed and considered in the analysis.

The groups of indicators are broken down to express more detailed indicators (see Table 4.2.3):

Table 4.2.3. Detailed indicator definition A to Z.

Indicator class	Indicator
1. Water quality	A. Chemical Ingredients
	B. Quality summarized
2. Water quantity	C. Available quantity drinking water
	D. Availability to all customers
3. Water Production Effectiveness	E. By water authority
	F. Rainwater collection
	G. Private wells
	H. Recycled water
4. Assurance of future supply	I. Ground water reserves
	J. Safety from industrial damages
	K. Prevent infiltration of chemicals
	L. Active water quality monitoring
5. Quality of life	M. Spas, sport and entertainment (with water)
	N. Design of water features and landscapes
	O. Secure water levels for buildings, streets, channels etc.
	P. Water for satisfactory hygiene
6. Protection against flood damage	Q. Protection against backpressure & submerged sewers
	R. Protection against flooding
	S. Against rise of groundwater level/wet cellars
7. Prevention against environmental factors	T. Pollution of groundwater
	U. Pollution of rivers and lakes
	V. Pollution of soil
	W. Drop in groundwater level
8. Water as business factor	X. Attractiveness of the city for businesses
	Y. Attractiveness of the city for tourism
9. Water supply	Z. Service and efficiency

The list of indicators A to Z was asked in the context of three propositions:
(1) In your opinion, how good is in your city ... (with the choice of answers: very good, good, improvable, bad, very bad)?
(2) How important do you judge an improvement of ... (with the choice of answers: not at all important, little important, important, quite important, very important)?
(3) Would you accept an increase of the water price/costs, if there is an improvement in ... (with the choice of answers: no, up to 5%, up to 10%, up to 20%, over 20%)?

It is important to ask different stakeholders (e.g. private households, experts, utilities) the same set of questions. The analysis will then show the results in the following structured way, first as quality-importance matrix (see Figure 4.2.8).

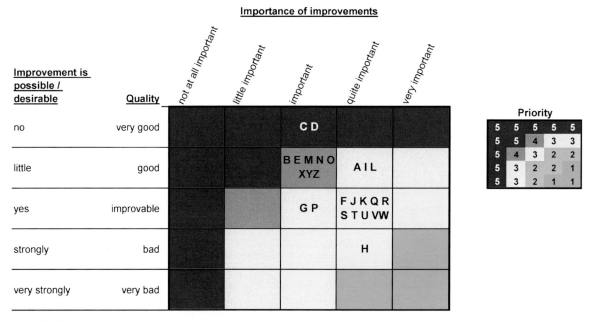

Figure 4.2.8. Quality-Importance matrix (example).

In this quality-importance matrix the vertical axis shows the quality as judged by the stakeholders (this example is taken from the results of the questionnaire survey among private households in the city of Rastatt). The better an item is judged, the less it needs to be improved. Vice versa, an item judged very bad needs very strongly to be improved. But also the improvement of the item should be considered as important. If the improvement of an indicator were felt to be of no importance, than it would not be attractive for any measures or spending of money to improve. Therefore the quality-importance matrix may serve to define priorities of actions. The small matrix at right representing the quality-importance weightings, serves to indicate (and change) the priorities (1 being of highest and 5 of lowest priority).

Having thus defined the priorities of measures (actions) it is questionable whether the willingness to pay for it corresponds to the priorities. The following matrix (see Figure 4.2.9) shows the results of a questionnaire survey. As it represents statistical mean values it is not surprising that the answers opted for are not of extreme values, possibly in contrast from what one would expect with a sample of 1 expert.

Priority		Willingness to pay more				
		no	up to 5%	up to 10%	up to 20%	over 20%
1						
2		F H	J K Q R S T U V W			
3		G P	A I L			
4		E M N O X Y Z	B			
5		C D				

Figure 4.2.9. Priority-Payability matrix (example)

From this willingness to pay matrix it can be seen that the stakeholders on average are willing to pay up to 5% more for water if some of the items ranked as second in priority are being improved. They would not be willing to pay as much for the improvement of water quality being ranked much lower (at priority 4), because the water quality is already good and the improvement is judged only as of (medium) importance. Consequently the stakeholders in charge of improvements should first of all look at improvements of the following indicators indicated in Table 4.2.4.

Table 4.2.4. Indicator list to be approved with regard to improvements.

Indicator class	Indicator
3. Water Production Effectiveness	F. Rainwater collection By water authority
	H. Recycled water
4. Assurance of future supply	J. Safety from industrial damageGroundwater reserves
	K. Prevent infiltration of chemicals
6. Protection against flood damage	Q. Protection against backpressure & submerged sewers
	R. Protection against flooding
	S. Against rise of groundwater level/wet cellars
7. Prevention against environmental factors	T. Pollution of groundwater
	U. Pollution of rivers and lakes
	V. Pollution of soil
	W. Drop in groundwater level

The next step in the analysis should then be to see whether the actual decision making or the designed action scenarios correspond to the priorities of the stakeholders. Among other aspects the action scenarios have been described by their influence on the indicators A to Z (see Figure 4.2.10).

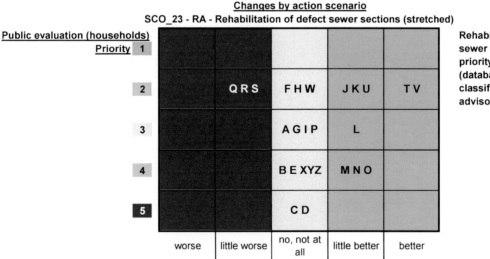

Figure 4.2.10. Comparison of objectives between water consumers and action scenarios (example).

This matrix shows very clearly that this action scenario will influence some of the indicators in a positive, some in a negative sense, and some not at all. In this case even two indicators (T and V) of high priority are changed to the better. Further analysis and reporting is done to show the effects in detail.

The SEESAW model calculates the costs of the action scenario and the increase of water price to cover the investment costs (see Figure 4.2.11). This can easily be compared with the expressed willingness to pay more for water. The economic calculation sheet allows evaluation of different business strategies such as expanding ("stretching") the investment over a longer or shorter period of time, or using different types of financing. Also, in further tables, it shows the effect on an average household budget when the water price is changed. It would be compared with average household incomes and minimum income situations, which might be treated differently. The following figure shows an example of a calculation over time.

From this example it can be seen (last line), that the investment-driven need for a water price increase is changing from below 1% now to about 5% in 5 years and 8.6% in 10 years. This is within the range that the private households indicated as being acceptable.

SCO_23 - RA - Rehabilitation of defect sewer sections (stretched)

		Now			+5 years				+10 years			0 - 5	0 - 10
		Invest	new	transf.	cumul.	Invest	new	transf.	cumul.	Invest	new	Total	Total
1 Investments and assets													
Investment expenditures	T-EUR	0				0				0		0	0
and / or yearly investments	T-EUR	1.100		1.100	5.500	1.200		1.200	11.500	0		6.700	11.500
Total investments	T-EUR	1.100		4.400	5.500	1.200	6.700	4.800	11.500	0	11.500	6.700	11.500
Assets (investment - depreciation)	T-EUR		1.073	4.345	5.418	6.618	6.452	4.740	11.192	11.192	10.912		
Depreciation time	years	40			40				40				
2 Costs													
Depreciation	T-EUR/Y		28	55	83		165	60	280		280	96	154
Maintenance costs	T-EUR/Y		0				0				0	0	0
Costs	T-EUR/Y		0	0,0%			0	0,0%			0	0	0
Capital costs	T-EUR/Y		33				199				336	116	184
Cost savings	T-EUR/Y		0				0				0	0	0
Total costs	T-EUR/Y		61				364				616	212	338
Consumption from public system	Mm3/Y		2,456				2,456				2,456		
3 Price of drinking water for households												Avg.	Avg.
Avg. drinking water price (incl. black water) for households	EUR/m3		2,76	0,5%			2,83	0,5%			2,90	2,79	2,83
Calc. Change of price	EUR/m3		0,02				0,15				0,25	0,09	0,14
New price (because of scenario)	EUR/m3		2,78				2,98				3,15	2,88	2,97
Price change (because of scenario)	%		0,89%				5,2%				8,6%	3,1%	4,8%

Figure 4.2.11. Action Scenario Calculation (example of rehabilitation measures stretched over a period of 10 years).

4.2.3.3 Data Acquisition, Input Data and Reports

The SEESAW Model helps directly in the process of information acquisition. It provides the user with input forms and questionnaires showing the information needed and the practical forms to collect and enter the data. It can be used by geology and engineering professionals and by socio-economists. The data is collected once and used by all.

In many situations it is useful to gain more information than is used for the structured evaluation of the quality-importance matrix previously described. For example, domestic users may be asked not only about their present water supply (e.g. with wells or rainwater collection) but also their personal preferences of water usage. Businesses as large water consumers or potential contaminators may be asked to explain their typical situation and preventive measures to protect the groundwater. For all this the questionnaires are more detailed for eventual use than the needed minimum.

The stakeholder interviews follow a schema similar to the analysis in an open discussion. At the end the interviewer may ask the stakeholder to fill in the questionnaire as well. The structure of the interviews is shown in Figure 4.2.12, and may also be used for introduction to the meeting, which is planned so as to allow for a flexible and creative discussion. The SEESAW model, although not producing complete reports, does most of the required analysis and display of graphs and tables, and these can be used for the sustainability report or other decision-oriented reports for stakeholders.

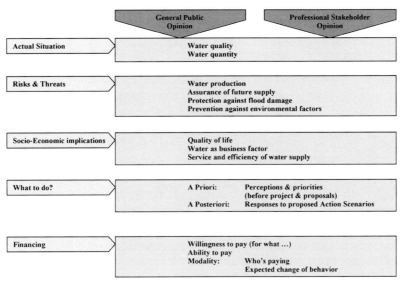

Figure 4.2.12. Guide to stakeholder interviews.

4.2.4 Decision supporting AISUWRS Deliberator

The AISUWRS Deliberator is a software tool showing the assessment of stakeholders for each scenario split into the relevant decision criteria. Based on the Deliberation Matrix methodology and the software 3D cube® developed during the European project GOUVERNe, the AISUWRS Deliberator presents the three important dimensions (action scenarios as alternatives, indicators as decision criteria, and stakeholders) in two-dimensional matrices being easy to understand and to communicate (as shown in Figure 4.2.13). The Deliberation Matrix concept and software was formulated and developed by FUTUREtec GmbH and C3ED in the context of the GOUVERNe Project, «Guidelines for the organisation, use and validation of information systems for Evaluating aquifer Resources and Needs», which is funded under Key Action 1 (Sustainable Management and Quality of Water, RTD Priority 1.1.3 - Operational management schemes and decision support systems) of the Energy, Environment And Sustainable Development theme of the European Commission's Fifth Framework Programme. The present application within the AISUWRS Project extends its use as a tool for the presentation of scenarios in a multi-stakeholder forum where a complex array of criteria need to be considered.

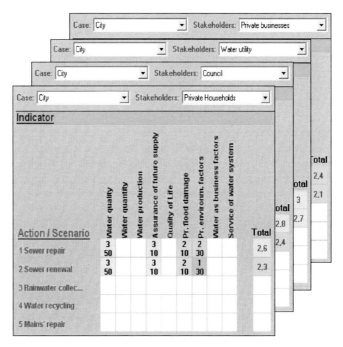

Figure 4.2.13. Stakeholder perspectives revealed by use of deliberator cube (adapted from the GOUVERNe project).

The AISUWRS Deliberator is designed for use by moderators and technical experts assembling the individual deliberations of a heterogeneous group of stakeholders and decision makers. Their assessment is systematically recorded for each alternative action scenario and decision criterion. The criteria are weighted with percentages whose scale may be chosen freely in each case, but which adds to 100% for the sum of indicators.

Having finished the individual assessment of each scenario, the AISUWRS Deliberator shows the results for all action scenarios and stakeholders. The numbers represent the weighted values of the criteria (indicators), emphasized by traffic light colours.

The example shown in Figure 4.2.13 serves to demonstrate the AISUWRS Deliberator tool. In each case the assessments of the action scenario alternatives are typically kept confidential, so in this example all values and deliberations are fictionalised and serve just for demonstration purposes. The following Figure 4.2.14 shows the input screen for deliberation of the alternatives by one stakeholder or a group of stakeholders. This step of the deliberation process should take place either by individual assessment or as a consensus during a group discussion.

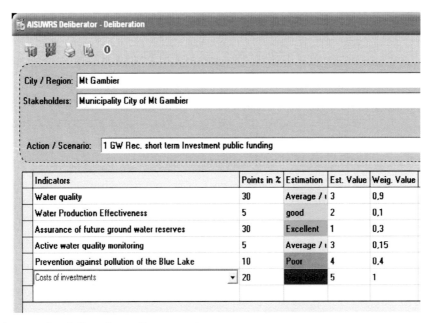

Figure 4.2.14. Input screen for deliberating action scenarios.

In this example, taken from the Mount Gambier study, deliberation takes place for each of the following action scenarios being alternative ways to implement the scenario "Greywater Recycling":

(1) GW Rec. short term investment, public funding
(2) GW Rec. investment over 10 years, public funding
(3) GW Rec. investment open time frame, private funding,
(4) GW Rec. investment public subsidy over 10 years, private funding

Instead of these alternative sub-scenarios one could as well discuss and compare completely different action scenarios under evaluation. In the next step, the deliberations of the alternatives by one stakeholder are compared (see Figure 4.2.15.

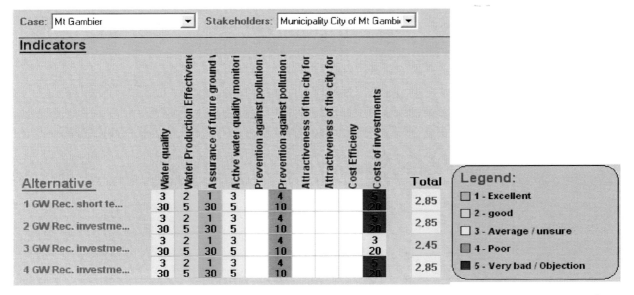

Figure 4.2.15. Comparison of the deliberations on alternative investment scenarios by the stakeholder "Municipality".

Not all indicators have been used for deliberation by the stakeholders. Each coloured field in Figure 4.2.15 shows the value (above) and the weight in % (underneath). At the end of each line the column "Total" shows the weighted sum of the deliberations of each indicator. Thus, in this example the stakeholder would prefer scenario No. 3, because in this case he would not have to cover the costs.

In the next step the deliberations of the different stakeholders are compared (see Figure 4.2.16. In this matrix each field represents the value of the column "Total" from the previous matrix. Comparing the deliberations of the different stakeholders it becomes evident that two stakeholders would strongly object to the scenario No. 3, preferred by at least 3 other stakeholders (please bear in mind that the values are fictionalized for the puposes of this example).

The AISUWRS Deliberator does not contain an algorithm to calculate a ranking of the alternatives under consideration. Instead it is designed to serve as a discussion tool visualizing the different deliberations and providing an opportunity to discuss whether a consensus could be reached or not. Thus the AISUWRS Deliberator is a useful tool to support participatory processes in the development of water resources policy, management and governance (Rehm-Berbenni, 2004).

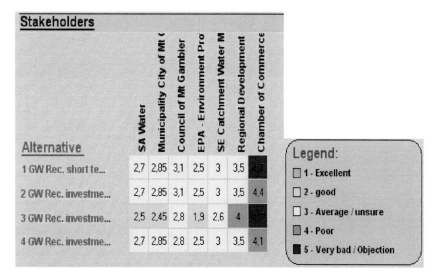

Figure 4.2.16. Synopsis of the different stakeholders' deliberations.

4.3 EXECUTIVE SUMMARY OF THE MODEL APPLICATIONS IN THE CASE STUDY CITIES

4.3.1 AISUWRS Sustainability Analysis for the City of Rastatt

For Rastatt, overall groundwater abstraction has exceeded the recharge of the 11 km² city area. Especially in 2004, with utility abstraction of app. 2.5 Mm³/y and assumed private abstraction of app. 1.7 Mm³/y, the low recharge of 1.1 Mm³/y in 2004 would indicate a strong disequilibrium. However, the catchment zone of the water works relates to a much larger area and the groundwater levels have been stable during the last two decades. Even in the long term, the average groundwater levels have not decreased significantly since 1913 (Wolf et al 2005b) and an equilibrium between the abstraction, recharge, and lateral groundwater flow has been found. This background information, which shows the city footprint in the broader context of the aquifer system, was not directly factored into the calculations, as the methodology relates only to the city footprint, but was used when drawing conclusions from the results.

Groundwater quality generally corresponds to the WHO drinking water recommendations. However, during recent years certain chemical parameters have increased locally, with trends that run counter to the aims of the EU Water Framework Directive 2000/60/EC to protect uncontaminated groundwater sources, avoid further contamination and improve already contaminated sites. Protection strategies could not be retrieved at the data collation stage, however legal regulations are quite strict in Germany and it can be assumed that protection and public promotion is at a relatively high standard compared to other countries in the EU. The Group I indicators generally leave a positive picture.

In Rastatt, the utility produces approximately 64% to 65% of the total societal water demand, which is quite high compared to badly supplied areas. Approximately 99% of the population is connected to the water supply, while private abstraction is mainly by the large industrial users. These are small in number and easily located, so much of the private abstraction can be followed up by the administration. For the Rastatt water supply utility, water production has slightly decreased since 1990. More than 90 % of the water produced during the last two decades went into consumption, and less water is wasted during the production process in recent times – only between 1995 and 2000 has water efficiency slightly decreased. The excess per capita drinking water production has stayed well below 8 m³/capita/y and shows a generally decreasing trend with only a short term high in 2000. Throughout the last two decades, the energy efficiency of the production process stayed constant: 3 m³ of water could be produced per kWh. The annual per capita energy consumption for the treatment process decreased slightly, as production decreased. Water losses in the network system have stayed well below 10% and show a very positive trend towards 4% in recent years. This leaves a generally satisfactory picture of the Group II Indicators.

For Rastatt, so far no data have been retrieved for a consumption trend analysis over time, but c. 110 l/capita/day have been estimated for domestic usage, i.e. in the low to normal range for western European cities.

Within the wastewater treatment sphere in Rastatt, treatment effectiveness with regard to BOD has been more or less constant during the last two decades at 98 to 99%, which is typical for biological treatment plants. Effectiveness in nitrogen (N) treatment considerably increased from 61% in 1990 to 79% in 2004 and P treatment increased from 67% in 1990 to 97% in 2004. The absolute loads that entered the water environment after treatment however strongly increased for BOD from app. 25,000 kg/y in 1995 to c. 46 tonnes/y in 2000 and 31 tonnes/y in 2004. N loads to the receiving water bodies decreased slightly when compared to the values from 1990, however rose again during the recent years (2004). P loads to the receiving water bodies nowadays (2.6 tonnes/y in 2004) are well below the loads in 1990 (15.3, tonnes/y). During the last two decades, less energy input is needed in order to reduce one kg of BOD, N and P load and efficiency has increased. The annual per capita consumption of chemicals and energy has however increased slightly since 1990, resulting from scale effects.

Rastatt does not so far face a major problem with regard to groundwater, water supply, consumption or wastewater services.

The socio-economic analysis does not indicate major problems with the sustainability of the urban water system. The household questionnaire survey showed some problems to be solved, but the discussion with water system experts would probably lead to a more subtle result with only few indicators being of priority. The matrix in Figure 4.3.1 shows the priorities of changes derived from the household survey and from expert opinions.

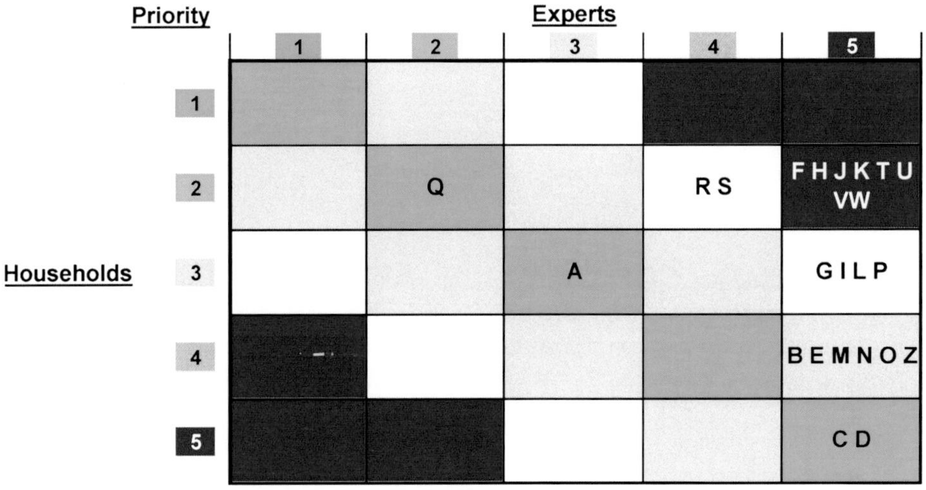

		Experts				
		1	2	3	4	5
Households	1					
	2		Q		R S	F H J K T U V W
	3			A		G I L P
	4					B E M N O Z
	5					C D

Figure 4.3.1. Comparison of the priority ranking by the public (households) and water experts in Rastatt (1 = high priority).

In this matrix the items with highest priority are listed in Table 4.3.1. As a result it shows clearly that both groups see two indicators of equal and relatively high priority: Q - Protection against backpressure & submerged sewers, and A – Water quality – Chemical ingredients. The latter refers to the hardness of water, which has no influence on health, but is a constant annoyance to the public.

The public opinion (from the household questionnaire survey) seems to agree that the water price, already being quite high, could be raised by up to 5% to cover the cost of improvement of these items. More results of this survey are shown above. As an overall trend, the technical consultees see less need for action than the households. There is no case in which the households are not aware of a risk that is perceived to be of relevance by the experts.

Table 4.3 - 1. Agreement between public and expert opinion for Rastatt.

Indicator class	Indicator
1. Water quality	A. Chemical Ingredients
3. Water Production Effectiveness	F. Rainwater collection
	H. Recycled water
4. Assurance of future supply	J. Safety from industrial damage
	K. Prevent infiltration of chemicals
6. Protection against flood damage	Q. Protection against backpressure & submerged sewers
	R. Protection against flooding
	S. Against rise of groundwater level/wet cellars
7. Prevention against environmental factors	T. Pollution of groundwater
	U. Pollution of rivers and lakes
	V. Pollution of soil
	W. Drop in groundwater level

4.3.1.1 Action Scenarios for the City of Rastatt

As the general public is satisfied overall with the current water management in Rastatt, the alternative scenarios calculated with the AISUWRS do not solve items of great urgency to local decision makers at the moment. For instance the encouragement of increased rainwater infiltration by the introduction of split tariffs is already planned in neighbouring cities, but is not envisaged in Rastatt.

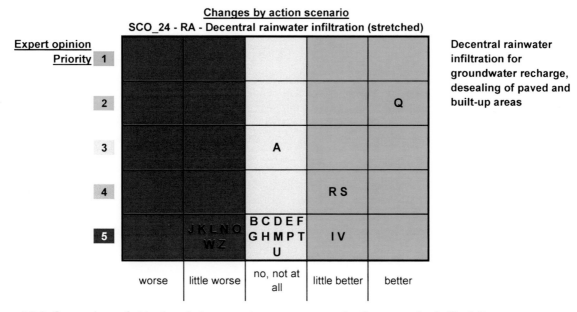

Figure 4.3.2. Comparison of objectives between water consumers and action scenarios in Rastatt.

Figure 4.3.2 depicts the impact of the rainwater infiltration scenario as expected by the researchers and the priority of these items from the perspective of the water system technical experts among the stakeholder in Rastatt. Only for items with lowest priority in the expert rating, the researchers expect a minor deterioration with respect to the current management practice. On the other hand, high priority items such as Q (protection against backpressure / reduction of hydraulic overload) will be improved by the scenarios. As the total investment of this scenario would be about 40 million euros (if modifications to the existing housing stock are carried out) it is not likely someone would be willing to envisage undertaking this completely at the moment. Nevertheless, the introduction of infiltration facilities is currently encouraged in new development areas where they can be integrated at low cost into the design plans.

Other scenarios like greywater reuse or total rehabilitation of the sewer system, which were also calculated with the AISUWRS models are deemed impossible due to the associated implementation costs.

From the socio-economic point of view it can be concluded that no major problems with sustainability exist in the city of Rastatt, and that overall the public is satisfied with the current system.

4.3.2 AISUWRS Sustainability Analysis for the Town of Mt Gambier

The water quantity available for Mount Gambier has always been sufficient, but in the last 10 years the withdrawal exceeded the annual recharge. Due to the direct hydraulic connection between the Blue Lake and the Gambier Limestone, it can be assumed that groundwater levels have fallen as well. Thus it seems necessary to reduce water consumption, which is already limited.

While the water quality of the lake is within the national guideline values and the WHO drinking water guidelines, trends indicate an increase in certain parameters, e.g. for nitrate. Mount Gambier is aware of the risk of pollution sources, as up to 35% of lake recharge is due to stormwater recharging the Gambier Limestone aquifer via drainage boreholes, and water managers aim via management plans to minimize hazards as much as possible. However, inspection and monitoring e.g. of private discharge boreholes is limited.

SA WATER supplies almost 100% of the population. The utilities production accounts for >80% of the total societal demand. Consumption from the public system covers from 55% (1995) to 62% (2005) of the societal water demand. All other demand is met via private, more scattered wells.

Domestic demand differs between app. 250 l/capita/d and 100 l/capita/d. Taking into account the total societal demand and all supply sources, total water demand would have decreased from 576 l/capita/d in 1995 to 500 l/capita/d in 2005 for total water usage (all usage such as commercial, industrial, agricultural etc. as estimated from public plus private abstraction figures).

BOD loads in the raw wastewater have been increasing, nitrogen (N) loads slightly decreasing during the last decade. Approximately 80% of the population is connected to the sewer system. Data on sewer exfiltration and loads from the septic tanks leaking into the aquifer system could not be estimated. Primary and secondary treatment results in an increasing effectiveness of 99% for BOD removal, > 90% for N treatment and 23% - 43% for P removal during the last decade. Neither energy recovery from biogas nor N and P recycling is currently taking place. So far Ecosan concepts are not under consideration. Stormwater is not collected in the sewer system but infiltrated directly into the aquifer via drainage boreholes where the sole process is some suspended solids removal via silt traps.

To conclude the environmental aspects, the urban water system in Mount Gambier shows some problems with regard to the water level drop and quantitative groundwater balance, and a future risk of increased contamination of the aquifer system and the Blue Lake.

With regard to future development, on a long-term basis little change of population and water needs for industry and agriculture is expected. However, behavioural change may reduce the per capita consumption of water, helping to stabilise the production and demand cycle without drastically influencing either quality of life or standard of living. Also the water quality is predicted to stay at a high level, favouring ensuring a good sanitary and public health environment for the city. Even a moderate increase of the water price would not affect the standard of living. However, a fall in employment opportunities in the region due to more restrictive water regulations would damage the economic situation of the population.

Based on the present research, the stakeholders interviewed in Mount Gambier are worried by the unknown level of risk inherent in direct stormwater infiltration to the subsurface and into the aquifer. Future (preventive) measures to improve the stormwater treatment and drainage depend to a large extent on this estimation of the risks. If major investment were needed, the municipality would depend upon financial transfers from other public sectors.

In summary, the questionnaire survey expressed the priorities and thus the concern of public opinion as shown in the matrix in Figure 4.3.3.

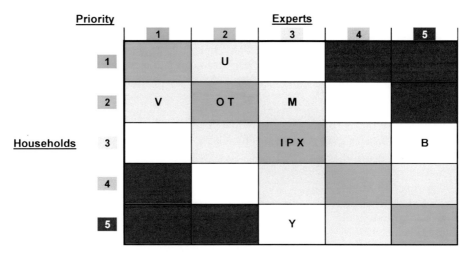

Figure 4.3.3. Priority ranking by the public (households) and water experts for Mt Gambier (1 = high priority).

Technical consultees and the public agree to a large extend on the priorities in the two highest ranks as indicated in Table 4.3.1:

Table 4.3.1. Agreement between public and expert opinion for Mt Gambier.

Indicator class	Indicator
7. Prevention against environmental factors	U. Pollution of rivers and lakes
	V. Pollution of soil
	T. Pollution of groundwater
5. Quality of life	O. Secure water levels for buildings, streets, channels etc.
	M. Spas, sport and entertainment (with water)

The indicators of highest priority express concern about possible damage in the future or man made aspects of quality of life, both of which can be influenced by society, so it is quite understandable how much the institutions involved care about water sustainability and the awareness and acceptance of the population.

4.3.2.1 Action Scenarios for the City of Mount Gambier

The originally proposed technical scenario "sewer rehabilitation" did not have the highest relevance for the decision makers and has mainly been undertaken to test the application of the AISUWRS procedures and modelling tools.

Measures of greywater recycling would decrease the volume of wastewater to be treated and thus the exfiltration from sewer systems by approximately 1/6. Water demand for drinking water production would decrease for c. 0.5 Mm^3/y according to the scenario modelled. Contaminant loads that enter into the subsurface water environment would be reduced (e.g. N loads in 2020 without greywater recycling 24,761 kg/y would be reduced by greywater recycling by 1461 kg/y).

However - compared to the stormwater problem – it has little relevance: the effects on water saving are small, but the costs are high. The additional costs of water to be covered by the price of water would raise the price of water by 75% over the next 5 years or to about 100% over the next 10 years. From the indicators of highest priority none will be covered by this scenario. Only the assurance of future supply of water by saving the ground water reserves (I), ranked as priority 3, will be enhanced.

4.3.3 AISUWRS Sustainability Analysis for the City of Ljubljana

For Ljubljana, assuming the reliability of the correctness of estimates, the abstraction for the city of Ljubljana has been decreasing from 1990 with c. 60 Mm^3/y to 40 Mm^3/y in 2004. Recharge values corresponding to the same catchment area were until October 2005 not available. However, an estimate of annual available groundwater for pumping has been given, showing that abstraction for the whole of Ljubljana is c. 37 % of the available groundwater volume.

During the last five years the water tables have been oscillating, with neither a general decreasing nor increasing trend, and groundwater quality is within the range required by national and international drinking water standards. However, certain chemical parameter concentrations have increased which may be markers for a more pervasive increase in groundwater pollution which the EC Water Framework Directive 2000/60/EC requires member states to try to mitigate. Legally binding regulations with regard to groundwater protection do exist, however, inspections, monitoring and strict implementation of the regulations is deficient.

The utility supplies water almost to 100% of the population, however water losses are very high and only around 50 – to 65 % of the water produced can actually be consumed by the population: 43% of the water in the network system was estimated to have been lost in 1990, 46% in 2000 and 36% in 2004. These values show a positive trend in reduction but at a very high level of losses. Utility production even exceeds the total societal water demand by far, which is assumed to include consumption from private wells of app. 5 Mm³/y. The water utility produces per year app. 154% (2000) to 127% (2004) of the total societal water demanded, however the water that is actually consumed by the population from the public system accounts for only 86% (1990) to 82% (2004) of the societal demand. (Comment: All data on private wells had to be estimated).

Production processes show a decreasing per capita consumption of energy from 1990 (87.1 kWh/capita/y) to 2004 (48.5 kWh/capita/y). Consumption of chemicals per capita and year has been increasing from 1990 to 2004. The per capita water consumption in Ljubljana has decreased during the last 14 years from 367 l/capita/d in 1990 to 282 l/capita/d in 2004 (including all consumption).

Contaminant loads that infiltrate into the aquifer from leaky sewers are unknown but with estimated exfiltrating volumes of 500.000 m³/y, these could be estimated to approximately 20,500 kg/y nitrogen (N) and 4,500 kg/y phosphorus (P). Wastewater treatment is primary (settlement), and no biological or chemical treatment is in place. Treatment effectiveness is low and the absolute loads that enter the ecosystem even after treatment are very high, especially for BOD (around 9 million kg/y) and N (more than 1 million kg/y). It has however been reported that biological treatment processes were planned for 2005. In Ljubljana, neither energy recovery from biogas nor N or P recycling is currently under consideration.

The urban water system in Ljubljana shows some problems with regard to water losses from production and with regard to the reduction of contaminant loads e.g. via wastewater treatment. As the reduction of BOD, N and P during treatment processes is very limited at the moment, the influence of sewer exfiltration is assumed to be minor when compared to the general loads entering the water environment after primary treatment. Furthermore, in Polje, only 70% of the households are covered by piped wastewater services, and loads entering the subsurface from septic tanks might be significant.

The socio-economic analysis shows great interest in water related topics among the stakeholders, recent incidents of chemical contamination having increased awareness of the potential threats. This is expressed in the questionnaire survey as shown in Figure 4.3.4.

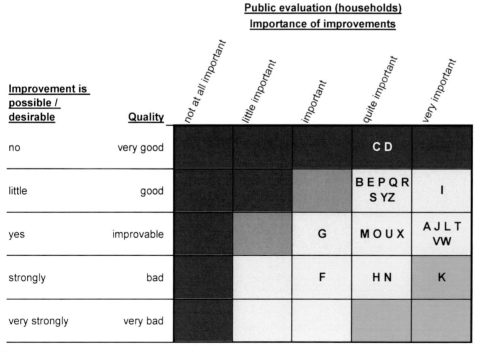

Figure 4.3.4. Quality-Importance matrix for Ljubljana.

The contributors to this analysis make a distinct statement that the highest priority lies with the indicators in Table 4.3.2, of which K is seen as the main concern:

Table 4.3.2. Indicators of highest priority for Ljubljana.

Indicator class	Indicator
1. Water quality	A. Chemical Ingredients
3. Water Production Effectiveness	F. Rainwater collection
	H. Recycled water
4. Assurance of future supply	J. Safety from industrial damage
	K. Prevent infiltration of chemicals
	L. Active water quality monitoring
5. Quality of life	M. Spas, sport and entertainment (with water)
	N. Design of water features and landscapes
	O. Secure water levels for buildings, streets, channels etc.
7. Prevention against environmental factors	T. Pollution of groundwater
	U. Pollution of rivers and lakes
	V. Pollution of soil
	W. Drop in groundwater level
8. Water as business factor	X. Attractiveness of the city for businesses

4.3.3.1 Action Scenarios for the City of Ljubljana

The scenario "sewer rehabilitation" did not have the highest relevance for the decision makers and has mainly been calculated in order to demonstrate the application to the model and DSS methodologies tool.

While the scenario of septic tank connection to the wastewater service in Ljubljana and the rainwater infiltration in Ljubljana was considered to be of higher relevance, within the limited study area chosen for the AISUWRS model chain, the septic tank scenario is of lower relevance and results of a model calculation cannot be extrapolated to the scale of the whole city. Nevertheless, it is interesting to compare the effects of this scenario with the assessment of the stakeholders' opinion, as shown in Figure 4.3.5, where the priority of the highest ranked indicator (K) is being improved by the action scenario 'Connection of landblocks to sewerage network'. The costs of this limited investment would not increase the cost structure significantly.

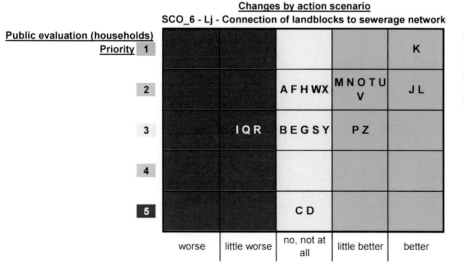

Figure 4.3.5. Comparison of objectives between water consumers and the action scenario in Ljubljana.

4.3.4 AISUWRS Sustainability Analysis for the City of Doncaster

The Doncaster groundwater situation is influenced by steadily decreasing annual withdrawal by the utility, since abstraction was limited to 83 Ml/day, resulting in abstraction rates of below 70 Ml/day by the late 1990s from the well sites east of Doncaster. Private abstraction could only be approximately estimated from licensed permits to be c.13 Ml/day in 2004/2005, but is actually likely to be very substantially less. Assuming that recharge data from the In_Geo

Forms of the SEESAW model and withdrawal data from Yorkshire Water take into account the same catchment area, the ratio of annual abstraction for Doncaster versus annual recharge would be c. 56 %. In general, groundwater levels are at shallow depths. Locally, water levels have been reported to have risen during the last decade, whereas other regions such as Hatfield Moor suffer from falling groundwater levels. Therefore, no clear conclusion on ground water availability in quantity can be made on a general scale. Also the water quality of the groundwater in the general vicinity of Doncaster shows a patchy distribution, ascribed to local confinement by Quaternary superficial deposits or by the Triassic Mudstones to the east of the urban area, and to depth stratification. Low concentrations of agricultural pesticides have been widely detected showing increasing anthropogenic influence. Instances of petroleum hydrocarbons contamination from leaking tanks and spillages have been reported. Nitrates are locally a water quality problem exceeding the EU Water Quality Guidelines. In addition, rising groundwater levels might contribute to higher mobilization of contaminants already in the unsaturated zone.

Yorkshire Water being the supplying company and the Environmental Agency as the regulator are both aware of the problems and risks of groundwater contamination. They both carry out intensive monitoring programmes. Inspection and monitoring of private abstraction wells is so far very limited.

In Doncaster, the utility supplies water to almost 100 % of the population. Approximately 10 % of the water consumed by Doncaster is surface water imported from further north under a regional transfer facility, while 90 % is locally abstracted groundwater. Taking into account water losses of 10-12% in the network (c. 3Mm^3/y), the water abstracted locally in the Doncaster well field accounts for 85% of the estimated societal water demand in the area. This results in a high effectiveness to meet the societal demand and private production is thus thought to be minor. It is observed that estimated water delivery from outside (surface waters from the north) is in the same range as the reported water losses through leakage.

Efficiency calculations could not be conducted. Being a private company, data availability from Yorkshire Water was limited due to commercial sensitive. The numbers given for water actually consumed from the public system relates only to the study area of Bessacarr Cantley. Thus no conclusions can be made with regard to the total for Doncaster. It has been reported in the In_Util Form that daily per capita water consumption (in Bessacarr) increased from 138 l/capita/day in 2002/2003 to 144 l/capita/day in 2003/2004 and decreased again to 135 l/capita/day in 2004/2005 Reasons for this are unknown and have not been stated. As no related figures for connected population could be obtained, scale effects cannot be calculated, as official statistical population figures do not correspond to the figure for population connected to the water supply that could be obtained for 2004/2005.

The total annual societal water demand has been estimated to be c. 30 Mm^3/y in 1997/1998, decreased to c. 26 Mm^3/y in 2001/2002 and slightly increased again to 28.8 Mm3/y in 2004/2005. Stated daily per capita domestic consumption rates indicate a total domestic consumption of app. 13 Mm^3/y and more than 15 Mm^3/y consumption by other users than domestic, thought to be mainly industrial, commercial, community services and agricultural. Data reliability is however questionable.

As waste water services in the Doncaster area split between among Yorkshire Water and Severn Trent Water the analysis of indicators from Group IV could not be carried out due to lack of data availability. For the same reason, the indicator on recovery and recycling at the service utility (Group V) could not be calculated for Doncaster.

The problem of system boundaries that was referred to in section 5.2.2 is especially a problem in Doncaster, where application of the AISUWRS computer models was restricted to a small part of town namely Bessacarr-Cantley, however statistical data were available mostly for bigger reference systems, and where administrative and infrastructural boundaries complicate the picture. Furthermore, the data for indicator calculation is usually compiled from various sources but the compilation often needs to rely on the same background figures, for example that a common set of population statistics is used throughout. Where this is not possible, and there were severe data shortages for the Doncaster example, conclusions should be suitably tentative.

With regard to future development, on a long-term basis little change of population and water needs for industry and agriculture is expected. However, behavioural change may reduce the per capita consumption of water, helping to stabilise the production and demand cycle without drastically influencing either quality of life or standard of living. Also the water quality is predicted to stay at a high level, favouring ensuring a good sanitary and public health environment for the city. Even a moderate increase of the water price would not affect the standard of living. However, a fall in employment opportunities in the region due to more restrictive water regulations would damage the economic situation of the population.

In summary, the questionnaire survey expressed the priorities (1 being the highest) and thus the concern of the public and expert opinion as shown in the matrix in Figure 4.3.6.

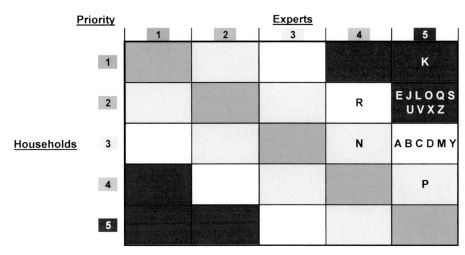

Figure 4.3.6. Priority ranking by the public (households) and water technical consultees in Doncaster.

From this diagram the most important actions to be taken from the indicator list described in section 4.2.3.2 are listed in Table 4.3.3.

Table 4.3.3: Most important actions to be concluded for Doncaster (derived from Table 4.2.3).

Indicator class	Indicator
4. Assurance of future supply	K. Prevent infiltration of chemicals
5. Quality of life	N. Design of water features and landscapes
6. Protection against flood damage	R. Protection against flooding

The water technical consultees and the public representative did not agree on most of the priorities. This suggests that if a major technical/strategic policy needed to be endorsed by the public, the deliberation process described previously would be helpful.

4.3.4.1 Action Scenarios for the City of Doncaster

The originally proposed technical scenario "sewer rehabilitation" did not have the highest relevance for the decision makers and has mainly been calculated in order to demonstrate the application of the model train.

When compared to the base case, the scenario of rainwater harvesting for further use for garden irrigation and toilet flushing resulted, according to the AISUWRS computer modelling exercise, in a reduction of the water supply demand of c. 152.000 m^3/y for the 4 km^2 of Bessacarr-Cantley. The reduction in mains water demand was shown to be 24% compared to the base case. Abstraction of groundwater could then be reduced accordingly. The impact on groundwater levels on the city at large was slight, from a maximum of 0.32 m at the centre of Bessacarr-Cantley, diminishing quickly towards its outskirts, and the impact on groundwater recharge was similarly modest at below 5% compared to the base case, and could be counterbalanced by rebound of the water table from reduced abstraction due to reduced water supply demand, a feature which cannot be calculated within the scenario modelling exercise.

From the socio-economic point of view the rainwater harvesting scenario does not seam to be economically justifiable as proposed because domestic consumers do not have an incentive to invest in the necessary equipment, either with a flat rate water consumption tariff or metered water charges, because the price of water is still low. If there were to be a political debate on increasing rainwater harvesting, it would be necessary to consider investment subsidies to households.

5
AISUWRS Urban Water Resources Toolbox – a brief summary

Leif Wolf[1], Brian Morris[2], Peter Dillon[3], Joanne Vanderzalm[3], Joerg Rueedi[4], Stewart Burn[3], Steven Cook[3]

[1] Department of Applied Geology (AGK), University of Karlsruhe, Germany
[2] British Geological Survey (BGS), Wallingford, UK
[3] CSIRO, Melbourne, Australia.
[4] Robens Centre for Public and Environmental Health, EIHMS, University of Surrey, Guildford, UK.

5.1 INTEGRATING GROUNDWATER INTO URBAN WATER MANAGEMENT

This book demonstrates the major relationships and links between urban drainage, urban water supply and urban groundwater resources. For instance, the diversion of urban surface runoff into the local aquifer can lead to rising groundwater levels, and the rehabilitation of sewers below the water table can stop the groundwater drainage function of the sewer system, causing flooded cellars and other underground structures. In addition, groundwater deterioration can be caused by leaky sewer systems. If cities are to become self-sustaining, then the continued use of urban groundwater for the urban water supply is critical. For this to be a practical policy, aquifer water quality needs to be maintained at a sufficiently high standard to minimise treatment requirements, which can only be achieved through the result of a well-established and enforced protection policy.

But the prediction of the effects on aquifer water levels and groundwater quality in the urban environment is not a trivial exercise. Recharge sources (and their corresponding water qualities) are many, and vary both in space and time. Cities combine open spaces that may be little changed from their previous rural setting (parks and other open spaces) with those that have been radically developed and may in some cases have been subject to a succession of land uses from rural to residential to industrial to residential again. This is a complex environment to work in and an understanding of processes associated with water and contaminant transport throughout the system is indispensable for success. Analysis has to be multidisciplinary, involving branches of civil engineering (structural, materials, sanitary, distribution, hydraulics), land use planning (town and landscape design), computing science (database and GIS analysis, system modelling), earth science (soil science, hydrogeology) and water quality (chemistry, microbiology). Additionally, individual components can be difficult and time-consuming to measure and are also subject to significant variability and uncertainty which only extensive long-term monitoring can attempt to quantify. Most importantly, the available data need to be utilised in models to provide results that policy makers in water utilities, urban government and regulatory agencies can use to predict trends and evaluate alternative strategies for water use and waste disposal.

In order to predict the medium and long-term response of urban aquifers, numerical groundwater models or other means of assessing risk to water supplies are urgently needed for most cities that utilise groundwater for their potable water supply. Such groundwater models are traditionally used for delineating the catchments of water supply wells and boreholes or for the assessment of single contaminated sites. The application of these models to urban water management is often hindered by the lack of knowledge of the complex urban groundwater recharge patterns and the unfamiliarity of town planners with these tools. The AISUWRS model chain now links models, which describe the passage of water through the urban infrastructure (UVQ), estimate leakage from the pipe network (NEIMO) and then through the unsaturated zone (SLeakI/POSI, UL_FLOW) using either numerical groundwater models or risk assessment methods.

Once the model framework has been set up for a city, it can easily be updated in the future and it can be used for other purposes like planning of local remediation measures in the vicinity of individual contaminant spillages.

5.2 GENERIC CONCEPT FOR THE ASSESSMENT OF URBAN GROUNDWATER RESOURCES AND SYSTEMS IN OTHER CITIES

Achieving a sound assessment of the sustainability of the urban water cycle involves activities much broader than the modelling task alone. An overview of the generic concept for the assessment of urban water resources is shown in Figure 5.2.1. The initial stage of the background study should provide a concise list of the available documentation on environmental, technical and social issues connected to the use of water in the city. This needs to be reviewed in the light of critical points and known bottlenecks. An assessment of the quantitative and qualitative status of ground and surface waters needs to be performed based on existing data. Besides the usual question of violation of water quality guideline values also the spatial and temporal trends need to be addressed: is one part of the city more affected than another? Is there an adverse trend that could signify major problems in the future? In order to receive a first picture it is sensible to compare water quality upstream, within and downstream of the city area.

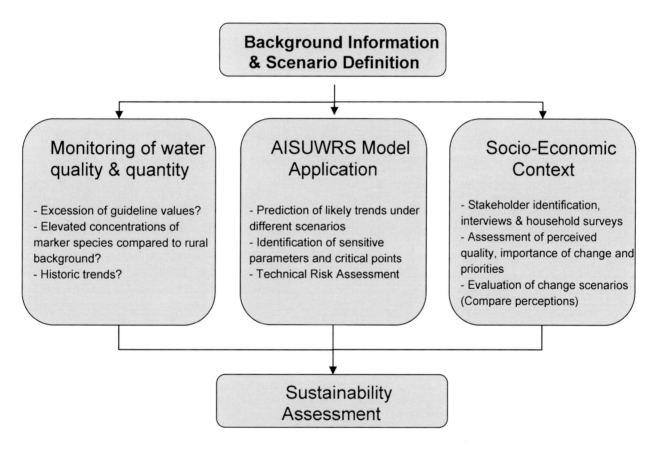

Figure 5.2.1. Generic elements for the assessment of urban water systems and resources.

Beyond the city boundary itself, possible implications of regional or global changes may impact the water system and must be identified. For example, an integrated framework like the DPSIR-approach (Drivers, Pressure, State, Impact, Response) should be used. The "drivers" and "pressures" may be of external origin like climate change or macroeconomics but also include migration and population change, demand change, evolving legal and regulatory frameworks, economics and finally the limitations of the natural system. Zaadnoordnijk et al. (2004) define pressures as the agents that potentially stress the environment. According to EEA (1999) they fall into three main categories: (1)

emission of chemicals, waste, radiation, (2) use of environmental resources and (3) area of land used. Urban groundwater is patently subject to the first two pressures.

Ideally, an expert panel would now assemble a list of possible technical or societal responses (e.g. greywater reuse) to the identified challenges. Based on this information, a first set of scenarios would be set up as a basis for further discussions.

The next step is the identification and consultation of stakeholders not only to keep key players informed but also to map their position on the different future scenarios and to discover potential shortcomings and problems. This may be achieved by a series of personal interviews, household surveys or by means of internet based techniques. Following the stakeholder consultation the final definition of scenarios that are to be modelled needs to be done.

As the AISUWRS tools are able to incorporate a very broad range of parameters and processes which may require time consuming data preparation, it is recommended to set a first target for the required precision even at this relatively early stage of the assessment. It may be that a first-pass approximation will achieve the desired strategic objective, or allow the number of scenarios to be reduced and refined for a subsequent, more in-depth, assessment.

The scenario simulations are performed with the individual models described in detail in Chapter 2 of this book, with the examples given in Chapter 3 to provide guidance on the practical problems encountered during preparation of input data and execution of the model codes and interpretation of results. Additional hints concerning the UVQ model can be found in the more extensive reports on the model setup in the case study cities (Rueedi & Cronin 2005, Souvent & Moon 2005, Klinger et al 2006).

In order to attempt cost-benefit analysis the costs associated with different water management scenarios also need to be evaluated. The results of the scenario simulations and the socio-economic studies can be presented to the local public and other stakeholders in a final deliberation workshop to discuss and sort out competing issues and finally to define the optimal future strategies and actions. This process is described in Chapter 4.

5.3 INNOVATION IN SOFTWARE DEVELOPMENT

5.3.1 UVQ

Research on the modelling of water and mass fluxes has nowadays become a recognized necessity for the successful development of urban environments. For example, contemporary research on megacities addresses the interactions between settlement structures, urban growth, different management systems and the urban water and solute balance. With the UVQ model, a well-tested tool is now available, which integrates all major elements of the water cycle from source to sink.

While there are several models devoted to urban water balance modelling (Mitchell and Diaper, 2005), typical representations of the urban water cycle consider the man-made and natural systems as separate entities. Within these two systems, modelling approaches generally only concentrate on one or the other aspect of the water cycle. UVQ integrates these networks into a single framework to provide an integrated view of the water cycle, using simplified algorithms and conceptual routines to provide a holistic and integrated view (Diaper & Mitchell, this book).

Urban Volume Quality (UVQ) is a conceptual, daily time step, urban water and contaminant balance model that simulates an integrated urban water system. UVQ estimates the contaminant loads and the volume of water flowing through the water system, from sources (e.g. precipitation or mains inflow) to discharge points (e.g. drainage systems or groundwater). The model has been developed to provide a means for rapidly assessing the impacts of conventional and non-conventional urban water supply, stormwater and wastewater development options on the total water cycle. UVQ is a lumped deterministic hydrological model and does not use finite element analysis techniques. The conceptual approach to simulating the urban water system permits significant flexibility in the manner in which water services are represented and provides the ability to mimic a wide range of conventional and emerging techniques for providing water supply, stormwater and wastewater services either to an existing urban area or to a site which is to be urbanised. Thus for instance, either piped sewerage or on-site sanitation or a combination of both can be accommodated.

UVQ may be regarded as the most advanced of the AISUWRS models as it is the result of a longer development process which started some years ago and went on throughout the present project. Its application in European cities led to the development of a special snow package and the incorporation of septic tank disposal in the range of options. Within the AISUWRS concept, special attention was placed on the interaction with the underlying groundwater and a separate set of output files is now generated, either for direct use in groundwater models or for further automated processing in the decision support system (DSS).

5.3.2 NEIMO (Network Exfiltration and Infiltration Model)

The problem of leaky sewers has been the subject of widespread research interest over the last decade. Numerous studies were performed at both laboratory and field scale, with sometimes rather diverse results. However, no model was available for the upscaling to larger networks. Most approaches rely on a simple extrapolation based on pipe length (e.g. Dohmann et al. 1999) and the application of CCTV-data into the predictions (Wolf et al. 2003, Wolf & Hötzl 2005) has been only recently tried. The AISUWRS project has responded to the lack of dynamic models to enable the replication of the complexity of large sewer networks by developing the Network Exfiltration and Infiltration Model (NEIMO). NEIMO is a deterministic model designed to provide water authorities with quantitative and qualitative data on leakage from wastewater networks to satisfy the information requirements of existing or future legislation. Based on the results of the UVQ-model, NEIMO calculates the water level in the sewer pipe and applies it to a Darcy-based prediction of exfiltration and infiltration. The most sensitive input parameters for exfiltration are the defect size and position and the hydraulic conductivity and the thickness of the colmation layer. For towns with an incomplete coverage of CCTV inspection records, generic curves are available which relate the defect size to pipe age, diameter and material. For the calculation of the infiltration rate NEIMO requires the mean depth of the sewer pipe below the water table. As a contribution to a scientific discussion of the results it must be stated that also NEIMO is not able to avoid the significant uncertainties that govern contemporary research on sewer leakage. However, it is the most accurate extrapolation of Darcy-based exfiltration estimates known to date and can easily be adapted to include future research results.

5.3.3 SLeakI/POSI

Only a small quantity of water in the urban water system is actually resident in the unsaturated zone. Nevertheless almost all water recharging the aquifers needs to pass through it. The unsaturated zone and especially the soil layer is a major protective shield for groundwater, providing vital scope for attenuation processes (adsorption, biologically mediated transformations, complexation, dispersion, filtration, hydrolysis, precipitation) to either eliminate or reduce the contaminant load reaching the saturated aquifer. Due to the complexity of the processes involved and the relevance of even small-scale heterogeneities, modelling the fate of components that leak from sewers through the unsaturated zone has received only little attention. Within AISUWRS, two models were developed which describe the contaminant transport in the unsaturated zone using analytical solutions. The challenge was to obtain a sufficiently fast algorithm to calculate water flow and contaminant transport for the numerous open areas and sewer leaks.

The developed models distinguish between point and spatial sources: Infiltration beneath unsealed urban surfaces is described using the POSI model (Public Open Space Index). It assumes that infiltration occurs over a sufficiently large area and absence of lateral subsurface flow (such as in perched aquifers above the regional water table) such that the boundary effects can be neglected and the flow pattern is one-dimensional. On the other hand, the leakage from point sources is described using SLeakI (Sewer Leak Index). The necessary consideration of three dimensions around a point source was achieved with a fast Gaussian quadrature that gave an accurate evaluation of the integral of moisture content over the depth of the unsaturated zone and hence provides an estimate of the minimum residence time. Each model allows up to two soil type layers.

Both programs offer the possibility for user defined attenuation or decay rates and hence may be adjusted to a wide range of substances. With the program comes a set of default parameters that are based on existing literature references including functions for hydraulic conductivity and moisture with respect to suction head for a range of soil types. SLeakI and POSI are integrated into the DSS which automates the exchange with UVQ and NEIMO but they will also function as standalone versions, either in batch mode or using an interactive graphical user interface..

5.3.4 UL_Flow

The estimation of travel times through the unsaturated zone is a key element in the quantification of any attenuation process. These processes were investigated in greater detail to supplement the model results of SLeakI/POSI and to find alternative solutions especially with regard to the transient and non-linear characteristics of unsaturated zone flow phenomena. As it is not possible, due to the current computational limits, to apply a numerical solution of the Richards equation for the strongly heterogeneous system of an urban area, it was of importance to develop analytical solutions and to test them in comparison with more complex approaches. From this background, the UL_FLOW model was developed. UL_FLOW uses a one-dimensional steady state solution and a simple volume balance for quantifying transient water flow and conservative tracer residence time in layered unsaturated soil profiles from an areal infiltration source (e.g. open space, large infiltration basins) to the water table. Considering steady state in a one-dimensional system helped simplify the problem because it defines the volumetric outflow as equal to the inflow due to continuity, since no intermediate storage is available. While the solution itself is steady state, the transient behaviour of the urban recharge is depicted by performing the calculation for a time series consisting of an appropriate number of steady state models. Within the AISUWRS project, the effect of various time step lengths (daily, monthly, yearly) on the travel time was assessed. It was concluded that yearly time steps are not appropriate but that at least monthly time steps are required for an adequate approximation.

Using the UL_FLOW model it is very simple to route a conservative tracer through the whole profile and then compute the cumulative tracer residence time for the entire profile. In comparison with the numerical computation of solute transport in unsaturated soils using HYDRUS1D (Simunek et al. 2005), this could be computed rapidly. The groundwater recharge and tracer residence time obtained by UL_FLOW were found to be in accordance with the respective results obtained by HYDRUS1D and indicate that it is sensible and appropriate to employ the fast analytical UL_Flow approaches to a city area instead of the time consuming numerical simulation. However none of the solutions cited above can account for preferential flow paths, in which even shorter residence times must be assumed after strong rain events.

5.3.5 AISUWRS DSS

The Decision Support System (DSS) developed within AISUWRS is the integrating software that supports the selection and comparison of predefined urban water scenarios. This is achieved by linking the individual models together to allow them to be run as one entity, rather than as individual stand-alone models. The DSS comprises a graphical user interface that enables the user to set key parameters for each of the models, to run the model chain and finally to view a summary of the modelling outputs.

The DSS also manages the event day approach, which has been developed to allow for the consideration of different rain events in the Network Exfiltration and Infiltration Model while keeping computation time within manageable limits (see Chapter 2.6). The AISUWRS DSS is linked to a Microsoft Access® database that stores the output of each model as well as the input files and parameters used in each run. The direct linking of the database content to geographical information systems was tested but is not included in the standard version.

A key function of the DSS is to combine the outputs from the unsaturated zone models (SLeakI and POSI) to produce the groundwater output file that is then written to the database. This file records the contaminant load and water quantities reaching the aquifer for each day in the DSS run. The database stores this information in a format that can then be used as an input for a number of commercially available groundwater transport models, such as Feflow and Modflow/MT3D or within the risk assessment framework developed for karstic aquifers.

5.3.6 AISUWRS Deliberator and the SEESAW tool

For the socioeconomic analysis it was necessary to collect information from a broad audience and to build a framework to display and compare the judgements of the individual stakeholders.

The SEESAW Model provides guidance in the acquisition of relevant data from the different stakeholders in the urban community and prepares the key indicators used for the sustainability assessment. SEESAW is an application in Microsoft Excel® that has been chosen for its flexibility and adaptability to the ever changing situations under study. The SEESAW Model is available in English and German, while other languages may easily be added by adapting the language table. It comprises input forms (e.g. for water utilities, consultants, research institutions, households, businesses), service tables (for different languages, scenario administration and statistical analysis) and finally output and reporting tables with the assessment of the actual situation and the different scenarios under different aspects (e.g. changes in the cost of living).

The AISUWRS Deliberator is a software tool showing the assessment of stakeholders for each scenario split into the relevant decision criteria. Based on the Deliberation Matrix methodology and the software 3D cube, developed during the European project GOUVERNe, the AISUWRS Deliberator presents the three important dimensions (action scenarios as alternatives, indicators as decision criteria, and stakeholders) in two-dimensional matrices that are easy to understand and to communicate (as shown in chapter 4). The Deliberation Matrix concept and software was developed and formalised in the context of the GOUVERNe Project, «Guidelines for the Organisation, Use and Validation of information systems for Evaluating aquifer Resources and Needs». It has been adapted for application within the AISUWRS Project as a tool for the presentation of scenarios in a multi-criteria, multi-stakeholder perspective.

The use of the Deliberator is a good example of cross fertilisation between EC-supported research projects, with a tool developed in one project adapted for use in a different context but similar function by research teams in another.

5.4 RISK ASSESSMENT APPROACH FOR PROTECTING DRINKING WATER SUPPLIES IN A KARSTIC AQUIFER

Sources and loadings to a karstic aquifer from sewer leaks and stormwater drainage wells were determined using UVQ, NEIMO, and SLeakI/POSI. However in a karstic system with compound faulting, solute transport modelling using Feflow and Modflow/MT3D was considered unreliable and an alternative module was inserted to determine the risks to the drinking water supply emanating from these sources of hazards.

This module is a risk based assessment consistent with the Hazard Analysis and Critical Control Point (HACCP) framework for managing drinking water systems in Australia (NHMRC and NRMMC, 2004) (Vanderzalm et al. 2004). This approach takes contaminant concentrations (hazards) from the urban water contaminants and assesses the potential for their attenuation in the saturated zone prior to extraction for water supply, and thus the overall impact on the quality of the water supply. This approach can be used in a qualitative sense, where quantitative input data are poor, but can also accommodate greater complexity when input data allow the use of a quantitative risk assessment utilising Microsoft Excel® and/or @Risk® software. Furthermore an initial risk assessment can be used to prioritise the contaminants or hazards that pose the greatest risk for more rigorous analysis.

Aside from source concentrations, attenuation co-efficients (dilution, solubility limits, half-lives) and saturated zone residence time are required to evaluate the impact of hazards on the quality of the water supply. Where the groundwater discharges into an open water body, attenuation in the receiving body is also assessed. A range of residence times can be incorporated to accommodate the variable flowpaths in a dual porosity system. These estimates are then validated with field data, such as an applied SF_6 tracer test or the positive detection of x-ray contrast media.

5.5 EXAMPLES OF INNOVATIVE METHODOLOGIES EMPLOYED IN THE FIELD STUDIES

While the field studies and monitoring programmes in the four case study cities were essential as a validation to the AISUWRS modelling exercises, the field studies also produced substantial new scientific knowledge, especially in the provision of "hard facts" and new monitoring strategies. Representative for a larger number of research results, the following results are of special interest:

- A range of pharmaceutical residues was found in seepage water as well as in urban groundwater bodies (Rastatt).
- Iodated X-ray contrast media were found in both towns where they were sampled (Rastatt & Mt Gambier). Their use as marker species was described.
- A concise set of sewer leakage rates was measured with different packer systems (double packer, asset packer tests) in Rastatt and Melbourne.
- A test site has been set up at Rastatt-Kehler Strasse, offering the first long-term monitoring of quantity and quality of sewage exfiltration under the real operating conditions of a public sewer network.
- A set of focus groundwater monitoring wells was drilled in the direct vicinity of defective sewers, enabling the assessment of worst-case conditions (Rastatt).
- An innovative construction method was utilised to build multilevel piezometers in Doncaster and this permitted quantitative residence time estimation of discrete depths over more than 40m of the upper part of the saturated aquifer using the environmental tracers CFC-11 and CFC-12 and SF_6.
- Microbiological investigations showed widespread occurrence of faecal indicators. In Doncaster, these faecal indicators, and to a lesser extent enteric viruses were found at depths of 60 m below the surface, indicating a fast transport, which was confirmed by the use of the environmental CFC and SF_6 tracers.
- A specially constructed urban lysimeter in Ljubljana allowed the monitoring of seepage processes below pavements and railroad tracks in the city area.
- Innovative new passive sampler devices were successfully employed in Mount Gambier to monitor the polycyclic aromatic hydrocarbon (PAH) loadings in recharge entering the karst zone via stormwater drainage boreholes.
- Sorption characteristics for boron on Mount Gambier Limestone were obtained in laboratory studies.
- An example of the application of SF_6 as an artificial applied tracer in a karstic aquifer system was provided in Mount Gambier.
- An extensive set of results from the sewage sampling campaigns in all case studies is available.

Summing up, the field investigations were able to show the urban impact on groundwater resources in the case study cities. In all case studies, the impact of the city without the consideration of catastrophic events (e.g. oil spillages) does not pose a major problem to the groundwater resources in the case study cities. Violations of health guideline values were only found for microbiological indicators. This remarkable conclusion is a testimony to the resilience of the subsurface and especially to the efficacy of the various attenuation processes, demonstrating that informed management of the urban water system offers real scope for the continued use worldwide, for water supply purposes, of aquifers underlying cities.

5.6 SOCIO-ECONOMIC CONTEXT OF URBAN WATER MANAGEMENT

The personal interviews with decision makers and the public at large in all four case study cities revealed quite divergent sensibilities and perceptions which were the result of different social, cultural and economic circumstances. In areas with a known water scarcity problem like Australia, public awareness was very high compared to the humid European settings. In the European case studies, different priorities between the experts and the public were observed. In almost all cases, the European experts saw less need for action than the public and acted somewhat defensively with regard to the current system. In Mount Gambier the perceptions were distributed more evenly (see chapter 4).

Economic constraints were frequently mentioned during the discussions with the stakeholder and cost-benefit analysis regarding the protection of groundwater would have been very welcome in the discussion. However, most of the alternative measures considered in the AISUWRS project (e.g. sewer rehabilitation) are more suited to reduce a risk in the long term than to show an immediate effect on public spending and the establishment of cost-benefit relationships for risks with low frequencies is often not feasible and can be misleading.

While a proper stakeholder involvement is a critical element in successful sustainability assessment, a strong tendency to stick to known and well-established solutions was observed. Consequently, the application of scenario simulation tools, which are able to explore alternative concepts at low cost, is seen as a major stimulus for innovation in the cities' urban water management.

5.6.1 How sustainable is the water management in the case study cities?

In general, the model results from all four case study cities confirm that no major problems with respect to water quantity are expected under the current water management options. On the contrary, it would even be possible to make more intensive use of groundwater in Doncaster, Ljubljana and Rastatt. With increasing competition for finite groundwater resources in Mt Gambier a fine balance has already been reached and it is not recommended to increase the abstraction without additional water management measures. The effect of increased artificial rainwater infiltration was calculated for the Rastatt case study but had negligible impact on groundwater levels. The most significant impacts on groundwater levels were observed for the worst-case climate change scenario calculated for Doncaster, resulting in a decline of up to 8 m. A similar scenario applied to Mount Gambier, which in combination with a reduction in recharge from the significantly larger rural capture zone for Blue Lake, would also be expected to have a major impact on groundwater levels and a further study has commenced to evaluate the historical effect of climate change. Population and demand change scenarios likewise had only minor impact on groundwater levels.

With regards to water quality, the picture is much more complicated due to the broad array of substances which have to be considered and their correspondingly wide range of behaviours and characteristics. The AISUWRS project mainly modelled key indicators like chloride and boron for demonstration purposes. These are relatively conservative inorganic species (not readily attenuated) that are widespread but not highly toxic except at very high concentrations. So they are unlikely to exceed the limits set in national drinking water guidelines. Nevertheless, as rising trends have been predicted for these indicators, these may provide a warning that similar trends may occur also for trace substances with much lower action thresholds.

The studies in the four cities also confirmed that the main threats to urban groundwater systems are still dictated by point source events like accidental contaminant spillages, or malfunctioning industrial installations and old contaminated sites, where the most prominent substances to look for would be halogenated hydrocarbons, polycyclic aromatic hydrocarbons and petroleum by-products. The effect of diffuse contamination from the new and different routes of recharge in the urban environment seems to be rather muted and may be no more than that found in the farmed rural area outside the city, albeit for a different range of substances, with different rates of degradation and levels of toxicity.

In addition to these events, the groundwater resources in all four cities were affected by the constantly present sources of sewer leakage and infiltrating road runoff. Pharmaceutical residues were screened for in Rastatt and Mount Gambier and found to be widespread in aquifers also used for drinking water production. Given the persistence of some pharmaceutical residues it is only a question of time until they also reach the water supply wells. The presence in raw groundwater of these trace compounds would warrant a more rigorous risk assessment of water supply. Likewise, urban groundwater in the case study cities was affected by the microbiological load associated with the exfiltrating wastewater. Standard parameters like E.coli regularly exceeded the guideline limits in samples taken from shallow urban groundwater observation wells. More sophisticated screenings conducted in Doncaster revealed that bacteria and viruses can be transported down to depths of 60 m b.g.l. in consolidated aquifers where fracture flow may play a role.

In Doncaster, the effect on contaminant indicator trends in terms both of total load and absolute concentrations is moderate in all of the scenarios. This is partly due to the relatively modest applied load, and partly to dilution effects, both from precipitation-derived recharge and from pore water with a low background concentration (effective porosity 15%). This suggests that as a receptor, the aquifer system beneath the study area may be relatively resilient against many dissolved–phase contaminants. Except for nitrate, background concentrations are low in terms of potability norms and the system is hydraulically well damped.

Ljubljana's town planning vision for the future is to move industry from the city centre to industrial zones on the borders of the city, generally downgradient of major water supply wellfields. Also a recognized water quality problem for Ljubljana city water supply is not the urban or industrial impact but the agriculture in surrounding periurban areas around the city. The initial results from this study suggest that residential land uses in an urban area may have significantly smaller impact on the groundwater than agriculture or industry, especially if these districts have complete foul sewer coverage. In this respect, use of sustainable urban development systems like on-site infiltration of roof runoff and improved sewer control and standards could lead to considerable improvements in groundwater quality.

In Rastatt, significant differences with respect to the loads to the aquifer could be noted in the different scenarios. Comparison between the sewer focus group of wells and the corresponding urban background group of selected hydrochemical indicators demonstrated the diffuse pollution originating from the combined effects of a large number of leaks. However, the high flow velocities within the main aquifer lead to a strong dilution of the contaminant input. While it was predicted that increases in the marker species would be observed even kilometres downstream of the city, the increases are too small to cause an environmental concern and may in reality be difficult to separate from short term variations and non-urban sources. With regards to sewer leakage, the predicted concentrations remained below the measured concentrations and this highlights the necessity to include also the sewer network beneath private properties and the effects of fertilizers and organic rich solid wastes. The rainwater infiltration scenario calculations showed that the city might go further along this path without running into the problem of rising groundwater levels. The results from Rastatt once again demonstrate the importance in some hydrogeological settings of an effective groundwater protection policy for the hinterland recharge areas, in this case the Black Forest Mountains, as these areas provide the vital dilution (via throughflow) that mitigates the effects of intra-urban sources of pollution.

In Mount Gambier the drinking water supply from the Blue Lake will always remain comparatively vulnerable due to the karstic nature of the aquifer system that feeds it. Besides spillages connected to single events like road accidents the infiltration of stormwater is a possible threat to the groundwater. Nevertheless Cook et al. (this book) conclude that the current water attenuation processes, provided by the saturated zone and in-lake processes such as the annual carbonate precipitation cycle, are sufficient for protection against potential contamination from the urban area. The employed HACCP model indicates the relative gain or loss in factors of safety with respect to water quality targets for proposed intervention scenarios. To ensure sustainable management it is recommended that future programmes for stormwater monitoring should continue to include total and soluble Al, Cr, Cu and Zn. Furthermore the regular monitoring programme of the Blue Lake should include more organic species like polycyclic aromatic hydrocarbons and pesticides (especially atrazine). Regarding stormwater infiltration a pre-treatment of stormwater at the wellhead would be advisable for protection of groundwater quality. This could prevent a significant portion of metals and organics from entering the groundwater system and would allow for their permanent removal from the system.

5.7 WHAT DOES THIS MEAN FOR OTHER CITIES?

Basically, every city is unique in its hydrogeological and infrastructural setting and so is a case unto itself. It would for example, be pointless and misleading to extrapolate conclusions from these examples (medium sized cities, well served by drainage infrastructure, located in moderate recharge climate) to larger agglomerations sited in semi-arid areas on less resilient aquifers. Strongly changing groundwater levels occur in megacities worldwide and the corresponding damage to urban infrastructure shows that improper management of the total urban water system can have strong economic and ecological consequences (Morris et al. 2003, Howard & Gelo 2002). Furthermore, more dramatic results can be expected in regions where wastewater is directly used for irrigation (Morris et al. 2003).

It is, therefore, recommended to apply similar scenario modelling tools for all major cities that currently, or intend to, rely on groundwater supplies in order to facilitate the thinking on total urban water and solute cycles and to demonstrate the linkages between the individual actions to decision makers and stakeholders. Of course the application to the context of very large cities and those where stormwater and sewage systems are integrated and/or canalised requires additional simplifications but especially in these cities the sustainability assessment is crucial due to their large resource demand.

5.8 COMPLEMENTARY INTERNATIONAL RESEARCH ACTIVITIES

Clearly the holistic view on the urban water cycle is more complex than the processes considered within the AISUWRS project. Fortunately, however, AISUWRS was part of a whole research cluster on integrated water management, CITYNET. This cluster involves projects for the detailed rehabilitation planning of drinking water networks like CARE-W as documented in Saegrov (2005). Of special importance to the sewer component of the water cycle are the projects CARE-S (Computer Aided Rehabilitation of Sewer Systems) and APUSS (Assessing Performance of Urban Sewer Systems). CD4WC (Cost effective development of urban wastewater systems for Water Framework Directive compliance) and DAYWATER (An Adaptive Decision Support System for the Integration of Stormwater Source Control into Sustainable Urban Water Management Strategies) are suited to address the broader picture as seen from the engineering point of view. However, for further reading on urban groundwater issues, it is recommended to consult Chilton et al. (1997) and Lerner (2004).

5.9 OUTLOOK & FUTURE AVAILABILITY

The key outcome of the AISUWRS project is the establishment of a software framework with user interface (AISUWRS DSS), which incorporates all major parts of the urban water and solute balance. Once set up for a case study, it demonstrates the causal connections in urban water management and allows the fast comparison of different scenarios. Within the project, scenarios of climate change, demand change, decentralised rainwater infiltration, greywater reuse and different strategies for sewer rehabilitation were assessed.

Apart from the DSS front end, all process models are also available as stand alone versions, along with separate manuals or documentation. The accessibility of the software products as shareware is foreseen. To date, information concerning the availability can be found at the website www.urbanwater.de, which will be maintained after the project ends.

While the established framework allows qualitative comparisons of the scenarios and indicates probable responses to a specific management option, much effort will also be required in future urban water research to increase the reliability of the quantitative results. AISUWRS implicitly acknowledges that there are areas where much needed research can improve the situation and looks forward to the next stage. The AISUWRS system now can be used for a sensitivity analysis of the individual parameters and can serve as a guideline for future research directed to the most critical points.

References and further readings

ABS (Australian Bureau of Statistics) (2001): Community Profile Series 2001 Census – Mount Gambier LGA, Cat No. 2001.0.

Adams, C.E.Jr et al (1999): Wastewater Treatment. Environmental Engineers Handbook: Crcnetbase 1999. CRC Press.Liu DHF and Liptak BG(Eds).

AISUWRS (2006): Network Exfiltration and Infiltration Model (NEIMO) Manual. Ver. 2. Available at: http://www.urbanwater.de/ results/publications/NeimoV2.pdf.

Allen, D.J., Brewerton, L.J., Coleby, L.M., Gibbs, B.R., Lewis, M.A., Macdonald, A.M., Wagstaff, S.J., Williams, A.T. (1997): The Physical Properties of Major Aquifers in England and Wales. British Geological Survey Technical Report (WD/97/34).

Almeida, M.C., Butler, D. and Friedler, E., (1999): At-source domestic wastewater quality. Urban Water, 1: 49-55.

Amer, A.M., Sherif, M.M. and Masuch, D. (1997): Groundwater rise in Greater Cairo; cause and effects on antiquities. In Groundwater in the urban environment (eds J. Chilton et al.), Vol. 1: Problems, Processes and Management, pp. 213-217. Balkema, Rotterdam.

Amick, R.S. and Burgess, E. (2003): Exfiltration in sewer systems. Accessed 10.07.2004, available at: http://www.epa.gov/ORD/NRMRL/Pubs/600R01034/600SR01034.pdf.

Anderson, M.P. and Woessner, W.W. (1992): Applied groundwater modelling. Simulation of flow and advective transport. Academic Press Limited. London, UK.

ANZECC and ARMCANZ (2000): National water quality management strategy; Australian and New Zealand guidelines for fresh and marine water quality, ANZECC and ARMCANZ, Australia.

Arcadis Trischler & Partner (1999): Kanalisations-Einzugsgebiete "Links der Murg", unpublished report, Civil Engineering Office Rastatt, Germany.

ARQ (Australian Runoff Quality) (2003): Proceedings of the Australian Runoff Quality Symposium, Albury, Australia, 16-17 June 2003, Institute of Engineers, Australia.

ATV-M149 (1999): Zustandserfassung, - klassifizierung und -bewertung von Entwässerungssystemen außerhalb von Gebäuden n. - Gesellschaft zur Förderung der Abwassertechnik e.V. (GFA). St. Augustin, Germany.

Australian Bureau of Statistics (ABS) (2004): see URL: http://www.abs.gov.au/Ausstats/abs@.nsf/0/.

Barrett, M.H., Hiscock, K.M., Pedley, S., Lerner, D.N., Tellam, J.H., French, M.J. (1999): Marker species for identifying urban groundwater recharge sources: a review and case study Nottingham. UK. Water Res. 33, 3083-3097.

Berger, C. and Lohaus, J. (2005): Zustand der Kanalisation in Deutschland - Ergebnisse der DWA-Umfrage 2004. KA-Wasserwirtschaft, Abwasser, Abfall.

Blackwood, D.J., Ellis J.B., Revitt D.M. and Gilmour D.J. (2005): Factors influencing exfiltration processes in sewers, Water Science and Technology, 51(2), 147-154.

BLMC (Blue Lake Management Committee) (2001): The Blue Lake Management Plan, BLMC.

Böhm, E., Hiessl, H. and Hillenbrand, T. (1999): Auswahl und Bewertung von Techniken zum nachhaltigen Umgang mit Wasser in Neubaugebieten. - Arbeitsstudie in Zusammenarbeit mit dem Bundesministerium für Verkehr, Bau- und Wohnungswesen, Fraunhofer Institut für Systemtechnik und Innovationsforschung (ISI), 81 S.; Karlsruhe, Germany.

BOM (Bureau of Meteorology) (2005): Climate of Mt Gambier, available at: http://www.bom.gov.au/weather/sa/mtgambier/climate_and_history.shtml.

Boughton, W. C. (1990): Systematic Procedure for Evaluating Partial Areas of Watershed Runoff. Journal of Irrigation and Drainage Engineering 116: 83-98.

Boughton, W.C. (1966): A Mathematical Model for Relating Runoff to Rainfall with Daily Data. Civil Engineering Transactions, CE8 (1) 83-97.

Brečko, V.M. (1999): Landscape vulnerability of Ljubljana's most important water source/Pokrajinska ranljivost najpomembnejšega vodnega vira Ljubljane. Acta Geographica 39.

Brooks, R.H., Corey, A.T. (1964): Hydraulic properties of porous media. Hydrology Paper No. 3, Colorado State University, Fort Collins, Colorado.

Brown, I.T., Rushton, K.R. (1993): Modelling of the Doncaster Aquifer. School of Civil Engineering, University of Birmingham, UK.

Burke, J.B., Moench M.H. (2000): Groundwater and Society: Resources, Tensions and Opportunities. UN-DESA New York

BWK (2003): Nutzungskonflikte bei hohen Grundwasserständen - Lösungsansätze. ISBN 3-936015-15-5, BWK Bund der Ingenieure für Wasserwirtschaft, Abfallwirtschaft und Kulturbau.

Carsel, R.F., Parish, R.S. (1988): Developing joint probability distributions of soil water retention characteristics. Water Resour. Res. 24, 755-769.

Chapman, D. (1996): Water quality assessments: a guide to the use of biota, sediments and water in environmental monitoring. Second edition. UNESCO/WHO/UNEP publication. (London: E & F N Spon).

Chapman, T. (1998): Stochastic modelling of daily rainfall: the impact of adjoining wet days on the distribution of rainfall amounts, Environmental Modelling and Software 13, 317–324.

Chilton J. et al. (1997). Eds. Proceedings of the XXVII IAH Congress on groundwater in the Urban Environment, Nottingham, UK 21-27 Sept. 1997. Vol 1: Problems, Processes and Management. Balkema, Rotterdam, The Netherlands.

Christova-Boal, D., Eden, R. and McFarlane, S. (1996): An investigation into greywater reuse for urban residential properties, Desalination 106, 391-397.

Chung, C. and Gu, R. R. (2003): Estimating time-variable transformation rate of atrazine in a reservoir, Advances in Environmental Research, 7, 933-947.

Codex Alimentarius Commission (CAC) (2003): Recommended international code of practice: General principles of food hygiene, CAC/RCP 1-1969, Rev. 4-2003, p. 31.

Cook, S., Vanderzalm, J., Burn, S., Dillon, P., Correl, R., Vogel, M., Rosemann, S., (2005): Deliverable D12: Mount Gambier Field Investigations Report and Deliverable D27: Mount Gambier set-up of urban water model. Available at: http://www.urbanwater.de.

Cronin, A.A., Rueedi, J., Morris, B.L. (2005): The effectiveness of selected microbial and chemical indicators to detect sewer leakage impacts on urban groundwater quality. In Proceedings of the 10th International Conference on Urban Drainage, Copenhagen/Denmark, 21-26 August 2005.

Cytoculture (2004): Important New Site Remediation Protocol: Monitored Natural Attenuation of Contaminated Groundwater Technical Update for CytoCulture Clients. Accessed 10.07.2004, available at: http://www.cytoculture.com/MNA.htm.

De Bénédittis J., Bertrand-Krajewski J.-L. (2005): Infiltration in sewer systems: comparison of measurement methods. Water Sci. Technol. 52 (3), 219–227.

Delleur, J. W. (1999): The Handbook of Groundwater Engineering, Editor in Chief J.W.Delleur, Boca Raton, CRC Press, 1999.

Denmead, O.T. and Shaw, R.H. (1962): Availability of Soil Water to Plants as Affected by Soil Moisture Content and Meteorological Conditions. Agronomy Journal Vol. 54 (5): pp. 385-389.

DeSimone, L. A., Howes, B. L. and Barlow, P.M. (1997): Mass balance analysis of active transport and cation exchange in a plume of waste-water contaminated groundwater. Journal of Hydrology 200, 228 -239.

Diaper, C. and Mitchell, V.G. (2005): UVQ User Manual, CMIT Report No. 2005-282.

Diersch, H-G, (2005): Feflow 5.2 user manual, WASY Institute for Water Resources Planning and Systems Research Ltd., Berlin. http://www.wasy.de.

Dillon, P.J. (1988): An evaluation of the sources of nitrate in groundwater near Mount Gambier, South Australia. CSIRO Div Water Resources Series No 1, Canberra, 68pp.

Diskin, M. H. and Simon, E. (1977): A Procedure for the selection of objective functions for hydrologic simulation models. Journal Of Hydrology 34: 129-149.

Dizer, H., Grützmacher, G., Wiese, B., Szewzyk, R., and J. M. Lopez-Pila, J.M. (2003): Elimination of Coliphages by slow sand filtration. Accessed 08.06.2004, available at:
http://www.kompetenz-wasser.de/engl/downloads/NASRI-Workshop/Poster1_Becken.pdf.

Dizer, H., Grützmacher, G., Bartel H, Wiese, H.B., Szewzyk R. and López-Pila, J.M. (2004): Contribution of the colmation layer to the elimination of coliphages by slow sand filtration Water Sci. Technol. 50 (2), 211–214.

Dohmann, M., Decker, J., Menzenbach, B.: (1999): Untersuchungen zur quantitativen und qualitativen Belastung von Boden-, Grund- und Oberflächenwasser durch undichte Kanäle. In: Dohmann, M. (Hrsg.): Wassergefährdung durch undichte Kanäle – Erfassung und Bewertung; Springer-Verlag, Berlin Heidelberg, 1999.

Domenico, P.A. and Shwartz, F.W. (1998): Physical and chemical hydrogeology. John Wiley & Sons, Inc. New York, USA. (2nd ed)

Doncaster Metropolitan Borough Council, 1994. Sandy Lane Drainage Area Study, Doncaster Metropolitan Borough Council, Doncaster.

Ebeling, H. (1991): Kleiner Stadtführer durch Rastatt, 59pp., G. Braun, Karlsruhe, Germany.

EEA (1999): Groundwater quality and quantity in Europe – Environmental assessment report No 3. In: S. Nixon (Ed.): European Environmental Agency, Luxembourg. ISBN 92-9167-146-0.

Eiswirth, M., Hötzl, H.: (1997): The impact of leaking sewers on urban groundwater. In: Chilton, J. et al. (ed.): Groundwater in the Urban Environment. Vol. 1: Problems, Processes and Management, Balkema, Rotterdam, pp 399-404.

Eiswirth, M., Hoetzl, H., Burn, S., Gray, S. Mitchell, V.G. (2001): Contaminant loads within the urban water system - Scenario analyses and new strategies. In Seiler, K.P., Wohnlich, S. (eds): New approaches to characterising groundwater flow, Balkema, Vol 1, pp 493-498.

Eiswirth, M. (2001): Hydrogeological factors for sustainable urban water systems. In: Current problems of hydrogeology in urban areas, urban agglomerates and industrial centres, Proceedings of NATO Advanced Science Workshop, Baku Azerbaijan, June 2001 NATO Science Series IV, Vol 8, Kluwer Dordrecht.

Eiswirth, M. (2002): Balancing the contaminant fluxes within the urban water system - options for sustainable urban water resources. Postdoctoral lecture qualification (in German), University of Karlsruhe 2002, 266 pp.

Eiswirth, M, Wolf, L., and Hötzl, H. (2002): Balancing the contaminant input into urban water resources. In: Procs XXXII IAH and VI ALHSUD Congress on groundwater and human development. Mar del Plata 21-25 October 2002. (Mar del Plata, Argentina: UNIV National De Mar del Plata).

Eiswirth, M., Held, I., Wolf, L., Hötzl, H. (2003): AISUWRS work package 1, Background Study Rastatt, Applied Geology Karlsruhe, Germany. Available at: www.urbanwater.de.

Eiswirth, M., Wolf, L., & Hötzl, H. (2004): Balancing the contaminant input into urban water resources. Environmental Geology, 46, 2, 246 - 256.

EKVO: Eigenkontrollverordnung (2001): Verordnung des Ministeriums für Umwelt und Verkehr über die Eigenkontrolle von Abwasseranlagen, Beuth-Verlag GmbH, Germany.

Elkington, J. (1998): Cannibals With Forks: The Triple Bottom Line of 21st Century Business, Stony Creek, CT: New Society Publishers, 1998.

Ellis, J.B. (2001): Sewer infiltration/exfiltration and interactions with sewer flows and groundwater quality – Proc. Interurba II, Lisbon, Portugal, 19.-22.

Ellis, J.B., Revitt, D.M., Lister, P., Willgress, C. and Buckley, A. (2003): Experimental studies of sewer exfiltration. Water Science and Technology, 47, 61-67.

Emmett, A. J. (1985): Mount Gambier stormwater quality, Department of Engineering and Water Supply Report 84/23.

Environment Agency (2001): Water resources for the future: a summary of the strategy for England and Wales, Environment Agency of England and Wales, Bristol, England.

Erichsen, A.C., Andersen, K.K., and Mark, O. (1999): Sulphide occurrence in sewer networks – a new topic in MOUSE trap. 3rd DHI software conference. Accessed 11.06. 2004, available at:
http://www.dhisoftware.com/book/materials/book/Sulphide%20Occurence%20in%20Sewer%20Networks1.pdf.

Federer, C.A., (1979): A Soil-Plant-Atmosphere Model for Transpiration and Availability of Soil Water. Water Resources Research, 15 (3) 555-562.

Foster, S.S.D., Morris, B.L. and Lawrence, A.R. (1993): Effects of urbanisation on groundwater recharge, Proceedings of the ICE International conference on groundwater problems in urban areas, June 1993, Thomas Telford, London.

Foster, S.S.D., Lawrence, A.R. and Morris, B.L. (1998): Groundwater in urban development: assessing management needs and formulating policy strategies. World Bank Technical Paper, No. 390 (Washington DC: World Bank).

Freeze, R.A. and Cherry, J.A. (1979): Groundwater. Prentice Hall, Englewood Cliffs NJ, USA

Fuchs, S., Hahn, H. H., Roddewig, J., Schwarz, M., Turkovic, R. (2004): Biodegradation and Bioclogging in the Unsaturated Porous Soil beneath Sewer Leaks Acta Hydrochimica et Hydrobiologica 32(4), 277-286.

Gardner, W.R. (1958): Some steady-state solutions of the unsaturated moisture flow equation with application to evapotranspiration from a water table. Soil Sci. 85, 228-232.

Gaunt, G.D. (1994): Geology of the country around Goole, Doncaster and the Isle of Axholme. British Geological Survey, London

Gleeson, C., Gray, N., (1997): The coliform index and waterborne disease, Problems of microbial drinking water assessment, ISBN 0419 21870 X, E & FN Spoon Publishers London.

Goebel, P., Stubbe, H., Weinert, M., Zimmermann, J., Fach, S., Dierkes, C., Kories, H., Messer, J., Mertsch, V., Geiger, W.F., Coldewey, G.W. (2004): Near-natural stormwater management and its effects on the water budget and groundwater surface in urban areas taking account of the hydrological conditions, Journal of Hydrology 299 (2004), p267-283.

Gordon, C. and Toze, S. (2003): Influence of groundwater characteristics on the survival of enteric viruses J Applied Microbiology 95, 536-544.

Gould, S., Baur, R., Burn, S., Cooke, J., De Silva, D., Sigurd, Hafskold, S.L., Herz, R., Le Gat, Y., Konig, A., Lindemann, G., Orman, N., Saegrov, S., Vollertsen, S. and Williams, W. (2004): Sewer Future Assessment Models – Literature review of existing models. CARE-S WP2 – Description and Validation of Structural Conditions, Task 2.2.3, April 2004, available at: http://www.care-s.unife.it/

Gray, S. and Becker, N. (2002): Contaminant flow in urban residential water systems, Urban Water 4, 331-346.

Gray, S., Becker, N. and Booker, N. (1999): Contaminant Balance. - CSIRO Urban Water Program Working Paper, T1-2.

Grischek T., Foley, A., Schoenheinz, D. and Gutt, B. (2001): Effects of interaction between surface water and groundwater on groundwater flow and quality beneath urban areas. In: Current problems of hydrogeology in urban areas, urban agglomerates and industrial centres, Proceedings of NATO Advanced Science Workshop, Baku Azerbaijan, June 2001 NATO Science Series IV, Vol 8, Kluwer Dordrecht.

Gyasi-Agyei, Y. (2005): Stochastic disaggregation of daily rainfall into one-hour time scale, Journal of Hydrology 309, 178 – 190.

Hagendorf, U. (1996): Forschungsergebnisse zur Bewertung der Dichtheit von Kanälen. awt-abwassertechnik, 11-16.

Hagendorf, U. (2004): Gefährdungspotential undichter Abwasserkanäle - Risiko für Boden und Grundwasser? Seminarband Wasser - Reservoir des Lebens, FLUGS Fachinformationsdienst Lebenswissenschaften, Umwelt und Gesundheit, http://www.gsf.de/flugs/seminarband1.phtml.

Hallström, I. (2005): Lessons learned from an outbreak of Giardi Lamblia in the Bergen Water Supply System, Brescia, Italy.

Hamblin, A.P. and Greenland, D.J. (1972): Mineralogy of soils from the holocene volcanic area of Southern Australia, Australian Journal of Soil Research 10, 61-79.

Härig, F. and Mull, R. (1992): Undichte Kanalisationssysteme - die Folgen für das Grundwasser. gwf Wasser - Abwasser, 133, 196-200.

Härig, F. (1991): Auswirkungen des Wasseraustauschs zwischen undichten Kanalisationssystemen und dem Aquifer auf das Grundwasser, PhD Thesis, University of Hannover, Germany.

Hassett, D.J., Pflughoeft-Hassett, D.F. and Heebing, L.V. (2003): Leaching of CCBs: Observations from over 25 years of research. International Ash Symposium, 2003. University of Kentucky Paper 76.

Heaney, J.P., Wright, L. and Sample, D., (2002): Chapter 3: Sustainable Urban Water Management. Innovative Urban Wet-Weather Flow Management Systems. US EPA, Cincinnati, Ohio.

Held, I. Klinger, J., Wolf., L. and Hoetzl, H. (2006): Direct measurements of exfiltration in a sewer test site under operating conditions. Matthias Eiswirth Memorial Volume - IAH Special Paper Series, published by Balkema/Taylor and Francis.

Herczeg, A.L., Leaney, F.W.J., Dighton, J.C., Lamontagne, S., Schiff, S.L., Telfer, A.L. and English, M.C. (2003): A modern isotope record of changes in water and carbon budgets in a groundwater-fed lake: Blue Lake, South Australia, Limnology and Oceanography, 48(6): 2093-2105.

Herczeg, A.L., Richardson, S.B. and Dillon, P.J. (1991): Importance of methanogensis for organic carbon mineralisation in groundwater contaminated by liquid effluent, Applied Geochemistry 6, 533-542.

Hill, A.J., Lawson, J.S. (in press): Geological setting for the groundwater resources of the lower South East, SA Department of Water, Land and Biodiversity Conservation.

Howard, K.W.F., Gelo, K. (2002): Intensive Groundwater Use in Urban Areas: the case of Megacities. In: Liamas, R. and Custodio, E. (eds) Intensive use of groundwater. Challenges and opportunities p.35-58 Balkema Publishers, The Netherlands.

Howard, P.H., Boethling, R.S., Jarvis, W.F., Meylan, W. M. and Michalenko, E. M. (1991): Handbook of environmental degradation rates, Lewis Publishers, USA.

IPCC: International Panel on Climate Change (2001): Climate change synthesis report, Cambridge University Press.

IRGO (2002): Poročilo o monitoringu podzemne vode na vplivnem območja vodnega vira Pivovarne Union, d.d. ('Report on groundwater monitoring of source area of Union Brewery wells'). Internal report.

Jackson, C.R. (2004): User's manual for the groundwater flow model ZOOMQ3D. British Geological Survey Internal Report IR/04/140.

Jamnik, B., Bračič-Železnik, B., Loose, A., Janković, M. (2001): Groundwater as a Source of Drinking Water and its Management in City of Ljubljana. In: The Quality of Underground Water as the Source of Drinking Water, Workshop in Maribor, Slovenia, November 2001.

Jenkins, J. (2005): The Humanure handbook. Accessed 19.05.2005, available at: http://www.weblife.org/humanure/chapter3_7.html.

Jones, R., Page, C. and Sims, G., (2001): OzClim Version 2.0.1, CSIRO Atmospheric Research Climate Impact Group and International Global Change Institute New Zealand.

Karpf, C. and Krebs, P. (2004a): Application of a leakage model to assess exfiltration. Int. Conference on Urban Drainage Modelling, Dresden 2004.

Karpf C. and Krebs P. (2004b): Sewers as drainage systems - quantification of groundwater infiltration, Proceedings of Novatech 2004, 5th Int. Conf. on "Sustainable techniques and strategies in urban water management", Lyon, France, 6-10 June 2004.

Keitz, S.v. (2002): Handbuch der EU-Wasserrahmenrichtlinie Erich Schmidt, Berlin.

Klinger, J. (2003): Erstellung eines zweidimensional-stationären Grundwasserströmungsmodells für den urbanen Raum Rastatt. Unveröff. Diplomarbeit, Lehrstuhl für Angewandte Geologie, Universität Karlsruhe (MSc Thesis, in German).

Klinger, J., Schaefer, M., Wolf, L. (2006): AISUWRS workpackage 10: UVQ report, Applied Geology Karlsruhe, Germany, available at: www.urbanwater.de.

Klinger, J., Wolf L., Hötzl H. (2005): New modelling tools for sewer leakage assessment and the validation at a real world test site. – 6th International Conference of European Water Resources Management (EWRA), 7-10. Sept. 2005, Menton, France.

Knuteson, S. L., Whitwell, T. and Klaine, S. J. (2002): Influence of plant age and size on simazine toxicity and uptake, Journal of Environmental Quality, 31, 2096-2103.

Komarova, T., Bartkow, M. E., Carter, S., Vanderzalm, J. and Müller, J. F. (2005): Field evaluation of passive samplers: monitoring polyaromatic hydrocarbons (PAHs) in stormwater, in Proceedings of the 5th Annual Health and Medical Research Conference, 3-4 November, 2005, Queensland.

Kookana, R.S. Correll, R.L. and Miller, R.B. (2005): Pesticide impact rating index (PIRI)- A pesticide risk indicator for water quality. Water, Air, Soil Pollution: Focus 5(1-2): 45-65.

Koske, T.J. (2005): Composting and the Carbon:Nitrogen Ratio. Accessed 19.05.2005, available at: http://www.agctr.lsu.edu/en/lawn_garden/home_gardening/Composting+and+the+CarbonNitrogen+Ratio.htm

Kovar, K. (ed.), (2003): Calibration and reliability in groundwater modelling: a few steps closer to reality; proceedings of the ModelCARE 2002; conference held in Prague, Czech Republic, 17 - 20 June 2002 / ed. by K. Kovar - Wallingford, Oxfordshire: IAHS, 2003. - X, 525 S.;

Krothe, J.N., Garcia-Fresca, B., and Sharp Jr., J.M. (2002): Effects of urbanisation on groundwater systems. 45 in Procs XXXII IAH and VI ALHSUD Congress on groundwater and human development. 21-25 October 2002. (Mar del Plata, Argentina: Univ National de Mar del Plata).

Kuehlers, D. (2000): Instationäre Strömungsmodellierung im Einzugsgebiet Wasserwerk Rheinwald. Eine Studie zu Methodik und Machbarkeit. Unveröff. Diplom-Arbeit, Lehrstuhl f. Angewandte Geologie, Universität Karlsruhe (unpublished MSc Thesis, in German).

Lamontagne, S. (2002): Groundwater delivery rate of nitrate and predicted change in nitrate concentration in Blue Lake, South Australia, Marine and Freshwater Research 53, 1129-1142.

Lamontagne, S. and Herczeg, A. (2002): Consultancy report for South Australian Environment Protection Authority, CSIRO.

Lawrence, A.R., Morris, B.L., Gooddy, D.C., Calow, R., and Bird, M.J. (1997): The study of the pollution risk to deep groundwater from urban wastewaters: project summary report. British Geological Survey Technical Report WC/97/15, BGS Keyworth UK.

Lawson, J., Love, A. J., Aslin, J. and Stadter, F. (1993): Blue Lake hydrogeological investigation progress report no. 1 – Assessment of available hydrogeological data, Department of Mines and Energy Report book 93/14.

Leaney, F.W.J., Allison, G.B., Dighton, J.C. and Trumbore, S. (1995): The age and hydrological history of Blue Lake, South Australia, Paleogeography, Paleclimatology, Paleoecology, 118, 111-130.

Lerner, D.N., Issar, A.S. and Simmers, I. (1990): Groundwater Recharge: A Guide to understanding and estimating natural recharge. IAH International Contributions to Hydrogeology Vol 8. (Hannover, Heise).

Lerner, D.N. (2004): Urban Groundwater Pollution, International Contributions to Hydrogeology, 24, Balkema Publishers, ISBN 9058096297.

LfU (Landesanstalt für Umweltschutz Baden-Württemberg) (1996): Großräumiges Grundwassermodell Oberrheingraben zwischen Basel und Karlsruhe. Demonstrationsvorhaben zum Schutz und zur Bewirtschaftung des Grundwassers des deutsch-französisch-schweizerischen Oberrheingrabens (LIFE-Projekt). – Abschlussbericht, Landesanstalt für Umweltschutz Baden-Württemberg, Abteilung 4 – Referat 42, 166 S.; Karlsruhe.

LfU (Landesanstalt für Umweltschutz Baden-Württemberg) (2002): Vorkommen von Pharmaka und Hormonen in Grund-, Oberflächengewässern und Böden in Baden-Wüttemberg, Landesanstalt für Umweltschutz-212.

Love, A.J., Herczeg, A.L., Armstrong, D., Stadter, F. and Mazor, E. (1993): Groundwater flow regime within the Gambier Embayment of the Otway Basin, Australia: evidence from hydraulics and hydrochemistry, Journal of Hydrology 143, 297-338.

Lundin, M. (1999): Assessment of the Environmental Sustainability of Urban Water systems. Chalmers University of Technology, Department of technical Environmental Planning: Göteborg, Sweden.

Lundin, M., Molander, S. and Morrison, G.M. (2002): Indicators for the development of sustainable water and wastewater systems. Chalmers University of Technology, Department of technical Environmental Planning: Göteborg, Sweden.

Magara, Y., Tabata, A., Kohki, M., Kawasaki, M. and Hirose, M. (1998): Development of boron reduction system for sea water desalination, Desalination 118, 25-34.

Makepeace, D.K., Smith, D.W. and Stanley, S.J. (1995): Urban stormwater quality: summary of contaminant data, Critical Reviews in Environmental Science and Technology, 25(2), 93.139.

Miller, R., Correll, R., Dillon, P., Kookana, R., (2002): ASRRI: A predictive index of contaminant attenuation during aquifer storage and recovery', in P.J. Dillon (editor) Management of Aquifer Recharge for Sustainability - Proceedings of the 4th International Symposium on Artificial Recharge of Groundwater, ISAR-4, A. A. Balkema Publishers, Amsterdam, pp. 69-74.

Miller, R., Correll, R., Vanderzalm, J. and Dillon, P. (2006): Modeling of movement of contaminants from sewer leaks and public open space through the unsaturated zone to the watertable. CSIRO Report CMIS 05/41.Misstear, B., Bishop, P., White, M.E.D., Anderson, G. (1996): Reliability of sewers in environmentally vulnerable areas, CIRIA Project Report 44, London, 1996, 122 pp.

Mitchell, V.G. and Diaper, C. (2005): UVQ: a tool for assessing the water and contaminant balance impacts of urban development scenarios. Water Science and Technology 52(12) pp 91-98.

Mitchell, V.G., Diaper, C., Gray, S. and Rahilly, M. (2003): UVQ: Modelling the movement of water and contaminants through the total urban water cycle. Proceedings of 28th International hydrology and water resources symposium, Wollongong, NSW.

Mitchell, V.G., Mein, R.G. and McMahon, T.A. (2001): Modelling the Urban Water Cycle. Journal of Environmental Modelling and Software Vol. 16(7) pp 615 – 629.

Mitchell, V.G. and Diaper, C. (2005): UVQ User Manual, CSIRO Urban Water, CMIT Report No. 2005-282.

Mitchell, V.G., Diaper, C., Rahilly, M., Del'Oro, E. (2004): Urban Water & Contaminant Balance Model (UVQ), User Manual and Documentation. CSIRO Manufacturing and Infrastructure Technology.

Mitchell, V.G. (2005): Aquacycle User Manual, CRC for Catchment Hydrology, Monash University.

Mitchell, V.G. and Maheepala, S. (1999): Urban Water Balance Modelling, CSIRO Urban Water Program, Report T1-11, BCE 99/195.

Mohrlok, U. (2005): UL_FLOW 1.0 – Modelling pseudo transient vertical flow in unsaturated layered soil using a quasi steady state approach, Documentation and User Manual. Appendix to Groundwater Recharge Estimation in Case Study Cities, deliverable D14, Technical Report no. 820, Institute for Hydromechanics, Universität Karlsruhe.

Mohrlok, U. and Bücker-Gittel, M. (2005): Methodology for Modelling Groundwater Recharge in urban Areas. AISUWRS work package 6, deliverable D13, Report No. 814, Institute for Hydromechanics, Universität Karlsruhe.

Moore, J.E. (1979): Contribution of ground-water modelling to planning. Journal of Hydrology 43, pp. 121-128.

Morris, B.L., Rueedi, J., Cronin, A.A. and Whitehead, E.J. (2005): AISUWRS Work-package 4 Field investigations final report. British Geological Survey *Commissioned* Report, CR/05/028N BGS Keyworth England 130 pp.

Morris, B.L, Darling, W.G., Cronin, A.A., Rueedi, J., Whitehead, E.J. and Gooddy, D.C. (2006): Assessing the impact of modern recharge on a sandstone aquifer beneath a suburb of Doncaster UK, Hydrogeology Journal in press, 2006.

Morris, B.L. Lawrence, A.R., Chilton, P.J., Adams, B., Calow R.C. and Klinck, B.A. (2003): Groundwater and its Susceptibility to Degradation: A Global Assessment of the Problem and Options for Management. Early Warning and Assessment Report Series, RS.03-3. United Nations Environment Programme, Nairobi, Kenya.

Mualem, Y. (1976): A new model for predicting the hydraulic conductivity of unsaturated porous media. Water Resour. Res. 12, 513-522.

Munster, J. (2003): Research interests, Accessed 8.07.2004, available at: http://www.ic.sunysb.edu/Stu/jmunster/interest.htm.

Mustafa, S. and Lawson, J.S. (2002): Review of tertiary Gambier Limestone aquifer properties, lower South-East, South Australia, Department of Water, Land and Biodiversity Conservation Report 2002/24.

Muttamara, S. (1996): Wastewater characteristics, Resources, Conservation and Recycling 16, 145-159.

Nash, J.E. and J. V. Sutcliff 1970. River Flow Forecasting Through Conceptual Models Part 1 - Discussion of Principles. Journal Of Hydrology 10: 282-290.

NHMRC and NRMMC (2004): National water quality management strategy: Australian drinking water guidelines, NHMRC and NRMMC, Australia.

NHMRC and NRMMC (2005): National water quality management strategy; Australian drinking water guidelines, NHMRC and NRMMC, Australia.

Norman, W., MacDonald, C. (2003): Getting to the Bottom of "Triple Bottom Line", Business Ethics Quarterly, March 2003.

NRMMC and EPHC (2005): National guidelines for water recycling: Managing health and environmental risks, Draft for public consultation, October (2005): NMMRC and EPHC, Australia.

OECD (1998): Towards Sustainable Development: Environmental Indicators, Paris.

Oliver, Y.M., Gerritse, R.G., Dillon, P.J., Smettem, K.R.J., (1996a): Fate and mobility of stormwater and wastewater contaminants in aquifers: 1. Literature review, Centre for Groundwater Studies Report no. 67.

Oliver, Y.M., Gerritse, R.G., Dillon, P.J., Smettem, K.R.J., (1996b): Fate and mobility of stormwater and wastewater contaminants in aquifers: 2. Adsorption studies for carbonate aquifers, Centre for Groundwater Studies Report no. 67.

Osswald, J. (2002): Hydrogeologische und hydrochemische Zustandsbeschreibung Rastatts im Hinblick auf anthropogene Beeinflussung. Unpubl. MSc Thesis. Karlsruhe : University of Karlsruhe 2002; 169 p (in German).

Pfuetzner, B. (1995): Modellierung der Grundwasserneubildung im Raum Karlsruhe. - Gutachten des Büros für angewandte Hydrologie im Auftrag der Stadtwerke Karlsruhe GmbH ,(unpublished), (in German).

Pitt, R., Clark, S. and Field, R. (1999): Groundwater contamination potential from stormwater infiltration practices, Urban Water, 1, 217–236.

Planning, S.A. (2005): Population Projection Enquiry System. Accessed 17.03.2006, available at: www.planning.sa.gov.au/pop_land_monitor/enquiry_system_new.htm

Powell, K.L., Taylor, R.G., Cronin, A.A., Barrett, M.H., Pedley, S., Sellwood, J., Trowsdale, S.A., Lerner, D.N. (2003): Microbial contamination of two urban sandstone aquifers in the UK. Water Research, 37(2), 339-352.

Prestor, J. (1998): Raziskave za nadomestitev vodarne Hrastje – Raziskave na Ljubljanskem barju – I faza ('Researches for replacement of Hrastje water field – Researches of Ljubljansko barje - phase one'). GZL-IGGG (Geological Survey of Slovenia), Ljubljana. In: Mencej Z (1998) Črpalni poskus na globokih vodnjakih "vodarna Brest" (Pumping test on deep wells in Brest waterfield), Hidroconsulting, Dragomer.

Prudence (2005): Prediction of Regional scenarios and Uncertainties for Defining EuropeaN Climate change risks and Effects, Prudence project summary, see URL: http://www.prudence.dmi.dk/.

Python (2005): Accessed Oct. 2005, see URL: http://www.python.org.

Ramamurthy, L.M., Veeh, H.H. and Holmes, J.W. (1985): Geochemical mass balance of a volcanic crater lake in Australia, Journal of Hydrology 79, 127-139.

Rauch, W. and Stegner, Th. (1994): The colmation of leaks in sewer systems during dry weather flow. Water Sci. Technol., 30(1), 205–210.

Rausch, R., Schäfer, W., Therrien, R., Wagner, C. (2005): Solute transport modelling: An introduction to models and solution strategies, Gebr. Borntraeger Science Publishers, Berlin, Stuttgart, Germany.

Rehm-Berbenni, C., Druta, A. (2004) GOUVERNe – Guidelines for the Organisation, Use and Validation of Information Systems for Evaluating Aquifer Resources and Needs. In: Global Change and Ecosystems, Water Cycle and Soil-related aspects, Water Technologies: Results and Opportunities, European Commission DG Research, pp. 75 – 80.

Richards, L.A. (1931): Capillary conduction of liquids through porous mediums. Physics 1, 318-333.

Richardson, S. B. (1990): Groundwater contamination by cheese factory and abattoir effluent, Masters thesis, Flinders University of South Australia.

Rinaudo J-D., Benjamin, G., Loubier, S., Interweis, E. (2003): Economic assessment of groundwater protection, BRGM, Strasbourg, BRGM/RC-52323-FR.

Robinson, N.I. (2005): Steady state point source leakage from a sewer pipe to the water table. Gleammf Pty Ltd. Report written for CSIRO. Accessed 10.01.2006, available at: http://www.urbanwater.de/results/publications/robinson_point_source.pdf.

Rosemann, S. (2004): The protective function of the unsaturated zone and the monitoring of stormwater quality of Mount Gambier, South Australia, Diploma thesis, University of Karlsruhe.

Rueedi, J., Cronin, A.A., Taylor, R.G., Morris, B.L. (in review) Estimating sewer leakage from depth-specific hydrochemistry of urban groundwater. Water Research.

Rueedi, J., Cronin, A.A. and Morris, B.L.(2004): AISUWRS Work-package 4 Field investigations interim report. British Geological Survey *Commissioned* Report, CR/04/022N BGS Keyworth England 72 pp.

Rueedi, J., Cronin, A.A. (2003): Construction of 5 Depth Specific Groundwater Sampling Sites in Doncaster, UK. Robens Centre for Public and Environmental Health, Guildford.

Rueedi, J., Cronin, A.A., Morris, B.L. (2005a): Daily patterns of micro-organisms in the foul sewer system of Doncaster, United Kingdom. In Proceedings of the 10th International Conference on Urban Drainage, Copenhagen/Denmark, 21-26 August 2005.

Rueedi, J., Cronin, A. A., Morris, B.L. (2005b): AISUWRS workpackage 10, UVQ Report Doncaster, Technical Report, Robens Institute for Public and Environmental Health, Guildford, UK. Available at: www.urbanwater.de.

Rueedi, J., Cronin, A.A., Moon, B., Wolf, L., Hötzl, H. (2005c): Effect of different water management strategies on water and contaminant fluxes in Doncaster, United Kingdom. In: IWA World Water Congress, Marrakech, Morocco. Water Science and Technology, 52(9).

SA DEH (South Australian Department for Environment and Heritage) (2005): Volume of Stormwater and Effluent Discharged to the Marine Environment.

SA EPA (South Australian Environmental Protection Authority) (2003): Environmental Protection (Water Quality) Policy 2003 and Explanatory Report, SA EPA.

SA EPA (South Australian Environmental Protection Authority) (2006): Discharges to groundwater in urbanised areas of the lower South East of South Australia: Development of stormwater treatment guidelines, SA EPA.

SA EPA. (2003): Environmental Protection (Water Quality) Policy 2003 and Explanatory Report, South Australia Environment Protection Agency SA EPA.

SA Water (South Australian Water Corporation) (2004): SA Water Corporate GIS, SA Water.

Sacher, F., Gabriel, S., Metzinger, M., Stretz, A., Wenz, M., Lange, F.T., Brauch, H.-J., and Blankenhorn, I. (2002): Arzneimittelwirkstoffe im Grundwasser - Ergebnisse eines Monitoring-Programms in Baden-Württemberg. Vom Wasser, 183-196.

Sadler, P. and De Silva, D. (2005): Analysis and quantification of sewer pipeline defects. CMIT Doc. 2005-239. CSIRO, PO Box 90, Victoria 3149, Australia.

Sægrov, S. (2004): What is CARE-S?, 3. World Congress of IWA, Marrakech September 2004

Sægrov, S. (ed) (2005): Computer Aided Rehabilitation of Water Networks, IWA Publishing 2005; ISBN 1843390914.

Schaefer, M. (2006): Modelling of water and contaminant flows in the city of Rastatt under different climate and water management scenarios, Unpubl. MSc Thesis. Karlsruhe : University of Karlsruhe 2006 (in German).

Schmidt, L., Schultz, T., Correll, R. and Schrale, G. (1998): Diffuse-source nitrate pollution of groundwater in relation to land management systems in the south east of South Australia, PIRSA Technical Report no. TR266.

Schweinfurth, W., Reinhard, E., Rothenberger, G. (2002): Rastatt, Naturraum und Siedlung, in Der Landkreis Rastatt, Bd II, pp 341-391, Jan Thorbecke Verlag Stuttgart, Germany.

Semadeni-Davies, A., Lundberg, A., and Bengtsson, L. (2001): Radiation balance of urban snow: a water management perspective. Cold Regions Science and Technology Vol. 33(1) pp. 59-76.

Shepley, M.G. (2000): Notts-Doncaster Sherwood Sandstone Model-1998 Update, Version 2. Environment Agency Internal Report EA Olton, UK

SILO (2004): Point patched data set, available at: http://www.per.clw.csiro.au/silo.

Šimůnek, J., van Genuchten, M.Th., Šejna, M. (2005): The HYDRUS-1D Software Package for Simulating the One-Dimensional Movement of Water, Heat, and Multiple Solutes in Variably-Saturated Media, Version 3.0. Department of Environmental Science, University of California, Riverside, California.

Smallwood, C. (1998): United Nations Environment Program International Labour Organisation. International Program on Chemical Safety. Environmental Health Critieria 204 Boron. Available at: http://www.inchem.org/documents/ehc/ehc/ehc204.htm

Smedley, P.L., Brewerton, L.J. (1997): The natural (baseline) quality of groundwater in England and Wales. Part 2: the Triassic Sherwood Sandstone of the East Midlands and South Yorkshire. British Geological Survey Technical Report WD/97/52. BGS Keyworth England.

Smith, P.C. (1980): Groundwater pollution potential - Mount Gambier City Council rubbish dump, Report No. 2. 80/9, South Australia: Department of Mines and Energy South Australia.

Souvent, P. and Moon, B. (2005): UVQ-Report for Ljubljana, 24 pages. Available at: http://www.urbanwater.de

SRC (Syracuse Research Corporation) (2005): Physical Properties Database, SRC, available at: http://www.syrres.com/esc/physprop.htm.

star.ENERGIEWERKE (2002): Wasserweg Rastatt, Eine Begleitbroschüre mit vertiefenden Informationen (in German).

Swamee, P.K. (2001): Design of a sewer line. J. Envir. Engrg. 127(9), 776-781.

Treskatis, C. (2003): Saisonal auftretende bakterielle Befunde in Trinkwasserbrunnen eines dicht besiedelten Einzugsgebietes, IAH Workshop Grundwasserprobleme in urbanen Räumen, 31.7. - 1.8.2003, Karlsruhe. www.urbanwater.de

Turner, J.V., Allison, G.B. and Holmes, J.W. (1984): The water balance of a small lake using stable isotopes and tritium, Journal of Hydrology 70, 199-220.

UNCHS (1997): World Urbanization Prospects: The 1996 Revision. UN Dept of Economic and Social Affairs Population Division Report ST/ESA/SER.A/166, United Nations, New York.

United Nations Dept of Economic and Social Affairs, Population Division (1994): World Urbanization Prospects, 1994 Revision, UN New York.

United Nations Environment Programme (1996): Groundwater: a threatened resource. UNEP Environment Library, No.15. (Nairobi, Kenya: UNEP).

URBS PANDEN (2003): Website of EC funded URBS PANDEN (Urban Sprawl: European Patterns, Environmental Degradation and Sustainable Development) – Available at: http://www.pik-potsdam.de/urbs/

URL RS (2004): Pravilnik o pitni vodi ('Rules on drinking water'), March 2004.

URL RS (2005): Uredba o standardih kakovosti podzemne vode ('Decree on quality standards for groundwater'), November 2005.

URS (2000): Stormwater discharge monitoring, 1999, for Mount Gambier, URS, November 2000.

URS (2003): 2001/02 Mount Gambier stormwater discharge monitoring, URS, June 2003.

US Environment Protection Agency (US EPA) (1980): Onsite wastewater treatement and disposal system. Chapter 4: Wastewater Characteristics. EPA 625/1-80-012, US EPA, Washington.

USEPA (United States Environmental Protection Agency) (2005): Technical Factsheets, available at: www.epa.gov/OGWDW/dwh/t-soc/.

Van Genuchten, M. T. (1980): A closed form equation for predicting hydraulic conductivity of unsaturated soils Soil Sci Soc. Am. J. 44, 892- 898.

Veselič M., Pregl, M. (2002) Lizimeter v urbanem okolju na območju Pivovarne Union, d.d.: geološko in tehnično poročilo o izvedbi merilnih vrtin: št.: ip 365/2002. Ljubljana: IRGO, August 2002.

Vink, J.P.M. and Van der Zee, S.E.A.T.M. (1997): Pesticide biotransformation in surface waters: multivariate analysis of environmental factors at field sites, Water Research, 31(11), 2858-2868.

Vogel, M. (2005): Contaminant attenuation of urban water resources and systems in a carbonate aquifer – Gambier Limestone aquifer (South Australia)'. Project work Technical University of Berlin.

Vollertsen, J. and Hvitved-Jacobsen, T. (2003) Exfiltration from gravity sewers - a pilot scale study. Water Science and Technology, 47, 69-76.

Vollertsen, J. (2005): Department of Life Sciences, Aalborg University, Sohngaardsholmsvej 57, DK-9000 Aalborg, Denmark. Personal communication.

WASY (2002): WASY Software FEFLOW®. Demonstration Exercise. WASY Institute for Water Resources Planning and Systems Research Ltd., Berlin.

WASY (2005): Webpage: Feflow feature list, accessed 20.12.2005, see URL: http://www.wasy.de/deutsch/produkte/feflow/pdf/feflow_feature_list.pdf.

Waterhouse, J.D. (1977): The Hydrogeology of the Mount Gambier Area, Report of Investigations 48, Dept. of Mines South Australia.

Westerstrom, G. (1984): Snowmelt Runoff from Porson Residential Area, Lulea, Sweden, In Proceedings of the Third International Conference on Urban Storm Drainage, Goteborg, Sweden, June 4-8, Vol 1 pp 315-323.

Wolf, L. (2004): Rastatt Groundwater Flow Model, Proc. 3rd AISUWRS project meeting in MtGambier, March 2004. Available at www.urbanwater.de.

Wolf, L. (2005): Dept. of Applied Geology, University of Karlsruhe, Germany. Personal communication.

Wolf, L. and Hötzl, H. (in print): Upscaling of laboratory results on sewer leakage and the associated uncertainty. Matthias Eiswirth Memorial Volume - IAH Special Paper Series, published by Balkema/Taylor and Francis.

Wolf, L., Held, I., Eiswirth, M., and Hötzl, H. (2004): Impact of leaky sewers on groundwater quality. Acta hydrochim. hydrobiol. 32(4-5), pp. 361-373.

Wolf, L., DeSilva, D, Klinger, J, Moglia, M, Held, I, Burn, S., Sadler, P., Tjadraatmatdja, G., Gould, G. Eiswirth , M., Hötzl, H. (2005a). Leakage Rates - AISUWRS Deliverable D20. Department of Applied Geology Karlsruhe. Available at: www.urbanwater.de.

Wolf, L., Klinger, J, Held, I, Neukum, C., Schrage, C., Eiswirth, M. Hoetzl, H. (2005b). Rastatt City Assessment Report - AISUWRS Deliverable D9. Department of Applied Geology Karlsruhe. Available at: www.urbanwater.de.

Wolf, L. (2006): Assessing the influence of leaky sewer systems on groundwater resources beneath the City of Rastatt, Germany, , 170 p., Doctoral Thesis, University of Karlsruhe.

World Commission on Environment and Development (1987): Our common future, the World Commission on Environment and Development. Oxford University Press.

World Health Organisation (WHO) (2004): Guidelines for Drinking Water Quality, 3rd edition, available at: http://www.who.int/ water_sanitation_health/dwq/guidelines/en/index.html.

WSAA (Water Services Association of Australia) (2003): WSAA Facts, WSAA .

Zaadnoordnijk, W.J., Van den Brink, C., Van den Akker, C., Chambers, J (2004): Values and functions of groundwater under cities. In: Lerner, D. N. (ed.) (2004): Urban Groundwater Pollution, International Contributions to Hydrogeology, 24, Balkema Publishers, ISBN 9058096297.

Zheng, C. and Bennett. G. D. (1995): Applied contaminant transport: theory and practice. Van Nostrand Rheinhold. USA.